# People Skills for B

*People Skills for Behavior Analysts* provides a much-needed introduction to the people skills needed to succeed as a behavior analyst.

Divided into two primary parts—Foundational Skills and Specialized Skills—this book addresses an impressive breadth of people skills, focusing on intrapersonal and interpersonal skills, collaboration, consultation and training, leadership, and resource development. Relying on recent evidence-based practices and relevant literature tailored to meet the new BACB Task List, Professional and Ethical Compliance Code, and Supervised Independent Fieldwork requirements, the text includes contributions by leading figures from a wide variety of applied behavior analysis subfields to provide a truly balanced overview. The book delves into the literature from fields related to behavior analysis, such as counselling, psychology, graphic design, management, and education, and applies these perspectives to behavioral theories and principles to provide students, new graduates, and seasoned professionals with research, best practices, reflective questions, and practical techniques. From reflecting on one's practice, to learning essential therapeutic skills, running a great meeting, becoming a 'super' supervisor, and delivering a memorable presentation, all people skills are included in one place for the behavior practitioner.

This is a valuable resource for undergraduate and graduate students studying Applied Behavior Analysis (ABA), and will also appeal to recent graduates and behavior analysts looking to improve their existing skillsets.

**Carmen Hall**, PhD, BCBA-D is an assistant professor in the Master of Professional Education, ABA Program at Western University, London, Ontario, a professor in the Autism and Behavioural Science Graduate Certificate at Fanshawe College, Toronto, and co-owner and Clinical Director of Looking Ahead Clinical Services, delivering ABA services to children, youth, and adults with developmental disabilities and autism in Ontario.

**Kimberly Maich**, PhD (Cognition and Learning), OCT, BCBA-D, C Psych is a professor in the Faculty of Education at Memorial University of Newfoundland, St. John's, and an assistant professor (standing appointment) in the Master of Professional Education (ABA) at Western University, London, Ontario. She is a certified teacher, board-certified behavior analyst (doctoral), clinical, school, registered psychologist, owner of Exceptionally Yours Psychological Services, and co-owner and director of Insight Therapy and Assessment Services.

**Brianna M. Anderson**, MPEd, BCBA is a PhD student in the Child and Youth Studies program at Brock University, St. Catharines, Ontario. She is also an instructor in the Master of Professional Education (ABA) Program at Western University, London, Ontario, and professor in the Autism and Behavioural Science Graduate Certificate at Fanshawe College, Toronto. As a SSHRC-funded researcher, Brianna is currently evaluating behavior-based interventions for addressing obsessive-compulsive behaviors in children with developmental disabilities.

# People Skills for Behavior Analysts

Edited by Carmen Hall, Kimberly Maich and Brianna M. Anderson

NEW YORK AND LONDON

Designed cover image: © Getty Images

First published 2024
by Routledge
605 Third Avenue, New York, NY 10158

and by Routledge
4 Park Square, Milton Park, Abingdon, Oxon, OX14 4RN

*Routledge is an imprint of the Taylor & Francis Group, an informa business*

© 2024 selection and editorial matter, Carmen Hall, Kimberly Maich and Brianna M. Anderson; individual chapters, the contributors

The right of Carmen Hall, Kimberly Maich and Brianna M. Anderson to be identified as the authors of the editorial material, and of the authors for their individual chapters, has been asserted in accordance with sections 77 and 78 of the Copyright, Designs and Patents Act 1988.

All rights reserved. No part of this book may be reprinted or reproduced or utilised in any form or by any electronic, mechanical, or other means, now known or hereafter invented, including photocopying and recording, or in any information storage or retrieval system, without permission in writing from the publishers.

*Trademark notice*: Product or corporate names may be trademarks or registered trademarks, and are used only for identification and explanation without intent to infringe.

*Library of Congress Cataloging-in-Publication Data*
Names: Hall, Carmen, editor. | Maich, Kimberly, 1969– editor. | Anderson, Brianna M., editor.
Title: People skills for behavior analysts / edited by Carmen Hall, Kimberly Maich, and Brianna M. Anderson.
Description: New York, NY : Routledge, 2024. | Includes bibliographical references and index. |
Identifiers: LCCN 2023022160 (print) | LCCN 2023022161 (ebook) | ISBN 9781032292243 (hardback) | ISBN 9781032292236 (paperback) | ISBN 9781003300465 (ebook)
Subjects: LCSH: Persuasion (Psychology) | Influence (Psychology) | People skills for professionals.
Classification: LCC BF637.P4 P37 2024 (print) | LCC BF637.P4 (ebook) | DDC 153.8/52–dc23/eng/20230802
LC record available at https://lccn.loc.gov/2023022160
LC ebook record available at https://lccn.loc.gov/2023022161

ISBN: 978-1-032-29224-3 (hbk)
ISBN: 978-1-032-29223-6 (pbk)
ISBN: 978-1-003-30046-5 (ebk)

DOI: 10.4324/9781003300465

Typeset in Sabon
by Newgen Publishing UK

Carmen Hall dedicates the book to all of the clients, parents, educators, collaborators, and students over the years who have inspired her to make Applied Behaviour Analysis (ABA) accessible to all. Their endless discussions on how people skills paired with ABA made the most effective treatments inspired us to bring authors together to create this resource. We hope it will answer comments received over the years from new graduates: "I wish I was taught that!" As always, Carmen thanks her amazing family and her inspirational co-authors Kimberly Maich and Brianna Anderson for their support to make this dream a reality.

Kimberly Maich dedicates the book, as always, to her family, including her husband John of 32 years, and her wonderfully neurodiverse adult offspring, Robert, Grace, and Hannah, as well as their supportive spouses, Mark and Steven. Special dedications go out to her adorable grandson, Quincey, and her newest grandson, Caleb. This project would not have been possible without the vision and leadership of co-editor Carmen Hall and the dedication and collaboration of co-editor Brianna Anderson.

Brianna Anderson dedicates this book, first and foremost, to her former clients and their families. It was through their kindness and tenacity that she learned how to be a collaborative and compassionate clinician. Brianna thanks her students for opening her eyes to new ways of thinking and making true the saying "in learning you will teach, and in teaching you will learn." To her mentors, from whom she has learned so much, co-editors Carmen Hall and Kimberly Maich included: she truly wouldn't be here without their guidance and generosity. And, last, Brianna thanks her loving husband, who has selflessly supported her during this incredible journey.

# Contents

| | |
|---|---|
| *List of Contributors* | xi |
| *Foreword* | xv |
| TYRA SELLERS | |
| *Preface* | xvii |
| *Acknowledgments* | xxii |

**PART I**
**Foundational Skills** — 1

1 **Intrapersonal Skills** — 3
  1.1 The Importance of Intrapersonal Skills for Behavior Analysts — 4
     CARLY MAGNACCA, KENDRA THOMSON, AND LINDA MOROZ

  1.2 The Discerning ABA Practitioner — 16
     EMILY BURNS

  1.3 Self-Care for Behavior Analysts: A Guide to Eating Fewer Doughnuts and Taking More Naps — 27
     ALBERT MALKIN, KARL F. GUNNARSSON, AND KRISTINA AXENOVA

2 **Interpersonal Skills** — 49
  2.1 20 Key Behaviors for Rapport Building — 50
     ANA LUISA SANTO AND KIMBERLEY TAYLOR

  2.2 Cultural Competence in Communication — 77
     TASMINA KHAN

  2.3 A Critical Look at Autism — 88
     CHRISTOPHER PETERS

| | 2.4 | Accepting Feedback Makes the World Go 'Round: From Student to Practitioner<br>DON TOGADE AND MARINA JIUJIAS | 96 |
|---|---|---|---|
| | 2.5 | How to Teach Rapport Building Skills to Behavior Analysts<br>ANA LUISA SANTO AND KIMBERLEY TAYLOR | 109 |
| 3 | **Collaboration** | | 127 |
| | 3.1 | No More "Train and Train More": A Functional, Contextual Approach to Collaboration with Families<br>CRESSIDA PACIA, CIARA GUNNING, AOIFE MCTIERNAN AND JENNIFER HOLLOWAY | 128 |
| | 3.2 | Playing Nicely in the Interdisciplinary Sandbox: A How-to Guide on Effective Collaboration with Various Professionals<br>TAYLOR SLOBOZIAN | 143 |
| | 3.3 | Who's the King of the Castle? Collaborating with Educators and Schools—as a Behavior Analyst<br>KAREN MANUEL AND BRITTANY DAVY | 151 |
| | 3.4 | Overcoming the Aversive: Handling Difficult Conversations with Professionalism and Compassion<br>BRIANNA M. ANDERSON AND DANA KALIL | 159 |
| | 3.5 | Running Efficient and Effective Meetings with Collaboration and Compassion<br>OLIVIA NG AND KERRY-ANNE ROBINSON | 169 |

**PART II**
**Specialized Skills** 179

| 4 | **Consultation and Training** | | 181 |
|---|---|---|---|
| | 4.1 | Don't Blame the Mediator: Keeping Applied Behavior Analysis Doable for Best Possible Outcomes<br>KENDRA THOMSON, REGHANN MUNNO, LOUIS P. A. BUSCH, AND MAURICE FELDMAN | 182 |
| | 4.2 | Beyond the PowerPoint<br>CARMEN HALL | 192 |

4.3 Behavior Analyst as Changemaker: Science, Advocacy and Activism in Practice 203
LOUIS P. A. BUSCH, JAIME SANTANA, AND MARK A. MATTAINI

4.4 Using Movement Education, Motor Learning, and Phenomenology to Inform Teaching Strategies in Other Professional Contexts 217
MAUREEN CONNOLLY AND ELYSE LAPPANO

5 Leadership 229
5.1 "It's Not You—It's Us": Fostering a Behavior Analytical Supervisor/Supervisee Relationship 230
CÉLINE BOURBONNAIS-MACDONALD AND ALEXANDRA WHITE

5.2 People Skills for Behavior Analysis: Putting the *Super* in Your Supervisory Relationships 236
KELLY ALVES

5.3 Giving Performance Feedback that Makes a Difference 243
MICHAEL G. PALMER

5.4 Interviewing Strategies: Integrating Practices from Industrial/Organizational Psychology into Behavior Analysis for Successful Hiring 252
MICHAEL G. PALMER

6 Innovative, Creative Ways to Use ABA 261
6.1 Making Applied Behavior Analysis Accessible to Consumers 262
NICOLE NEIL AND ANASTASIA KLIMOVA

6.2 Telehealth to Implement Applied Behavior Analysis Treatment 280
NATALIE PAQUET CROTEAU AND ERICA FRANCO

6.3 "You're on Mute!" People Skills Behind the Screens and Behind the Scenes 289
ANDERS LUNDE AND KIMBERLY MAICH

6.4 Making a Document People Will Actually Read 299
CARMEN HALL

6.5 A Trauma-Informed Approach to Applied Behavior
    Analysis                                                      308
    BRIANNA M. ANDERSON AND DANA KALIL

6.6 People Skills of Behavior Analysts as Co-designers
    with Autistic Adults: Applied Behavior Analysis
    Integrated into Virtual Reality Game-Based
    Intervention Supports                                         316
    JAVIER ALEJANDRO ROJAS

*Index*                                                           327

# Contributors

**Kelly Alves,** EdD, BCBA, is a professor in the Behavioural Health Sciences Program at Seneca College, Toronto.

**Brianna M. Anderson,** MPEd, BCBA, is a PhD student in the Child and Youth Studies Program at Brock University, St. Catharines, Ontario, professor and coordinator of the Autism and Behavioural Science Graduate Certificate Program at Fanshawe College, Toronto, and lecturer in the Master of Professional Education (ABA) at Western University, London, Ontario.

**Kristina Axenova,** MA, is a PhD Candidate in the Clinical Psychology program at York University, Toronto.

**Céline Bourbonnais-MacDonald,** EdD, is an instructor at Western University, London, Ontario.

**Emily Burns,** EdD, BCBA, is an ABA Specialist for the Lambton Kent District School Board.

**Louis Busch,** MA, BCBA, is a community support specialist with Shkaabe Makwa's Workforce Development Program and a behavior therapist at the Centre for Addictions and Mental Health in Toronto.

**Maureen Connolly,** PhD, is a professor of Physical Education and Kinesiology at Brock University, St. Catharines, Ontario.

**Brittany Davy,** MA, BCBA, is a PhD candidate in the Health and Society Program, and clinical supervisor at GAIN Learning Centre.

**Maurice Feldman,** PhD, BCBA-D, is a professor emeritus in the Department of Applied Disability Studies at Brock University, St. Catharines, Ontario, and a clinical psychologist.

**Danielle Flood,** MA, BCBA, is an autism consultant at Kinark Child and Family Services.

**Erica Franco**, MADS, BCBA, has been working with children diagnosed with autism and their families in the Toronto area through regional providers and supports the Northern Ontario community through her private practice, N&E Behavioural Consulting.

**Karl F. Gunnarsson**, PhD, BCBA-D, is an assistant professor at the University of Iceland and National University Hospital, Reykjavík, Iceland.

**Ciara Gunning**, PhD, BCBA, is a lecturer in the Psychology (Behaviour Analysis) program at the University of Galway, Ireland.

**Carmen Hall**, PhD, BCBA-D, is an assistant professor in the Master of Professional Education at Western University, London, Ontario, professor in the Bachelor of Behavioural Psychology Program at Fanshawe College, Toronto, and clinical director of Looking Ahead Clinical Services in London, Ontario.

**Tasmina Khan**, MPEd, is a senior behavior therapist at Magnificent Minds.

**Jennifer Holloway**, PhD, BCBA-D, is CEO and senior psychologist at All Special Kids, and director of the PhD and MSc in Applied Behaviour Analysis Program at the University of Galway, Ireland.

**Marina Juijias**, MS, BCBA, is a professor at George Brown College, Toronto, and senior supervising therapist in her own private practice.

**Dana Kalil**, MA, is a research assistant at Brock University, St. Catharines, Ontario, with experience blending behavior therapy with recreational dance to facilitate improvements in children's self-coping strategies.

**Anastasia Klimova**, BSc, is a graduate student in the Master of Professional Education (ABA) Program at Western University, London, Ontario, and behavior therapist at Kerry's Place Autism Services.

**Elyse Lappano**, MA, is the manager of student information and services at Brock University, St. Catharines, Ontario.

**Anders Lunde**, MEd, is a special education teacher in British Columbia.

**Carly Magnacca**, MA, is a graduate student in the Clinical Developmental Psychology program at York University, Toronto.

**Kimberly Maich**, PhD (Cognition & Learning), OCT, BCBA-D, C Psych is a professor in the Faculty of Education at Memorial University of Newfoundland, St. John's, and a psychologist.

*List of Contributors* xiii

**Albert Malkin**, PhD, BCBA-D, is an assistant professor in the Faculty of Education at Western University, London, Ontario, within the Master of Professional Education (ABA).

**Karen Manuel**, MADS, BCBA, is the director and owner of GAIN Learning Centre.

**Mark Mattaini**, DSW, ACSW, holds an emeritus appointment at Jane Addams College of Social Work, University of Illinois at Chicago, and former director of the doctoral program.

**Aoife McTiernan**, PhD, BCBA-D, is a lecturer in the psychology (Behaviour Analysis) program at the University of Galway, Ireland.

**Linda Moroz**, MA, BCBA, is a behavior therapist at the Bethesda Foundation.

**Reghann Munno**, MADS, BCBA, is a behavior therapist / consultant at Pelham Psychotherapy and DS Consulting.

**Nicole Neil**, PhD, BCBA-D, is assistant professor and coordinator of the Master of Professional Education (ABA) Program at Western University, London, Ontario, and clinical director at Looking Ahead Clinical Services.

**Olivia Ng**, MA, BCBA, is clinical supervisor at Progressive Steps Training and Consultation, and a part-time faculty member in the Master of Professional Education (ABA) at Western University, London, Ontario.

**Cressida Pacia**, MA, BCBA, is a PhD candidate in the Applied Behaviour Analysis Program at the National University of Ireland.

**Michael G. Palmer**, PhD, BCBA-D, is director of the Certificate in Applied Behaviour Analysis Program at the University of New Brunswick, Fredericton.

**Natalie Paquet Croteau**, PhD, BCBA-D, is the owner of ABA Northern Services and Training.

**Christopher Peters** is a PhD Candidate in the PhD in Education (Educational Sustainability) at Nipissing University, North Bay, Ontario.

**Kerry-Anne Robinson**, MEd, BCBA, is the owner and clinical director of Progressive Steps Training and Consultation.

**Javier Alejandro Rojas**, MA, is an interdisciplinary PhD candidate at Memorial University of Newfoundland, St. John's.

**Jaime Santana**, MADS, BCBA, is clinical director at Santana Behavioural Services, Toronto.

**Ana Luisa Santo**, MA, BCBA, is a senior behavior therapist, director of Children and Youth, and Manager of Autism Services at Surrey Place in Toronto.

**Tyra Sellers**, PhD, BCBA-D, is the CEO of the Association of Professional Behavior Analysts and has held positions as an assistant professor at Utah State University, Logan, Utah, and as the director of ethics at the Behavior Analyst Certification Board.

**Taylor Slobozian**, OCT, MPEd (Field of ABA) supports community partners, family members, and learners of all ages diagnosed with Autism Spectrum Disorder (ASD) since 2016 in Northwestern Ontario with the focus being on adaptive, communication, functional, and school readiness skills.

**Kimberley Taylor**, MADS, BCBA, is a senior behavior therapist at Surrey Place in Toronto.

**Kendra Thomson**, PhD, BCBA-D, is an associate professor in the Department of Applied Disability Studies at Brock University, St. Catharines, Ontario.

**Don Togade**, PhD, BCBA-D, is a professor and coordinator of the Honours Degree in Behaviour Analysis Program at George Brown College, Toronto, and adjunct professor at the Chicago School of Professional Psychology.

**Alexandra White**, MADS, BCBA, is a senior therapist at bitKIDS Behaviour Consulting and part-time professor at Mohawk College and McMaster University, both in Hamilton, Ontario.

# Foreword

## Tyra Sellers

**Looking Ahead**

The health and success of the profession of behavior analysis are predicated on many factors. One important factor is the provision of effective supervision. Supervisors are responsible for ensuring that services provided are ethical and effective, that caregivers are engaged in meaningful collaboration throughout service delivery, and that individuals working toward certification develop effective skills to provide and supervise services. Each supervisor can permanently contribute to, or damage, the health and success of the profession, as many of the individuals they supervise will go on to have their own trainees and supervisees. Those trainees and supervisees will become supervisors with trainees and supervisees of their own—and so on. If supervisors pass along defective or deficient supervisory repertoires, at least some of their trainees and supervisors will subsequently pass those less-than-optimal skills sets along, and vice versa for passing along effective supervisory repertoires. In this way, supervision practices have lasting future impacts in our profession.

The criticality of effective supervision is enhanced by the rapid growth of the profession. For example, as of December 2022, over 50 percent of individuals certified at the BCBA level have been certified for five or fewer years. This growth is exciting, as it means an increased ability to provide important services to individuals and organizations. However, this also means that trainees and supervisees are likely being trained and supervised by individuals with newly acquired supervisory skills. Many individuals have received little explicit training on how to be an effective supervisor. In the time since the Behavior Analysts Certification Board (BACB) began certifying individuals in 1998, they have systematically increased requirements, including those related to supervision (e.g., increasing the percentage of fieldwork hours supervised, requiring supervision specific continuing education units and completing of specific training to qualify to supervise, prohibiting the supervision of trainees for one year post-certification unless under the guidance of a qualified Consulting Supervisor). The ethics standards guiding supervisory practices have also been enhanced regarding areas such as responsibility, diversity, and continuity of supervisory services.

Perhaps in response to the profession's growth, an increased understanding of the importance of effective supervisory repertoires, and additional requirements, over the past several years there seems to have been an increase in scholarly work and conference presentations related to supervision. Behavior analysts have written practice recommendation papers focusing on important supervision topics such as compassion, culturally responsive practices, self-assessment, and core professionalism skills. Several books and workbooks are now available to support those teaching supervisory skills in courses, during supervised fieldwork, and in the workplace, with a particular focus on novice supervisors. In fact, some practice guideline papers and books rightly focus on the trainee, as a sort of antecedent intervention encouraging them to take an active role in their supervision, outlining for them what they should expect from their supervisor and guiding them through practical and important skills such as soliciting and giving feedback. The resources have helped push discussion about the need for high quality supervision to the forefront of our profession, with researchers and professionals focusing on strategies for developing the critical repertoires for being a successful supervisor and for teaching those repertoires to others.

Effective, high-quality supervision is undoubtedly more effortful than low-quality supervision, just as is the case with low-quality clinical service. However, the return on investment is realized immediately for the supervisee, supervisor, clients, and caregivers. Importantly, the positive impact also reaches exceptionally into the future, as effective supervisory repertoires are passed along. This type of supervision—the type that can sustain a healthy and successful profession—is not born from a unilateral, transactional model. This type of supervision requires the active development and continual facilitation of a collaborative and mutually respected relationship. It is born from humility and the knowledge that we all have something to teach, and something learn from each other.

This book provides relevant guidance related to interpersonal skills, collaboration, consultation and training, and leadership. The content of this book adds to the body of scholarly work that acknowledges the importance of developing high-quality supervisory and leadership skills, and it is a breath of fresh air in the profession. It is a call to ensure that behavior analysts balance the application of their science with an approach that values and centers humanity and connectedness. In this balance there is beauty and hope that our profession will endure with professionals who are well equipped to make meaningful changes in the world in compassionate and purposeful partnership with consumers and with other professionals.

# Preface

*People Skills for Behavior Analysts* is a unique professional learning tool for continuing education, focusing on skills beyond the technical skills in the work that behavior analysts undertake on a daily basis. People skills are broad, diverse, and far-reaching personal attributes or character traits that contribute and generalize to areas such as therapeutic relationships—or simply how we get along with others—including clients, their parents (where applicable), our colleagues, and more.

Much of the past literature on this topic (Parlamis & Monnot, 2019) has used the term *soft skills*; however, like others who aim to retire the term, we feel that this is a misnomer. We believe that there is nothing soft about people skills: they are the strong and solid foundation that leads to clinical success. At the same time, if they are considered at all, they are often thought of as being secondary to technical skills (the so-called "hard skills"). The reader will see from this broad set of fascinating monographs that people skills are broadly influential in the moment-to-moment work of the behavior analyst. In fact, we can see the presence of people skills in the Ethics Code for Behavior Analysts (BACB, 2020) and the 5th Edition Task List (BACB, 2017). However, students often will tell us that no one taught them how to complete these soft skills—for example, how to adjust their body language, how to prepare for a presentation, or what specific language will assist them in building a relationship with their client. The goal of this book is to provide the "how-to" for all those times we were asked.

The use of person-first language (e.g., person with autism) and identity-first language (e.g., autistic person) is used interchangeably and at the discretion of individual chapter authors. While many chapters are written from the perspective of the medical model, highlights from the social model and from critical perspectives also are featured.

Contributors to this edited volume underscore and bring perspectives from various behavior analysts' clinical practices and research labs, and include professionals from other fields, such as education, social work, and psychology to encapsulate the various research and practices to build these skills. Contributing authors were encouraged to look into various literature beyond that of applied behavior analysis (ABA), including business, graphic design, psychology, counselling, and social work. Through capturing the research and best practices across fields, the authors made it specific to the practice of behavior analysis.

This book is divided into two sections: Foundational Skills (Part I) and Specialized Skills (Part II). Foundational Skills is comprised of 13 unique chapters focused on Intrapersonal Skills, Interpersonal Skills, and Collaboration. Specialized Skills, building on presented information and perspectives in the preceding section of foundational information, is comprised of an additional 15 chapters focused on Consultation and Training, Leadership, as well as Innovation and Creativity.

## Part I: Foundational Skills

### Chapter 1: Intrapersonal Skills

Chapter 1 examines skills required for intrapersonal communication skills—the conversations that we have with ourselves. Such skills, described in these monographs, involve thinking and analyzing, self-reflecting, and examining caring for oneself as a therapist:

1. The Importance of Intrapersonal Skills for Behavior Analysts
2. The Discerning ABA Practitioner
3. Self-Care for Behavior Analysts: A Guide to Eating Fewer Doughnuts and Taking More Naps

### Chapter 2: Interpersonal Skills

Chapter 2 looks at the skills required in the communication acts between two or more people—the interpersonal skills. These skills are necessary to form relationships with both colleagues and interdisciplinary teams as well as clients and their families. The skills required are both verbal and non-verbal and are complex and intertwined:

1. 20 Essential Interpersonal Skills for Mediator Rapport Building
2. Cultural Competence in Communication
3. A Critical Look at Autism
4. Accepting Feedback Makes the World Go Round: From Student to Practitioner
5. How to Teach Rapport-Building Skills to Behavior Analysts

*Chapter 3: Collaboration*

This chapter highlights a core component of the role of a behavior analyst: working with clients, families, and other service providers to ensure that the services implemented are in the best interest of the client and agreed upon by all parties. Carrying out collaboration with different backgrounds and approaches can sometimes be complicated. This chapter includes the specific skills in collaborating and working with others:

1 No More "Train and Train More": A Functional, Contextual Approach to Collaboration with Families
2 Playing Nicely in the Interdisciplinary Sandbox: A How to Guide on Effective Collaboration with Various Professionals
3 Who's the King of the Castle? Collaborating as a BCBA with Educators and Schools
4 Overcoming the Aversive: Handling Difficult Conversations with Professionalism and Compassion
5 Running Efficient and Effective Meetings with Collaboration and Compassion

## Part II: Specialized Skills

*Chapter 4: Consultation and Training*

Being a behavior analyst involves a great deal of consulting and training of others to implement strategies. Thus, how to conduct effective trainings, including consultations, small group trainings, or large group trainings, is needed for future behavior analysts.

Information on evidence-based strategies to conduct training is provided, including theory, practice, and research from popular resources in other related fields.

1 Don't Blame the Mediator: Keeping Applied Behavior Analysis Doable for Best Possible Outcomes
2 Beyond the PowerPoint: Becoming an Engaging Presenter
3 Behavior Analyst as Changemaker: Science, Advocacy and Activism in Practice
4 Using Movement Education, Motor Learning, and Phenomenology to Inform Teaching Strategies in Other Professional Contexts

*Chapter 5: Leadership*

Behavior analysts are responsible for overseeing clients and coaching and training those working with them. Thus, strong leadership skills are needed in conjunction with the clinical skills learned in graduate programs. Supervision skills cover discovering individual supervision philosophies, establishing guidelines, and boundaries at the outset, providing positive and corrective feedback, and effective ongoing leadership and mentorship. Lastly, hiring and maintaining employees through burnout and other factors in the field are explored.

1. It's Not You, It's Us: Fostering a Behavior Analytical Supervisor/Supervisee Relationship
2. Putting the Super in Your Supervisory Relationships
3. Giving Performance Feedback that Makes a Difference
4. Interviewing Strategies: Integrating Practices from Industrial/Organizational Psychology into Behavior Analysis for Successful Hiring

*Chapter 6: Innovative, Creative Ways to Use ABA*

As a field, ABA has been criticized for not being accessible to consumers (Critchfield et al., 2017). This issue has been attributed to increased jargon and lack of accessible information formats for consumers. Mainstream graphic design, web design, accessible standards, and media formats and trends also need to be taught to graduates to ensure that the resources and materials are up-to-date, accessible, and attractive to clients. Presenting the information in this matter will assist with dissemination and communication. The following monographs highlight these topics:

1. Making Applied Behavior Analysis Accessible to Consumers
2. Telehealth to Implement Applied Behavior Analysis
3. You're on Mute! People Skills Behind the Screens and Behind the Scenes
4. Creating Resources People Will Actually Read
5. A Trauma-Informed Approach to Behavior Analysis
6. Behavior Analysts' Skills for Co-Developing Virtual Reality Game-Based Technologies as Applied Behavior Analysis Programs for Autistic Adults

## References

Behavior Analyst Certification Board. (2017). *BCBA task list* (5th ed.). Author.
Behavior Analyst Certification Board. (2020). *Ethics code for behavior analysts.* https://bacb.com/wp-content/ethics-code-for-behavior-analysts/

Critchfield, T. S., Doepke, K. J., Kimberly Epting, L., Becirevic, A., Reed, D. D., Fienup, D. M., Kremsreiter, J. L., and Ecott, C. L. (2017). Normative emotional responses to behavior analysis jargon or, how not to use words to win friends and influence people. *Behavior Analysis in Practice*, *10*(2), 97–106. https://doi.org/10.1007/s40617-016-0161-9

Parlamis, J., & Monnot, M. J. (2019). Getting to the CORE: Putting an end to the term "soft skills". *Journal of Management Inquiry, 28*(2), 225–227.

# Acknowledgments

The editors would like to thank all the authors for their time and effort in pulling together transcripts within and outside of their clinical practices and research areas to make this book fulsome and varied. The number of topic areas that it spans is profound. We would also like to thank all of the students and research assistants who assisted the authors in the preliminary explorations and searching across literature databases to ensure that people skills across disciplines would be captured.

# Part I
# Foundational Skills

# 1
# Intrapersonal Skills

# 1.1 The Importance of Intrapersonal Skills for Behavior Analysts

*Carly Magnacca, Kendra Thomson, and Linda Moroz*

**What Are Intrapersonal Skill**

Intrapersonal skills can involve behaviors that are considered to be "within the self" (or private events; Skinner, 1953) and focus on one's own internal behaviors, including self-care, self-awareness, and self-regulation. Although there is some literature outlining how to use behavior analytic strategies to improve specific and operationally defined intrapersonal skills in the literature (e.g., Kalis et al., 2007; Shaffer et al., 2019), there is limited information on intrapersonal skills more broadly. Intrapersonal skills can be defined operationally and impacted by the same behavioral principles as public events, including the basic contingencies of reinforcement and punishment. For example, Skinner (1953) mentioned how private events, although inaccessible to outside observers, are impacted by the same contingencies of reinforcement and punishment as overt behaviors and can impact future private events and overt behaviors. Further, intrapersonal skills may also be shaped by interlocking contingencies given that they may involve the behavior of others (e.g., having a conversation with a co-worker about feeling overwhelmed at work (Borba et al., 2017; Glenn, 2004). Intrapersonal skills may also involve metacontingencies as they shape our covert behavior. For example, if self-care is promoted at a place of work, one may behave differently than if working at an organization where it is not valued or reinforced, and the organization may have staff retention issues (Glenn, 2004). Moreover, intrapersonal skills may also involve competing contingencies given that you may be interested in targeting multiple different instances of your own behaviors (e.g., health goals and career goals). In addition, other behavioral concepts such as rule-governed behavior, which can include instructions from others or one's own private events, may also impact intrapersonal skills, along with many other behavioral concepts. Although intrapersonal skills have not been largely discussed in the behavior analytic literature, for behavior analysts they are an essential skill that will impact the longevity and quality of one's career

DOI: 10.4324/9781003300465-3

and the support provided to others. Throughout the following section, we will explore a few specific intrapersonal skills, discuss the importance of these skills for behavior analysts, and review practical strategies that are supported by the literature to improve intrapersonal skills.

The intrapersonal skill that many are likely the most familiar with is self-care. The term self-care has become increasingly popular since the first self-care publication in 1946 (Reigel et al., 2021), especially throughout the COVID-19 pandemic. Although there have been many attempts to create an operational definition for the term, there has been little agreement (see Table 1.1). Throughout the self-care literature, some themes have been identified, including physical (e.g., engaging in physical activity), psychological (e.g., receiving counselling for one's mental health), spiritual (e.g., meditating), and support (e.g., consulting with a supervisor or colleague) aspects of self-care. Self-care may look different for everyone, but it consistently involves engagement in behaviors to increase or maintain one's well-being.

Another popular intrapersonal skill, especially in the business and management field, is self-awareness (Eurich, 2018). Just as with self-care, there are many ways self-awareness is defined in the literature. Sometimes self-awareness is defined as involving internal and external traits. Internal traits are defined as one's own perspective of their values and goals. Comparatively, external traits involve perspective taking, including taking into account how others view one's values and goals. Further, across fields, self-awareness is also used interchangeably with other terms such as self-knowledge and self-consciousness (Carden et al., 2022; Eurich, 2018). Despite the lack of clarity on the term, Carden et al. (2022) conducted a systematic review compiling and comparing different definitions to identify consistent themes. The authors found that, based on the current literature, the term can involve both interpersonal categories (e.g., focusing on others' perceptions and on your behaviors) and intrapersonal categories (e.g., awareness of own values and private events).

Although there are many other skills that are considered intrapersonal skills, the final skill discussed will be self-regulation. There are many other words used synonymously with self-regulation, including willpower, effortful control, self-control, and self-management (Murray et al., 2015). Self-regulation can be defined as managing behaviors and private events to engage in goal-directed behaviors (Murray et al., 2015). There are many overlapping domains of self-regulation, including cognitive (e.g., goal setting, self-monitoring, executive functioning), emotional (e.g., managing unpleasant private events), and behavioral (e.g., following rules, delay of gratification). These three domains interact in a bidirectional manner (Blair & Ursache, 2011).

*Table 1.1* Previous Definitions of Self-Care

| Self-Care | Definition |
|---|---|
| Pincus (2006) | "[What] one does to improve [the] sense of subjective well-being. How one obtains positive rather than negative life outcomes" (p.1). |
| Carrol, Gilroy, and Murra (1999) | "intrapersonal work, interpersonal support, professional development and support, and physical/recreational activities" (p.135). |

| Type of Self-Care | Definition |
|---|---|
| Physical | The physical part of self-care has been generally defined as including physical activity (Carroll et al., 1999), which in this context is defined as physical movement that uses energy and can be completed through exercise, sports, housework, and other daily tasks. Including physical exercise, which in this context is defined as physical movement that demands the use of energy and can be done through exercise, sport, home chores, and other activities done on a regular basis (Henderson & Ainsworth, 2001). |
| Psychological | Psychological self-care is obtaining one's own personal counselling (Coster & Schwebel, 1997; O'Connor, 2001). Personal counselling is described as psychological treatment for any sort of discomfort or incapacity (Norcross, 2005). |
| Spiritual | Given how flexible the interpretation of spirituality is, it is also essential to clarify it broadly when referring to self-care. A general definition of spirituality is a sense of the meaning of existence and the relationship one makes to this knowing (Estanek, 2006; Hage, 2006; Perrone et al., 2006) |
| Support | Connections and interactions that grow out of support systems, both personally and professionally, are part of the support element of self-care. Professional support is characterized as peer, colleague, and supervisor counseling and supervision, as well as continuing professional development (Coster & Schwebel, 1997; O'Connor, 2001; Stevanovic & Rupert, 2004). |

## Why Are Intrapersonal Skills Important?

Limited research has been conducted on the importance of intrapersonal skills for behavior analysts specifically; however, research conducted in other fields involving health care and service provision have indicated that these skills impact quality of work, stress management, psychopathology, and burnout. For example, clinical psychology graduate students' self-care practices, such as sleep hygiene, social support, and emotion regulation, were explored in relation to their perceived levels of stress. Myers and colleagues (2012) found that these self-care practices were significantly associated with perceived stress level and, therefore, recommended the inclusion of education on self-care practices within clinical training. Further, Pope and Vasquez (2016) discuss the potential consequences for clients when service providers neglect self-care, such as clinicians disrespecting clients, trivializing work, making more mistakes, lacking energy, using work as a tool to distract oneself from aversive private events, and losing interest in one's work. In addition, burnout has also been demonstrated to be associated with decreased quality of care for mental health care providers (Salyers et al., 2015) and increased ethical violations for counsellors (Everall & Paulson, 2004).

The direct impact of self-awareness on quality of work has been explored in some health care fields. For example, self-awareness has been described as an integral tool for nurses to develop an effective therapeutic relationship with their patients (Rasheed et al., 2019). Just as within the behavior analytic fields, the therapeutic relationship has been demonstrated to be critical for nurses, impacting the patient's physical and mental health and improving the quality of nursing care (Feo et al., 2017; Strandås & Bondas, 2018). Further, in the counselling field, self-awareness is strongly encouraged to gain a better understanding of how personal biases may impact quality of service, especially when working with diverse populations (Collins & Pieterse, 2007; Pieterse et al., 2013).

Self-regulation has been linked with increased overall well-being, along with academic achievement and socioeconomic success (Murray et al., 2015). Particularly for service providers, self-regulation can impact the quality of their work in a variety of ways. For example, Apgar and Cadmus (2022) described the importance of self-regulation practices for social workers to manage personal feelings that may arise when delivering an intervention. Given that self-regulation includes cognitive items such as self-monitoring, goal setting, taking perspective, executive functioning, problem-solving, and decision making, it is easy to consider how these factors would impact the work of a behavior analyst, such as quality of supervision, along with designing and delivering interventions. Further, poor self-regulation is associated with engagement in interfering and risk-taking behaviors, attention difficulties, and mental health concerns (Apgar & Cadmus, 2022).

Although limited research has explored how intrapersonal skills influence behavior analysts specifically, it is evident that these skills may directly impact many items included within the Behavior Analyst Certification Board Ethics Code for Behavior Analysts (Behavior Analyst Certification Board, 2020). For example, the first section, "Responsibility as a

Professional," can be impacted by intrapersonal skills, especially in terms of cultural responsiveness and diversity, setting appropriate boundaries, and being aware of one's personal biases (Wright, 2019). Similarly, intrapersonal skills impact the second section ("Responsibility in Practice") given the role of these skills impacting quality of practice, and the fourth section ("Responsibility to Supervisees and Trainees") impacting the quality of supervision.

### How to Improve Intrapersonal Skills?

Now, with a better understanding of what intrapersonal skills are and the importance of these skills as future behavior analysts, it is important to consider how to target these skills. The concepts of intrapersonal skills, including self-awareness, self-regulation, and self-care, are closely related to the six core processes of acceptance and commitment therapy/training (ACT). As a brief description of ACT, it is a modern behavioral intervention that targets psychological flexibility, which is defined as one consciously connecting with the present moment and persisting in behavior to accomplish value-directed goals (Bond et al., 2006; Hayes et al., 1999; Hayes et al., 2006; Zhang et al., 2018). ACT outlines six core processes that work in an interconnected manner to target psychological flexibility, including acceptance, defusion, present-moment awareness, self-as-context, values, and committed action (see Tarbox et al., 2020 for behavioral definitions of each of the processes). Each core process is designed to ameliorate a maladaptive psychological process that is thought to influence the emergence of psychological inflexibility, visually depicted using the hexaflex (see Figure 1.1).

Psychological flexibility has been demonstrated to be a protective factor for burnout and other psychological symptoms for a variety of populations, including for those working in the health care field (e.g., Frögéli et al., 2016). In addition, although there is limited research on ACT for behavior analysts, recent research suggests that ACT is an effective intervention to target burnout for individuals working in the health care field. For example, Towey-Swift et al. (2022) conducted a systematic review and narrative synthesis of controlled trials examining ACT to target burnout in health care staff and found that, following the ACT intervention, nine of the 14 studies included in the review reported decreased burnout scores. Further, Pingo et al. (2020) evaluated an ACT-based intervention for individuals working in the developmental service sector. The authors found that the intervention led to improvements in work performance, and participants self-reported levels of workplace stress, job satisfaction, and psychological flexibility remained rather stable.

*The Importance of Intrapersonal Skills for Behavior Analysts* 9

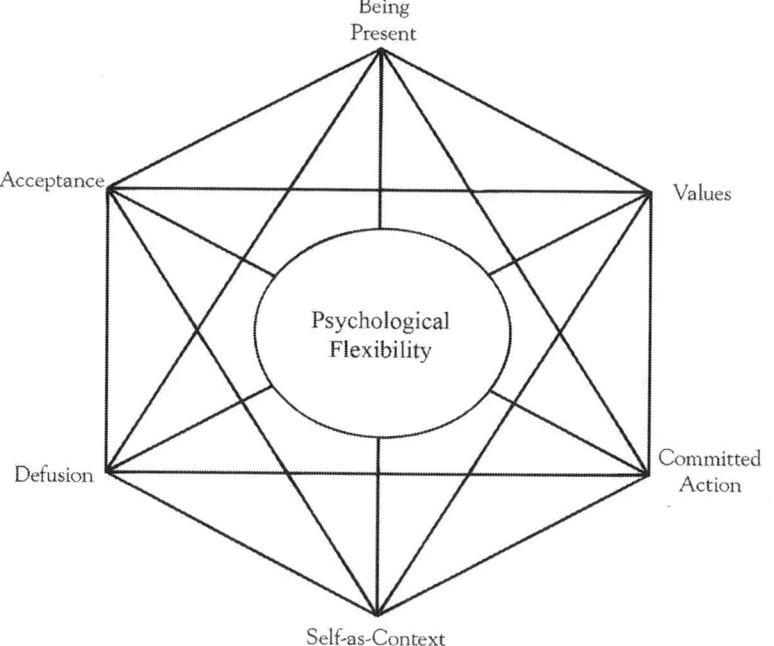

*Figure 1.1* The ACT Hexaflex.

In addition to ACT interventions, mindfulness interventions have also been demonstrated to have many health benefits, including reducing stress and preventing burnout. Similar to the ACT core process of present moment awareness, mindfulness involves attending to stimuli in the present moment, rather than attending to private events regarding the past or future (Little et al., 2020). Despite the limited research exploring the use of mindfulness-based interventions for behavior analysts, it has been demonstrated to prevent or be associated with decreased chances of burnout for psychologists (Benedetto & Swadling, 2014), educators (Abenavoli et al., 2013), and health care professionals (Asuero et al., 2014). Further, some fields such as psychotherapy encourage the incorporation of mindfulness practices within their clinical training to promote well-being and prevent burnout (e.g., Christopher & Maris, 2010).

Focusing on the self-awareness facet of intrapersonal skills, another strategy that may be helpful to include is reflective practices. Although reflective practices are not always taught, they are a helpful tool to consider incorporating. Taylor et al. (2019) describes different reflective practices that can be incorporated into your work as a service provider to improve specific skills, such as compassion. Moreover, although not behavior analytic specific, Koshy and colleagues (2017) outline how health care workers can effectively incorporate reflective practices to get the most out of the experience.

**Interview with a Behavior Analyst**

To provide a demonstration of intrapersonal skills applied in a professional and personal setting by an individual working in the field, we interviewed Linda Moroz, a Board Certified Behavior Analyst who has experience practicing ACT personally and professionally.

*CM: Tell Me a Bit about Your ACT Journey*

LM: In 2019, ACT was just a term that I heard tossed around, more so in the mental health community, but I really didn't have a good understanding of the intervention. I was interested in learning more about it, so when it was mentioned in meetings or if it came up with clients, I would have a better understanding. I found out that there was an ACT boot camp [for behavior analysts] being run in my area and received permission from my organization to attend. Originally, I was under the impression that it was a solely psychotherapeutic intervention but, throughout the training, I learned that ACT is used in more settings than just psychotherapeutic settings and it has a role in the applied behavior analytic stream as well, which I was thrilled about. At the training, I connected with others who were hoping to conduct ACT workshops for families of children with neurodevelopmental disabilities. During the first ACT workshop, I took on a smaller facilitation role and was profoundly touched with how impactful ACT was for the workshop participants. I wanted to learn more. Since that first ACT workshop, we have run a second and received a grant to run two more workshops through Kids Brain Health Network and the Brain Health Foundation. We have gotten such great feedback from families that it has been really helpful in encouraging them to have more self-compassion, think about their values and being present. I have also used it individually with families, such as using the Matrix activity with kids that I work with. The activity helps them to figure out what thoughts, feelings, and emotions they get stuck on, how they can continue to move forward, and what is important to them. As well, it has helped me to be more mindful and present when meeting with families, and they are sharing their story. It has helped me examine and reflect on my own biases and judgements, especially if I am asking them to do too much.

ACT has enhanced my ability to be more empathetic and compassionate with others.

*CM: How Do You Incorporate ACT into Your Personal Life, for Your Own Well-being?*

LM: Oh, in so many ways, for example becoming more mindful of being present. A lot of times, when I was with family and friends, I was thinking about the next thing that that I had to do or something from the past, rather than focussing on being with them in the moment. ACT has also helped with my own negative thoughts, feelings, and emotions. When things come up for me, I'm able to use ACT concepts to understand that these thoughts, feelings, and emotions are transient and to make a choice about the direction or action that I want to take in that moment. I can then pause to reflect on whether I am moving toward or away from my values and, even if it is moving away, consciously make that decision. It has helped develop more self-compassion and has helped me to reflect on being more supportive and accepting with my family and friends.

*CM: What Would You Recommend to People Pursuing Certification or New Behavior Analysts, to Improve or Promote Strong Interpersonal Skills?*

LM: I would recommend everybody take an ACT boot camp. ACT supports the development of those intra- and interpersonal skills that are so important in our field. When working with others and they feel heard and supported, it opens the door to helping them effectively move in a direction that results in meaningful change. ACT has had such a profound impact on my personal and professional life. I have found it extremely helpful to remind myself that everyone struggles with thoughts, emotions, and feelings, maybe in different ways. We all suffer, we all feel joy and the range of emotions. It has encouraged me to be more authentic and vulnerable when interacting with others. I would really encourage others to learn more about ACT because it can offer tools that can be invaluable throughout one's career. There are many things that I have learned and experienced throughout my career. Few have had as great an impact as ACT.

## Practicing ACT as a Behavior Analyst

Many different fields are qualified to deliver ACT, although the delivery format may differ. For example, social workers and psychologists are qualified to deliver ACT to treat psychological diagnoses in a psychotherapy format. However, behavior analysts (who have received ACT training) are qualified to provide ACT from a training perspective, such as providing caregivers of individuals with neurodevelopmental disabilities with ACT tools that they can implement into their life (as Linda described). Just as with any behavioral intervention, behavior analysts require comprehensive training in order to competently deliver ACT. As explained by Tarbox et al. (2020), ACT is within the scope of practice for behavior analysts but, without appropriate training, it is not within one's scope of competence. Therefore, if interested in incorporating ACT into one's behavior analytic practice, it is important to receive appropriate training and supervision. Further, as Linda mentioned, training in ACT may be helpful in many ways, including improving one's personal life or increasing compassion, which is an important skill as a behavior analyst and can enhance therapeutic relationships (Taylor et al., 2019).

## Conclusion

Although intrapersonal skills are not frequently discussed in the behavior analytic literature, they are essential skills for a behavior analyst to acquire given their impact on one's career and personal life. Other fields, such as nursing, social work, and clinical psychology, have demonstrated the importance of intrapersonal skills, especially their relation to the quality of care, for clients and supervisees, and impact on burnout. Further, intrapersonal skills can indirectly impact many items included within the Behavior Analyst Certification Board Ethics Code for Behavior Analysts (Behavior Analyst Certification Board, 2020). Therefore, we encourage behavior analysts to consider the role of intrapersonal skills, specific intrapersonal skills you hope to improve on, and how this will impact your work as a practicing clinician, researcher, and/or supervisor.

## References

Abenavoli, R. M., Jennings, P. A., Greenberg, M. T., Harris, A. R., & Katz, D. A. (2013). The protective effects of mindfulness against burnout among educators. *Psychology of Education Review, 37*(2), 57–69.

Apgar, D., & Cadmus, T. (2022). Using mixed methods to assess the coping and self-regulation skills of undergraduate social work students impacted by COVID-19. *Clinical Social Work Journal, 50*(1), 55–66. http://dx.doi.org/10.1007/s10615-021-00790-3.

Asuero, A. M., Queraltó, J. M., Pujol-Ribera, E., Berenguera, A., Rodriguez-Blanco, T., & Epstein, R. M. (2014). Effectiveness of a mindfulness education program in primary health care professionals: A pragmatic controlled trial. *Journal of Continuing Education in the Health Professions, 34*(1), 4–12. http://dx.doi.org/10.1002/chp.21211.

Behavior Analyst Certification Board. (2020). *Ethics code for behavior analysts.* https://bacb.com/wp-content/ethics-code-for-behavior-analysts/.

Blair, C. & Ursache, A. (2011). A bidirectional model of executive functions and self-regulation. In K. D. Vohs & R. F. Baumeister, *Handbook of Self-regulation: Research, Theory, and Applications* (p. 300–320). Guilford Press.

Bond, F. W., Hayes, S. C., & Barnes-Holmes, D. (2006). Psychological flexibility, ACT, and organizational behavior. *Journal of Organizational Behavior Management, 26*(1–2), 25–54. https://doi.org/10.1300/J075v26n01_02.

Borba, A., Tourinho, E. Z. & Glenn, S. S. (2017). Effects of cultural consequences on the interlocking behavioral contingencies of ethical self-control. *Psychological Record, 67*, 399–411. https://doi.org/10.1007/s40732-017-0231-6

Carden, J., Jones, R. J., & Passmore, J. (2022). Defining self-awareness in the context of adult development: A systematic literature review. *Journal of Management Education, 46*(1), 140–177. https://doi.org/10.1177/1052562921990065.

Christopher, J. C., & Maris, J. A. (2010). Integrating mindfulness as self-care into counselling and psychotherapy training. *Counselling and Psychotherapy Research, 10*(2), 114–125. http://dx.doi.org/10.1080/14733141003750285.

Collins, N. M., & Pieterse, A. L. (2007). Critical incident analysis based training: An approach for developing active racial/cultural awareness. *Journal of Counseling & Development, 85*(1), 14–23. http://dx.doi.org/10.1002/j.1556-6678.2007.tb00439.x.

Di Benedetto, M., & Swadling, M. (2014). Burnout in Australian psychologists: Correlations with work-setting, mindfulness and self-care behaviors. *Psychology, Health & Medicine, 19*(6), 705–715. http://dx.doi.org/10.1080/13548506.2013.861602.

Eurich, T. (2018). What self-awareness really is (and how to cultivate it). *Harvard Business Review*, 1–9.

Everall, R. D., & Paulson, B. L. (2004). Burnout and secondary traumatic stress: Impact on ethical behavior. *Canadian Journal of Counselling, 38*(1), 25–35.

Feo, R., Rasmussen, P., Wiechula, R., Conroy, T., & Kitson, A. (2017). Developing effective and caring nurse-patient relationships. *Nursing Standard, 31*(28), 54. http://dx.doi.org/10.7748/ns.2017.e10735.

Frögéli, E., Djordjevic, A., Rudman, A., Livheim, F., Gustavsson, P., & Frögögéli, E. (2016). A randomized controlled pilot trial of acceptance and commitment training (ACT) for preventing stress-related ill health among future nurses. *Anxiety, Stress & Coping, 29*(2), 202–218. https://doi.org/10.1080/10615806.2015.1025765.

Glenn, S. S. (2004). Individual behavior, culture, and social change. *The Behavior Analyst, 27*(2), 133–151. https://doi.org/10.1007/BF03393175.

Hayes, S. C., Luoma, J. B., Bond, F. W., Masuda, A., & Lillis, J. (2006). Acceptance and commitment therapy: Model, processes and outcomes. *Behavior Research and Therapy, 44*, 1–25. https://doi.org/10.1016/j.brat.2005.06.006.

Hayes, S. C., Strosahl, K. D., & Wilson, K. G. (1999). *Acceptance and commitment therapy: An experiential approach to behavior change*. Guilford.

Kalis, T. M., Vannest, K. J., & Parker, R. (2007). Praise counts: Using self-monitoring to increase effective teaching practices. *Preventing School Failure:*

*Alternative Education for Children and Youth*, 51(3), 20–27. https://doi.org/10.3200/PSFL.51.3.20-27.

Koshy, K., Limb, C., Gundogan, B., Whitehurst, K., & Jafree, D. J. (2017). Reflective practice in health care and how to reflect effectively. *International Journal of Surgery. Oncology*, 2(6), e20. http://dx.doi.org/10.1097/IJ9.0000000000000020.

Little, A., Tarbox, J., & Alzaabi, K. (2020). Using acceptance and commitment training to enhance the effectiveness of behavioral skills training. *Journal of Contextual Behavioral Science*, 16, 9–16. https://doi.org/10.1016/j.jcbs.2020.02.002.

Murray, D., Rosanbalm, K., Christopoulos, C., & Hamoudi, A. (2015). Self-regulation and toxic stress: Foundations for understanding self-regulation from an applied developmental perspective. OPRE Report #2015-21, Washington, DC: Office of Planning, Research and Evaluation, Administration for Children and Families, U.S. Department of Health and Human Services.

Myers, S. B., Sweeney, A. C., Popick, V., Wesley, K., Bordfeld, A., & Fingerhut, R. (2012). Self-care practices and perceived stress levels among psychology graduate students. *Training and Education in Professional Psychology*, 6(1), 55. http://dx.doi.org/10.1037/a0026534.

Pieterse, A. L., Lee, M., Ritmeester, A., & Collins, N. M. (2013). Towards a model of self-awareness development for counselling and psychotherapy training. *Counselling Psychology Quarterly*, 26(2), 190–207. http://dx.doi.org/10.1080/09515070.2013.793451.

Pingo, J. C., Dixon, M. R., & Paliliunas, D. (2020). Intervention enhancing effects of acceptance and commitment training on performance feedback for direct support professional work performance, stress, and job satisfaction. *Behavior Analysis in Practice*, 13(1), 1–10. http://dx.doi.org/10.1007/s40617-019-00333-w.

Pope, K. S., & Vasquez, M. J. (2016). *Ethics in psychotherapy and counseling: A practical guide*. John Wiley & Sons.

Rasheed, S. P., Younas, A., & Sundus, A. (2019). Self-awareness in nursing: A scoping review. *Journal of Clinical Nursing*, 28(5–6), 762–774. http://dx.doi.org/10.1111/jocn.14708.

Richards, K., Campenni, C., & Muse-Burke, J. (2010). Self-care and well-being in mental health professionals: The mediating effects of self-awareness and mindfulness. *Journal of Mental Health Counseling*, 32(3), 247–264. https://doi.org/10.17744/mehc.32.3.0n31v88304423806.

Riegel, B., Dunbar, S. B., Fitzsimons, D., Freedland, K. E., Lee, C. S., Middleton, S., Stromberg, A., Vellone, E., Webber, D. E., & Jaarsma, T. (2021). Self-care research: Where are we now? Where are we going?. *International Journal of Nursing Studies*, 116, 103402. https://doi.org/10.1016/j.ijnurstu.2019.103402.

Salyers, M. P., Fukui, S., Rollins, A. L., Firmin, R., Gearhart, T., Noll, J. P., Williams, S., & Davis, C. J. (2015). Burnout and self-reported quality of care in community mental health. *Administration and Policy in Mental Health and Mental Health Services Research*, 42(1), 61–69. http://dx.doi.org/10.1007/s10488-014-0544-6.

Shaffer, R. C., Wink, L. K., Ruberg, J., Pittenger, A., Adams, R., Sorter, M., Manning, P., & Erickson, C. A. (2019). Emotion regulation intensive outpatient programming: Development, feasibility, and acceptability. *Journal of Autism and Developmental Disorders*, 49(2), 495–508. https://doi.org/10.1007/s10803-018-3727-2.

Skinner, B. F. (1953). *Science and human behavior*. Macmillan.

Strandås, M., & Bondas, T. (2018). The nurse–patient relationship as a story of health enhancement in community care: A meta-ethnography. *Journal of Advanced Nursing*, 74(1), 11–22. http://dx.doi.org/10.1111/jan.13389.

Tarbox, J., Szabo, T. G., & Aclan, M. (2020). Acceptance and commitment training within the scope of practice of applied behavior analysis. *Behavior Analysis in Practice*, 1–22. http://dx.doi.org/10.1007/s40617-020-00466-3.

Taylor, B. A., LeBlanc, L. A., & Nosik, M. R. (2019). Compassionate care in behavior analytic treatment: Can outcomes be enhanced by attending to relationships with caregivers? *Behavior Analysis in Practice*, 12(3), 654–666. http://dx.doi.org/10.1007/s40617-018-00289-3.

Towey-Swift, K. D., Lauvrud, C., & Whittington, R. (2022). Acceptance and commitment therapy (ACT) for professional staff burnout: A systematic review and narrative synthesis of controlled trials. *Journal of Mental Health*, 1–13. http://dx.doi.org/10.1080/09638237.2021.2022628.

Wright P. I. (2019). Cultural humility in the practice of Applied Behavior Analysis. *Behavior Analysis in Practice*, 12(4), 805–809. https://doi.org/10.1007/s40617-019-00343-8.

Zhang, C., Leeming, E., Smith, P., Chung, P., Hagger, M., & Hayes, S. (2018). Acceptance and commitment therapy for health behavior change: A contextually-driven approach. *Frontiers in Psychology*, 8, 2350–2355. https://doi.org/10.3389/fpsyg.2017.02350/ful.

# 1.2 The Discerning ABA Practitioner

*Emily Burns*

Introduction

ABA practitioners working in clinics, therapy centers, schools, assisted living facilities, private homes, and other care settings are essential members of multidisciplinary teams who must collaborate well with clients, families, educators, and community partners to promote clients' skill development and well-being. The population served by Applied Behavior Analysis (ABA) practitioners is diverse, including clients from all different races and cultures, and with varying diagnoses and abilities. This diversity can be overwhelming to ABA practitioners as they must practice only within their area of competence (1.05; BACB, 2020). Working as an ABA practitioner requires one to enjoy constant learning and networking with other professionals, both within the field of ABA and beyond to ensure one is practicing responsibly. The Ethics Code for Behavior Analysts (1.06 and 1.07; BACB, 2020) encourages behavior analysts to actively engage in professional development to acquire skills related to cultural responsiveness and diversity as well as to evaluate their own biases and ability to address the needs of those that they serve. As an ABA practitioner serving a diverse population, it is imperative to be aware of one's personal biases and to take appropriate steps, where necessary, to resolve any interference these biases may have on one's work.

Bias and Clinical Judgment

Bias in general is not necessarily problematic, as you learned in the previous chapter. However, there are biases that we may be more prone to as people working in the field of ABA. Some helping professions rely more heavily on clinical impression, or the professional's accumulation of knowledge and skills over time and the analysis of subjective data to understand a client's presenting problem. As ABA practitioners, we are likely biased to use reliably collected data in making intervention decisions, or

DOI: 10.4324/9781003300465-4

perhaps to defer intervention planning until a more reliable trend in the data is observed (Moran & Tai, 2001). However, when ABA practitioners rely on clinical impression alone, bias can become problematic and is less likely to benefit the client. Relying on clinical impression alone may be considered unethical, as it will likely lead an ABA practitioner to make erroneous assessment and intervention decisions, and to practice inefficiently. The ABA practitioner's impression of the client they are working with involves verbal processes shaped by the presenting problem behavior and by the practitioner's diverse reinforcement history, which may skew the accuracy of their judgment. Even experienced ABA practitioners are vulnerable to faulty clinical judgment if relying on their clinical impression only. Discerning ABA practitioners strive to rout out potential biases from their practice and become aware of them. They are also well positioned when working within multidisciplinary teams to reduce the impact of prevalent racial, cultural, and ableist biases. ABA practitioners are keen observers and interpreters of behavior. Their position and expertize allow them the opportunity to have a significant positive impact on the direction teams take in support of the clients they work with. It is crucial for ABA practitioners to engage in and model frequent self-reflection. Compassionate ABA practitioners can do uncompassionate things if they are not astutely aware of their own behavior's impact on others. As professionals and as caring adults, ABA practitioners are obligated to be aware of their personal biases and to act in the best interest of the clients they serve.

## Types of Biases

ABA practitioners are at risk, similar to other helping professionals, of focusing too much on rehabilitation as it relates to the dichotomy of "normal" vs "abnormal" behavior (Cox, Villegas, & Barlow, 2018; Moran & Tai, 2001; Shemberg & Doherty, 1999; Shyman, 2016). Cox, Villegas, and Barlow (2018) define "normal" as "the specification of conduct by which individuals are then rewarded for conforming, or punished for nonconforming, to the specified conduct" (p. 283). ABA practitioners must consider if the focus on conformity to specified standards is necessary to improve the client's quality of life, or whether an effort to change society's accommodation of diverse abilities would have a greater impact on the client's quality of life. This question is likely best answered on a behavior-by-behavior basis. It seems difficult to justify changing a client's unique patterns of behavior if those behaviors do not harm anyone or limit the client's ability to function within their environments. The BACB ethical code requires that ABA practitioners defer to clients and or their caregivers to make decisions about target behaviors for change based on their personal values and goals (2.09; BACB, 2020; Shyman, 2016). ABA practitioners should strive to create acceptance and opportunities for all individuals with diverse abilities to ensure that each client they work with is provided the means to achieve quality of life as a respected member of society.

There are several potential biases that ABA practitioners should be aware of and work to overcome. These include but are not limited to pathology bias, confirmatory bias, hind-sight bias, over-confidence in clinical judgment, racial or cultural bias, and ableism. Pathology bias is especially relevant in settings where clients are influenced explicitly and implicitly to demonstrate pathology, such as psychiatric and managed care settings (Moran & Tai, 2001; Shemberg & Doherty, 1999). ABA practitioners are experts at environmental audits and should conduct these regularly in conjunction with functional assessments of problem behavior. Be sure to search beyond other's, and your own, initial hypotheses and consider relevant alternatives. Working within multidisciplinary teams who offer a variety of viewpoints, as well as regular review of relevant literature and intervention data, should help to reduce the likelihood of pathology bias negatively impacting client outcomes.

Confirmatory bias occurs when one seeks or remembers only information that substantiates their initial hypothesis, while not attending to data that refutes that hypothesis or to information that could offer alternative hypotheses (Moran & Tai, 2001). ABA practitioners may be less likely to attend to and more likely to forget observations that are not consistent with their original hypothesis. This selective-remembering problem may be improved with careful data tracking and review. While some members of the multidisciplinary team may feel your client's problem behaviors are stemming from something internally exceptional to them, you have a unique lens and the ability to steer the conversation to aspects of the environment and observational cues that may be impacting the client and their behavior. Sometimes there is something in the environment that leads to problem behavior, like coffee breath, or the buzzing of the florescent lights in conjunction with lagging skills, and not some ominous unknown pathology. Remember that diagnoses are verbal behavior in that they are based on language and labels, and therefore reinforced through the mediation of other people who are behaving in a way that has been shaped and maintained by a verbal environment (Skinner, 1986). As a member of a multidisciplinary team, you are part of that verbal environment. Share your ABA knowledge and expertize to create innovative ideas and advocate for your clients.

Pathology bias for ABA practitioners may look like an over-focus on behaviors of concern and giving little attention to the skills and positive behaviors a client demonstrates (Moran & Tai, 2001; Shemberg &

Doherty, 1999). When working with our clients and their support systems, it is important to consider not only the problems and why they might be happening, but also the skills and resources available to the client that could be better understood and utilized to benefit that client. As ABA practitioners, we typically receive referrals because things are not going well, someone's safety is at risk, and a solution to a problem is needed quickly. Others may already have hypotheses for why a challenging behavior is happening and for potential solutions before we even become involved. When faced with these types of situations, take a step back from what you are seeing and hearing about a client's problem behavior and look at them and their situation holistically. Put your ABA detective skills to work and discover the things that are going well. How might the skills and environmental factors that are allowing for that success be harnessed to create even better outcomes for your client and used to overcome the challenging behaviors?

Deterministic reasoning happens when an ABA practitioner or other member of the multidisciplinary team attempts to understand why a client demonstrates a particular behavior, and their hypothesis then overestimates the influence of specific stimuli on the client's responses and ignores the likelihood that the client's behavior is multiply determined and influenced by many environmental factors (Moran & Tai, 2001). In a school setting, this could look like a teaching team determining that a student always leaves for extended periods of time during second period to escape the challenging math lessons. Their solution then is to move the student to a first period math class, a time when the student is typically more engaged and may be more willing to work through the challenging math tasks. They move the math class and the student's performance in this subject improves; however, the student continues to leave frequently during second period. Upon further investigation, the teaching team learns that the cafeteria staff put out freshly baked chocolate chip cookies during second period, just ahead of lunch and the student had been escaping not only to avoid the fatigue associated with the challenging math lessons, but also to access the cookies before the cafeteria becomes too busy for the staff to cope with. People do not typically escape to nothing. Most behavior is multiply reinforced.

Often a four-term function of behavior assessment is too simplistic to fully appreciate why a behavior of concern is happening. Dr. Greg Hanley's approach, known as practical functional assessment, offers a solution to this problem. Practical functional assessment includes an open-ended interview with those who know the client best and an interview-informed synthesized contingency analysis (IISAC) (Rajaraman et al., 2022). Based on information gained from the interview, the practitioner strives to create a context in which the client is happy, relaxed, and engaged by meeting all their needs before gradually introducing the test condition to turn the behavior of concern on and off. This process maintains the client's safety and dignity while they build rapport with the practitioner and allows the practitioner to gain a more fully informed understanding of why the behavior of concern is happening. Practical functional assessment is defining an exciting new direction in the field of ABA. The literature speaks to its positive impact on client outcomes and treatment satisfaction (Jessel, 2022; Rajaraman et al. 2022). As a bonus, it likely also helps to reduce deterministic reasoning. As an ABA practitioner using this approach, you must remain curious when considering why behavior is happening and be creative with the potential solutions you offer, as multiple unmet needs will likely need to be addressed for it to be successful.

Hindsight bias occurs when an ABA practitioner learns of an outcome and then acts as if they predicted the likelihood of that outcome happening (Moran & Tai, 2001). Be mindful of hindsight bias, as research indicates that practitioners who construct limited causal explanations for behavior tend to be overly deterministic or externally caused (Moran & Tai, 2001). Had the teaching team in the previous example assumed, given the evidence of the student's math performance improving, that their initial hypothesis was complete and correct, they would have been less likely to discover the alternative reasons the student was leaving class and failed to create an effective intervention plan for his eloping behavior. Because they remained curious, they were able to address the eloping behavior by not only moving the math class to first period, but also by asking the cafeteria personnel to set aside a chocolate chip cookie for the student to pick up as he walks to his second period class if he remained in that class for the duration of the lesson the previous day. Now the eloping behavior is even more likely to decrease, as the student's needs are being met and the target skill, remaining in class for the duration of the lesson, is being effectively reinforced by capitalizing on the student's inherent motivation.

Like all helping professionals, ABA practitioners, especially those with extended experience in the field, run the risk of becoming overconfident in their clinical judgment and may be more prone to the biases listed above. It is important to remember that the behavior of diagnosing a client—or determining the function of their behavior without the aid of reliable, objective data—is intermittently reinforced when confirmed later by reliable data. Typically, when a practitioner's judgment is found to be correct, it is more easily remembered and is often socially reinforced by others. As a result, the practitioner may be tempted to use unaided clinical judgment more often and thus be more prone to the biases. This is dangerous. One might assume that the accuracy of their judgment improves with experience; however, the extent of professional training and experience and the predictive accuracy of clinical judgment are unrelated (Moran & Tai,

2001). Thus, the reliance on objective data collection and analyses is crucial. Informed clinical judgment allows for effective intervention planning and reduces the risk of harm to clients.

**Cultural Diversity**

ABA practitioners must understand and consider the cultural diversity of the population they serve to be safe and effective in their work (1.01; BACB, 2020). It is important to be thoughtfully diligent and humble while learning about cultural diversity (Arango & Lustig, 2022). Cultural bias is present when ABA practitioners interpret client behavior and design interventions based on standards and practices inherent in their own cultures without consideration of, or based on, inaccurate assumptions about the client's culture (Fong et al., 2016). This can lead to an incorrect assessment of the behavior of concern's function, ineffective treatment, and problematic and potentially harmful outcomes for clients. ABA practitioners are obligated to respect the dignity of the clients and families they work with (1.07 & 1.08; BACB, 2020). Cultural considerations matter ethically in the context of behavior analysis because the values and beliefs of clients and their caregivers are expressions of their dignity and right to self-determination. Cultural awareness is understanding one's own cultural values, preferences, and characteristics as well as those from other cultures. It is the ABA practitioner's responsibility to be aware of their cultural beliefs and how they compare and contrast with their clients in order to fulfill all commitments to them (1.06 & 1.07; BACB, 2020).

**Ableism**

Most people associate ABA with teaching and clinical treatment that helps people with neurodevelopmental disorders such as autism, intellectual and developmental disabilities, and attention deficit hyperactivity disorder (ADHD) (ONTABA, 2020). More recently ABA has been used effectively with other populations, including people living with mental health concerns, acquired brain injuries, movement disorders, addictions challenges, feeding disorders, and dementia. Given that ABA practitioners work frequently with those living with diverse cognitive realities, knowing about biases in this area is essential to our work. Ableism is

> the devaluation of disability that results in societal attitudes that uncritically assert that it is better for a child to walk than roll, speak than sign, read print than read Braille, spell independently than use a spell-check, and hang out with nondisabled kids as opposed to other disabled kids.
> (Hehir, 2002, p. 1)

The pervasiveness of ableist assumptions reinforces prejudices against people with disabilities and is detrimental to their educational attainment, employment outcomes, and overall well-being. It is challenging to live in a world that was not designed with differing abilities in mind. Education and clinical service providers make things worse by focusing inordinately on the characteristics of the person's disability. Changing the disability should not be the goal of service providers. Instead, their focus must be on encouraging their clients to develop and use skills and modes of expression that are most efficient for them. Language is a powerful means of shaping how people think about disabilities (Bottema-Beutel et al., 2021). Medicalized, deficit-focused language like "at risk for autism" should be replaced by neutral terms like "increased likelihood of autism." Our verbal behavior and that of those around us has a significant impact on how we see each other and interact in the world. ABA practitioners must collaborate openly with their clients when designing intervention plans to ensure that the strategies and outcomes of these plans align with the client's future vision for themselves (2.09; BACB, 2020). When working with clients who are nonverbal, be a keen observer throughout intervention planning and application. Allow your clients to vote with their feet (Hanley, 2022). If they are demonstrating verbally or nonverbally that they do not want to participate in the intervention activities you have planned, get imaginative and work with your client and those who know them best to produce strategies that not only engage them but also promote their self-determined outcomes. As compassionate ABA practitioners, we must respect our clients' "voices" and honor their right to self-determination. It is also our responsibility to ensure that others in our client's support network appreciate the client's right to self-determination and focus on the goals that are most meaningful to them. Autonomy is at the heart of ABA practitioner's obligations to their clients.

**Strategies for Reducing Personal Biases**

Even the most brilliant and compassionate ABA practitioners run the risk of maintaining personal biases that negatively impact client outcomes if the practitioner remains unaware of or underestimates their errors in clinical judgment (Moran & Tai, 2021). Self-reflection and collaboration with a trusted supervisor will help one to become more aware of personal biases and their impact. ABA practitioners should seek and remain open to frequent and systematic feedback from a supervisor, clients, and community partners. Regular review of data on intervention integrity, outcomes, and social validity will also help to reduce the impact of personal bias on the work of ABA practitioners and improve their rapport with those they serve.

Cultural competence is essential to the work of ABA practitioners. Beaulieux and Jimeinez-Gomez (2022) assert that cultural competence consists of three dimensions: awareness, knowledge, and skills. These key dimensions provide a framework for ABA practitioners to engage with as they do the work of reducing personal biases. Awareness requires self-reflection and self-assessment. ABA practitioners need to be able to differentiate and identify cultural variables that impact their behavior and respect and appreciate differences. They must also work to recognize their biased behavior and the impact it may have on the services they provide. Self-reflection and self-assessment will allow the ABA practitioner to assess the limits and scope of their professional competence and better understand areas for learning and personal growth.

Knowledge involves identifying and understanding the past and current treatment of minoritized people (Beaulieux and Jimeinez-Gomez, 2022). ABA practitioners should aspire to learn more about the cultural groups with whom they work. This will help practitioners to identify and appreciate why certain intervention strategies are not good options for particular clients. ABA practitioners require knowledge of the cultural groups they serve to reduce barriers that may prevent people from accessing and fully benefiting from the services they provide.

Interpersonal skills in conjunction with ABA expertize is necessary for delivering ABA services in a culturally responsive way. Much of the clinical training ABA practitioners receive focuses on the analytic technologies of our science; however, our ability to share our science requires that we communicate and collaborate well with non-ABA practitioners. The effectiveness of our services is largely determined by our ability to navigate interpersonal situations compassionately and with humility. Rohrer et al. (2021) have created a checklist they call the Compassionate Collaboration Tool to be used for self-evaluation, procedural fidelity evaluation, and as a framework for comprehensive training to ensure ABA practitioners demonstrate effective communication and collaboration with the clients and families they serve. Self-monitoring relationships with clients to prevent and disrupt the impact of personal biases is essential (Beaulieux and Jimeinez-Gomez, 2022) and tools like Rohrer et al. (2021) Compassion Collaboration Tool can help.

ABA practitioners must also practice self-compassion. Building cultural awareness, knowledge, and skills requires one to be open and vulnerable to mistakes. It is a practice that one must commit to and engage in persistently, even when it is difficult. Cultural awareness will help the ABA practitioner to respond to cultural cues, communicate effectively, and build rapport with clients and community partners. ABA practitioners should ask open-ended questions and listen actively to clients and team members. They should seek and incorporate feedback. Interventions that are culturally responsive are most effective and best received. ABA practitioners must navigate interpersonal situations with compassion and humility if we hope to effectively disseminate our science, grow our field, and collaborate well with others (Rohrer et al., 2021).

**Others Have Biases Too**

It can be difficult and uncomfortable to speak to others about your observations of their personal biases and cultural ignorance, and perhaps even more uncomfortable to receive this type of feedback from others However, sharing with others and staying open and curious is the first step to cultural awareness, identifying our biases, and improving our professional practice (Fong et al., 2016). As ABA practitioners, we are keen observers, and we think deeply about the functions of our own behavior and the behavior of those around us. When our data shows that a problem exists, we must act. Behavior skills training has been shown to be an effective technique for improving culturally responsive behavior analytic practices (Neely et al., 2019). Neely, Gann, Castro-Villarreal and Villarreal (2019) provided behavior skills training to five educators to improve their implementation of culturally responsive class-wide behavior management practices. The behavior skills training program involved an overview of evidence-based classroom behavior-management strategies, discussions of culturally responsive instructional and classroom-management practices, and an opportunity to practice these strategies in classrooms and receive individualized constructive feedback. Behavior-skills training provides a useful structure for supporting those with whom we work to learn more about, and better integrate, the skills associated with cultural responsiveness in ABA practices. Keeping in mind that culture encompasses race, socioeconomic class, age, religion, sexual orientation, ethnicity, disability, nationality, and geographic context, the learning and self-reflection required of ABA practitioners to behave in culturally responsive ways takes great care and commitment. Modeling self-reflection and self-compassion, and prompting our colleagues to do the same, will reduce the potential negative impact of personal biases on service delivery and promote better outcomes for clients.

**Conclusion**

Collaborating with diverse clients, families, and community teams is one of the many privileges of being an ABA practitioner. While collaborating well when the stakes are high is not always simple, ABA practitioners are well positioned to have a positive influence on the decision-making of the multidisciplinary teams with whom they work. Utilizing one's keen observation skills and ability to interpret and respond to others' behavior in culturally responsive ways will result in positive outcomes for clients. ABA practitioners who commit to a practice of continued self-reflection and openness to learning and feedback reduce the risk of their personal biases negatively impacting their work and relationships. Compassionate ABA practitioners honor their client's right to self-determination and put client values and autonomy at the center of their practice.

**References**

Arango, A., & Lustig, N. (2022). Ignorance and cultural diversity: The ethical obligations of the behavior analyst. *Behavior Analysis in Practice*, 1–17. https://doi.org/10.1007/s40617-022-00701-z

Beaulieu, L., & Jimenez-Gomez, C. (2022). Cultural responsiveness in applied behavior analysis: Self-assessment. *Journal of Applied Behavior Analysis*, 55(2), 337–356. https://doi.org/10.1002/jaba.907

Behavior Analyst Certification Board. (2020). *Ethics code for behavior analysts*. https://bacb.com/wp-content/ethics-code-for-behavior-analysts/

Bottema-Beutel, K., Kapp, S. K., Lester, J. N., Sasson, N. J., & Hand, B. N. (2021). Avoiding ableist language: Suggestions for autism researchers. *Autism in Adulthood*. https://doi.org/10.1089/aut.2020.001418

Cox, D. J., Villegas, A., & Barlow, M. A. (2018). Lost in translation: A reply to Shyman (2016). *Intellectual and Developmental Disabilities*, 56(4), 278–286. https://doi.org/10.1352/1934-9556-56.5.278

Fong, E. H., Catagnus, R. M., Brodhead, M. T., Quigley, S., & Field, S. (2016). Developing the cultural awareness skills of behavior analysts. *Behavior Analysis in Practice*, 9(1), 84–94. http://doi.org/10.1007/s40617-016-0111-6

Hanley, G. (2022). *Today's ABA: An effective and humane approach for addressing PB*. FTF Behavioral Consulting. https://ftfbc.com/courses/todays-aba-an-effective-and-humane-approach-for-addressing-problem-behavior/

Hehir, T. (2002). Eliminating ableism in education. *Harvard Educational Review*, 72(1), 1–33. https://doi.org/10.17763/haer.72.1.03866528702g2105

Jessel, J. (2022). Practical functional assessment. In J. B. Leaf, J. H. Chihon, J. L. Ferguson, & M. J. Weiss (Eds.). *Handbook of applied behavior analysis interventions for autism* (pp. 443–464). Springer. https://doi.org/10.1007/978-3-030-96478-8_23

Moran, D. J., & Tai, W. (2001). Reducing biases in clinical judgment with single subject treatment design. *The Behavior Analyst Today*, 2(3), 196. https://doi.org/10.1037/h0099930

Ontario Association for Behavior Analysis. (2020). *Professional practice in ABA series*. https://training.ontaba.org/wp-content/uploads/2020/10/4-Understanding-ABA_rev4_May31.pdf

Rajaraman, A., Hanley, G. P., Gover, H. C., Ruppel, K. W., & Landa, R. K. (2022). On the reliability and treatment utility of the practical functional assessment process. *Behavior Analysis in Practice*, 15, 815–837. https://doi.org/10.1007/s40617-021-00665-6

Rohrer, J. L., Marshall, K. B., Suzio, C., & Weiss, M. J. (2021). Soft skills: The case for compassionate approaches or how behavior analysis keeps finding its heart. *Behavior Analysis in Practice, 14*(4), 1135–1143. https://doi.org/10.1007/s40617-021-00563-x

Shemberg, K. M., & Doherty, M. E. (1999). Is diagnostic judgment influenced by a bias to see pathology?. *Journal of Clinical Psychology, 55*(4), 513–518. https://doi.org/10.1002/(SICI)1097-4679(199904)55:4<513::AID-JCLP13>3.0.CO;2-W

Shyman, E. (2016). The reinforcement of ableism: Normality, the medical model of disability, and humanism in applied behavior analysis and ASD. *Intellectual and Developmental Disabilities, 54*(5), 366–376. https://doi.org/10.1352/1934-9556-54.5.366

Skinner, B. F. (1986). The evolution of verbal behavior. *Journal of the Experimental Analysis of Behavior, 45*(1), 115. https://doi.org/10.1901/jeab.1986.45-115

# 1.3 Self-Care for Behavior Analysts
## A Guide to Eating Fewer Doughnuts and Taking More Naps

*Albert Malkin, Karl F. Gunnarsson, and Kristina Axenova*

Introduction

Picture this: a professional jerks their head and wakes up, they look at their screen and see a string of "RRRRRRRRRRRRRRRRRRRRRRRRR..." for many pages. They rub their eyes, look at the clock, it's 4:30 a.m. and they realize disappointedly, they've done it again! They've worked into the early morning hours; they crawl into bed, only to wake in a few hours to get their child ready for school and get themselves to work in the morning. This is not going to be an easy day. Worse yet, this is just like every day for the past three months. They tell themselves, "It'll get better when I get over this next hump." Realistically, the peaks and valleys of work can be more like little humps at the top of a mountain of heightened stress. If the above sounds familiar, the road to burnout, health, and mental health issues may be around the corner (or they have already arrived). The field has long understood that people supported by behavior analysts must have the autonomy to choose to take a load off, eat some doughnuts, and take a nap; unfortunately, professionals may not afford themselves the same personal freedoms (see Bannerman et al., 1990). Certainly, there is freedom to work long hours, too; in either case, it is important to understand the potential harms.

Why would anyone put themselves through such detrimental trials and tribulations? To earn more money? To learn? To help people? Unfortunately, it may not matter; if work takes precedence over other life domains (e.g., exercise, family, leisure, etc.), personal well-being is likely to suffer (Gisler et al., 2018). Consider the well-meaning aim of being a "great" behavior analyst. Being "great" might mean working many hours, being in harm's way in challenging situations, or taking on immense responsibility. It might also *not* mean eating healthy foods, sleeping enough, starting a hobby, and definitely not spending quality time with loved ones. Rigid adherence to the rule of "being great" may have the short-term benefit of desirable work outcomes, however, long-term

DOI: 10.4324/9781003300465-5

performance can suffer. Sleep-deprived, malnourished, and highly stressed people make poor decisions with poor outcomes; people who over-value short-term rewards are more likely to have poor health outcomes (Bickel et al., 2012). Being "great" may backfire; it might be harmful personally, to loved ones, and people served. To steer clear of significant maladies, and instead, thrive in the inherently difficult task of improving socially important behavior, behavior analysts must practice caring for themselves.

**Well-Being and Burnout—Personal Risk Factors**

Behavior analysts must heed the charge of delivering quality services that improve socially important behavior; this is no modest task. In other words, service recipient's lives are in our hands. The field appears to be living up to high expectations in some ways. In fact, the field appears to be booming; Board Certified Behavior Analysts (BCBAs) have approximately doubled in number since 2017 (Behavior Analyst Certification Board; BACB, n.d.); and the field has attracted significant monetary investment (for better or worse; see Garner et al., 2022). Meaning the field consists of a young workforce willing to help people live their best lives. Regrettably, the growth may require wading through some murky waters.

Many new behavior analysts may quickly move into supervisory roles and face pressures in clinical problem-solving and in relationships with colleagues and supervisors. Professionals may have large caseloads while learning on the job. This is not ideal: excessive work demands are positively correlated with burnout (e.g., Bottini et al., 2020; Dounavi et al., 2019). When faced with difficult situations, turning to peers and supervisors for support is recommended; unfortunately, ideal support may not be available to all. Brown (2021) found that nearly 80 percent of behavior analysts surveyed reported experiences in which supervisors pressured them to act unethically. In contrast, supervisor and collegial support have been found to increase job satisfaction and reduce burnout (Gibson et al., 2009). We are not suggesting that most behavior analysts do not receive adequate training and/or act unethically, but most will face difficulties. If exposure to these situations persists, the risk of burnout increases (Lizano, 2015; Maslach et al., 2001).

Briefly, burnout is a syndrome that is a response to chronic stress (Maslach et al., 2001). In work-related burnout, emotional exhaustion is a central dimension, characterized by feelings of being emotionally overextended and depleted; this precedes depersonalization, which is characterized by distancing oneself from others and developing cynical attitudes towards clients; thereafter, a reduced sense of personal accomplishment refers to feelings of ineffectiveness and reduced self-efficacy (Maslach & Jackson, 1981; Shaddock et al., 1998).

The need for inoculating oneself to stressful events and difficulties seems especially relevant to new behavior analysts—over 60 percent of new practitioners report moderate to high burnout (Plantiveau et al., 2018). This risk is compounded by populations served (i.e., Autism Spectrum Disorder [ASD], service providers are more likely to report experiencing burnout; Bottini et al., 2020). The fact that new behavior analysts currently make up approximately half of all behavior analysts *and* that most work with people with ASD is cause for concern (BACB, n.d.; Fiebig et al., 2020). In short, half of all behavior analysts have an elevated risk, while the risk is non-zero for others. Personality factors are also a risk: ABA professionals who report personality traits consistent with "neuroticism" are also likely to report burnout (Hurt et al., 2013).

An account of burnout that takes environment–behavior relationships into account is possible and required for behavior change. A history of limited contact with reinforcement, frequent exposure to aversive events, and verbally constructed rules may result in work-related tasks acquiring aversive functions. Stepping foot in a clinic, attending client sessions, or speaking to caregivers may all be perceived as hopeless and draining. Worse yet, events that appear related may also acquire similar stimulus functions (see Dymond & Rehfledt, 2000, and Belisle & Dixon, 2022, for descriptions of relating stimuli in this manner—i.e., transfer and transformation of stimulus functions, and relational dynamics, respectively).

Regardless of the conceptual underpinnings, burnout poses risks to personal and organizational well-being (Morse et al., 2012). Risks to personal well-being encompass health-related challenges such as chronic fatigue, recurrent flu symptoms, and colds (Cordes & Dougherty, 1993), memory issues (Peterson et al., 2008), heart disease (Johnson et al., 2005), and increased substance use (Lizano, 2015). Burnout in organizations can result in high turnovers (Emery & Vandenberg, 2010), absenteeism (Hastings et al., 2004), reduced commitment (Billingsley & Cross, 1992), compromised care (Burton, 2010; Figley, 2002) and, worst of all, abuse (Montaner et al., 2022; White et al., 2003). In short, burnout impacts individual workers, the workplace, and clients served.

**Self-Care Essentials**

Self-care is more meaningful than time for pleasure (e.g., mani-pedis, binge-watching TV, etc.). It entails "the ability to care for oneself through awareness, self-control, and self-reliance in order to achieve, maintain, or promote optimal health and well-being" (Martinez et al., 2021, p. 17). Well-being can be maintained via the following domains of self-care: physical (e.g., healthy diet, regular sleep, and exercise); psychological (e.g., striving for self-awareness and sound decision-making via mindfulness, journaling, counselling, etc.); emotional (e.g., use of humor, positive self-talk, social activities, etc.); spiritual (e.g., identifying meaning and purpose, via prayer, meditation, etc.; from a religious or secular standpoint); and professional (e.g., professional development, quality collaboration and supervision, setting limits and boundaries, and advocating for one's needs within an organization; Bloomquist et al., 2015; Collins, 2021; Jarden & Jarden, 2022; Rehfeldt & Tyndall, 2022).

To address the domains above it is suggested to seek out advice from public health and other evidence-informed sources, while steering clear of unsustainable fads. Sadly, even the best-laid plans may go awry; fortunately, behavior analysts are well-positioned to assess and intervene in these cases.

*The Function of Self-Care or Lack Thereof*

One's self-care can be another's self-indulgence. Identifying function is crucial. For example, mountain biking or a day at the beach sound lovely. Unfortunately, these reinforcing activities can be used to avoid unpleasant tasks; some might avoid time with an elderly parent, or an annual check-up in the name of self-care. Others might say "yes" to responsibilities at work or to coaching a soccer team to avoid uncomfortable social interactions. Others may seek out conference presentations for social approval. The scenarios are endless, just like the idiosyncratic functions of the behavior of verbally capable people (Belisle et al., 2017).

The contextual variables that evoke and maintain poor self-care need to be identified. This process requires attention to the context that occasions the behavior (e.g., what is happening when you say "yes" when you shouldn't, and don't say "no" when you should?). This must be done in the moment, to foster flexible and context sensitive decision-making (Sandoz et al., 2022). It is important to identify whether a decision is being made in the service of a good quality of life, or to access short- or long-term reinforcers. Behavior may be sensitive to the immediate context (i.e., "my child wants to play with me") or rules (i.e., "must be great, no time to play"); also, actions in those contexts may set the occasion for contacting consequences that are augmented by what is important (i.e., caring for yourself and your family; Sandoz et al., 2022). To gain this type of insight, it is suggested to use the worksheet in Figure 1.2 (adapted from Pistorello, 2013) to analyze the contingencies in the workplace and in personal circumstances, as a first step in changing behavior to increase self-care.

One of the best ways to begin to understand the influence of your experiences on your actions is to observe what happens during difficult times, what you do in response to them, and what the consequences are. Each time you find yourself struggling with thoughts and feelings, record the following:

1. Describe the event or situation that gave rise to your thoughts and feelings.

2. What showed up inside you – thoughts, emotions, feelings, memories, physical sensations?

3. How did you act in response to them?

4. If there were any consequences to your actions, what were they?

5. Did these consequences get in the way of anything that's important to you?

6. If you were acting in the service of your values (e.g., what's important to you) in that moment, what might you have done differently, if anything?

*Building awareness is the first step in changing behavior. Fill out this worksheet as many times as possible until your awareness grows and you can catch yourself getting in your own way as it happens. This worksheet can also help you identify what works for you so you can make better informed choices when responding to your discomfort or pain.*

*Figure 1.2* A Functional Analysis. *Adapted from Pistorello (2013).*

## *Self-Care and Job Crafting*

Recent findings by Slowiak and DeLongchamp (2022) indicate that behavior analysts who report higher rates of self-care and job crafting also report lower rates of burnout and higher rates of work-life balance and work engagement. The term *job crafting* refers to "self-initiated change behaviors that employees engage in with the aim to align their jobs with their own preferences, motives, and passions" (Tims et al., 2012, p. 173). In other words, job crafting involves increasing autonomy in accordance with values in the workplace, thereby changing one's relationship to work.

Slowiak and DeLongchamp (2022) provided excellent resources in Appendix A of their manuscript that assessed self-care and job crafting. It is recommend to also complete the Self-Care Assessment for Psychologists (Dorociak et al., 2017) and the Job Crafting Scale (Tims et al., 2012). If deficits are identified, we recommend creating behavioral targets for self-management by ranking the items based on the lowest score and the most important, personally.

Our initial example of a behavior analyst's goal of "being great" may identify the need to focus on life balance using Dorociak et al. (2017), with a low score on the item, "*I spend time with my family.*" To address this, one can implement an antecedent manipulation to review their plan for the day and schedule family time; this can be done during an established morning routine (i.e., making coffee). Adding this new behavior to a previously formed habit allows for a convenient time to collect and review data on successful family time accomplishments (see Milkman, 2021; Oster, 2022 for more ideas). After all, a "family meeting" is a "family meeting," whether at work or home and should not be missed. This framing may augment the function of family time from standing in the way of "being great," to a mandatory appointment.

The above is one of many potential strategies that can arise via assessing one's own behavior. This plan incorporates some steps from the Cooper et al. (2020) recommendations for self-management; a comprehensive plan should: (1) specify a goal and define the behavior to be changed, (2) self-monitor, (3) contrive contingencies that compete with natural contingencies, (4) go public with a commitment to change, (5) involve a self-management partner, and (6) continually evaluate the program and redesign it as necessary.

*When Verbal Behavior Gets in the Way*

For the most part, the strategies listed above involve direct contingencies. Unfortunately, in some cases, rigid adherence to rules may limit sensitivity to contingencies (Kissi et al., 2020); this can limit engaging in actions to increase well-being. Take for example, working late nights, or doing excessive research on a clinical problem because that is being the "best," even at the expense of personal health. Striving to be the "best," may also mean anxiety about *not* being the "best" (i.e., being an imposter). This can interrupt or paralyze work due to endless worry and unnecessary effort on meticulous details. Stimulus Control training (e.g., McGowan & Behar, 2013) can be applied (i.e., the Pomodoro technique of 25 minutes of work and a much-needed five-minute break, under the control of a timer). For many, this intervention will work, but additional methods may be needed to deal with intrusive thoughts. For example, the knowledge that work should be done during a 25-minute period, may not result in doing so. Excuses, self-doubt, self-judgment, and the like may stand in the way.

Procedures based on Acceptance and Commitment Training (ACTr) may be applied in the service of self-care. More specifically, as covered in the previous chapter (1.1):

> [T]he goal of ACT is to increase situationally appropriate decision-making through decreasing the impact of internal events that occur in the short term and increasing the impact of external future rewards. Successful implementation of ACT results in a shift from engaging in behaviors for immediate gratification to behaviors that serve the purpose of maintaining meaningful long-term outcomes.
> (Morrison et al., 2020, p. 601)

The ACT Matrix is an practical model that aides in discriminating experiences and actions as either moving toward or away from chosen values (i.e., smaller-sooner or larger-later rewards; see Figure 1.3; Polk & Schoendorff, 2014). Practicing this discrimination involves identifying (1) valued life directions, (2) thoughts and feelings, (3) patterns of overt behavior, and (4) goals and plans to move toward valued directions. Repeated practice in this discrimination task can decrease experiential avoidance and increase valued activities (Levin et al., 2017). Building skills that lead to flexible behavior can also be achieved using the more typical ACTr model that consists of six core interrelated processes (Hayes et al., 2006). Table 1.2 suggests a list of activities associated with each process to foster flexible behavior when dealing with inevitable struggles both on and off the job.

## Conclusion

Practitioners face a constant tug-of-war "between other-care and self-care" (Skovholt & Trotter-Mathison, 2014, p. 4). Self-care needs to be taught and reinforced early and often. Supervisors and instructors need

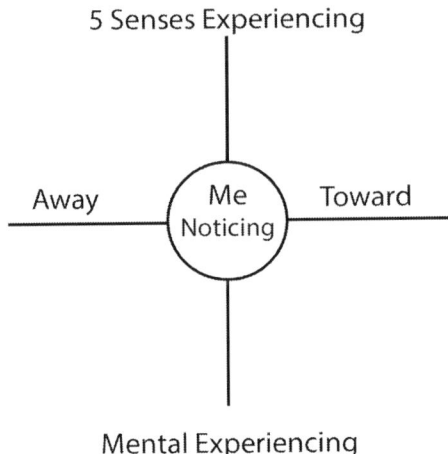

*Figure 1.3* The Matrix: 5 Senses Experience Image by Kevin Polk, PhD (ww.drkevinpolk.com). Reprinted with permission.

to model and incorporate self-care in their practices. The days of "no extensions under any circumstances" in education must be a thing of the past. If a student wants to spend time with a grandparent during their last few remaining days, a weekly quiz can wait; so can an Individualized Education Plan, or that hard-to-schedule meeting with a client's caregiver. Students and supervisees will appreciate the time caring for themselves and their families and will pay it forward to their students and supervisees. There needs to be a shift away from fanatical adherence to "contingencies" (i.e., meeting deadlines), and a shift toward humanizing practices. Acts of kindness toward others not only feel good, they may be a form of self-care; they will also occasion similar behavior in those you influence and in those they influence, and so forth.

Finally, deficits in self-care do not occur in a vacuum, but are shaped over time by culture. Sacrificing self-care to meet unreasonable workloads is likely part of the problem, and not part of the solution; behavior analysts must make systemic changes in the best interest of all those involved (Holland, 1978).

Table 1.2 Suggestions for Fostering Flexible Behaviour

| ACT Process | Description | Metaphor | Exercise |
|---|---|---|---|
| Acceptance | Acceptance means opening up and making room for painful feelings, sensations, urges, and emotions. We drop the struggle with them, give them some breathing space, and allow them to be as they are. Instead of fighting them, resisting them, running from them, or getting overwhelmed by them, we open up to them and let them be. (*Note: This doesn't mean liking them or wanting them. It simply means making room for them!*) | **The Struggle Switch**[1]<br><br>It can sometimes feel like there is a "struggle switch" in the back of our mind that is switched on as soon as we have an uncomfortable emotion or painful memory. When we struggle against these things, our emotions can get amplified, hence creating more anxiety and frustration. This amplification can impact and influence our life pulling us down or causing us to engage in self-defeating behaviors. When an uncomfortable emotion, painful memory or painful emotion shows up and we keep the struggle switch "off," we are intentionally deciding not to invest our time, energy and effort into struggling against it. Instead, we invest the pain, emotions or memories into doing meaningful life enhancing activities. | **Four Quick Steps to Acceptance**[2]<br>1. OBSERVE. Bring awareness to the feelings in your body.<br>2. BREATHE. Take a few deep breaths. Breathe into and around them.<br>3. EXPAND. Make room for these feelings. Create some space for them.<br>4. ALLOW. Allow them to be there. Make peace with them<br><br>Some people find it helpful to silently say to themselves, "I don't like this feeling, but I have room for it," or "It's unpleasant, but I can accept it." |

(*Continued*)

Table 1.2 (Continued)

| ACT Process | Description | Metaphor | Exercise |
|---|---|---|---|
| **Defusion** | Defusion, otherwise known as "cognitive defusion," means learning to "step back" and separate or detach from the our thoughts, images, and memories. Instead of getting caught up in our thoughts or being pushed around by them, we let them come and go as if they were just leaves floating down the stream, or clouds passing by in the sky. We step back and watch our thinking instead of getting tangled up in it. We see our thoughts for what they are—nothing more or less than words or pictures. We hold them lightly instead of clutching them tightly. | **Clouds in the Sky**[3] A common metaphor is to see your mind as a sky and your thoughts as the clouds passing you by. This describes how all of our thoughts are impermanent and changing. Like the popular saying, "this too shall pass"—new thoughts (or "clouds") will rise into our awareness, and then drift out of awareness and disappear. And if we give it enough time, every thought eventually dissipates and passes us by, just like a cloud in the sky. | **Repetition Techniques**[4] 1. The thought could be watched dispassionately, repeated several times out loud until only its sound remains, or treated as an external observation by giving it a shape, size, color, speed, or form. 2. Try not taking the thought literally by noticing the thought and saying "I am having the thought that X, Y, Z," noting that your thoughts are experiences that inevitably come and go. |
| **Self-As-Context (The Observing Self)** | In everyday language, we talk about the "mind" without recognizing that there are two distinct elements to it: the thinking self and the observing self. We are typically familiar with the thinking self: that part of us which is always thinking—generating thoughts, beliefs, memories, judgments, fantasies, plans, etc. But most people are unfamiliar with the observing self: the aspect of us that is aware of whatever we're thinking, | **Life is a Stage**[5] Imagine that your life is like a stage play. All of your experiences, thoughts, feelings, and behaviors are like actors on the stage. Our observing self is the part of us that can sit in the audience and watch the play unfold in front of us and around us, rather than getting caught up on the stage and chasing the characters (i.e., our | **The Observing Self Exercise**[2] The aim of this exercise is to increase awareness of "Self-as-Context," or the "Observing Self": that aspect of us which is experientially distinct from whatever event is being observed. The basic formula is: 1. Bring your attention to X (can include thoughts, feelings, sensations, urges, memories, your body, the roles you play, etc.). |

# Self-Care for Behavior Analysts

feeling, sensing, or doing in any moment. As you go through life, your body changes, your thoughts change, your feelings change, your roles change, but the "you" that's able to notice or observe all those things never changes. It's the same "you" that's been there your whole life.

*thoughts and feelings) around. Your observing self can also control the lights on the stage, choosing to turn down the lights on certain characters and turn up the lights on others. You can intentionally adjust your focus.

2. As you notice X, be aware that you are noticing it.
3. There is X, and there you are, observing it.
4. X changes.
5. However, the you that notices X does not change.
6. Once again, notice X, and be aware that you're noticing it. There's X, and there's you.

## Present Moment

Contacting the present moment means being psychologically present and consciously connecting with and engaging in whatever is happening in the present moment. It also means consciously paying attention to our here-and-now experience instead of drifting off into our thoughts (of the past or future), or operating on "automatic pilot."

### Dropping Anchor[6]

Imagine you are a boat, being tossed around at sea. The weather is rough, representing all the external storms or crises around you that you have no control over, along with the inner emotional storm you may be experiencing. Rather than be tossed around in every direction, by these rough seas, you DROP ANCHOR....
To steady your boat, steady yourself. Dropping your anchor will hold you steady until the storm passes; it will not stop the storm, the weather is still happening, however you're less affected by it. That's why we drop anchor; to be steady while the storm passes.

### How to Drop Anchor[6]

Use the acronym A-C-E:
A: Acknowledge your thoughts and feelings, and/or bodily sensations. One way to do this is to pause and notice what's showing up inside, in your inner world. Another way to do this is to say to yourself "I'm noticing I'm having the thought...", or "I'm noticing I'm feeling ....," or "I'm noticing a sensation of...." Acknowledging your inner experience at this point in time. While acknowledging your thoughts and feelings, you... move on to "C."

C: Come back into your body. You can do this in a variety of ways. If you're sitting; you might feel your feet on the floor (press them into the floor, or wiggle your toes), lengthen your spine, move your arms for a

(Continued)

Table 1.2 (Continued)

| ACT Process | Description | Metaphor | Exercise |
|---|---|---|---|
| | | | stretch, or drop your shoulders with a shrug. You can utilize breathing here as well—take a deep breath in through your nose, and out through your mouth (with or without a sigh), or do a few conscious breaths in a way you might have found helpful before. This step is about expanding your awareness, to be more than the thoughts and feelings that are arising; aware that you can also move your body, and control this movement. Now move on to "E." |
| | | | E: Engage in what you're doing; or with where you are. You might look around and really notice what you see in the room you're in; or notice what you are doing—if you're in a meeting, consciously engaging in the meeting—listen to what is being said, placing your attention on what you are doing in this moment. |
| | | | You can do this in as little as 30 seconds, or as long as 10 minutes |

| | | |
|---|---|---|
| Present Moment | Contacting the present moment means being psychologically present and consciously connecting with and engaging in whatever is happening in the present moment. It also means consciously paying attention to our here-and-now experience instead of drifting off into our thoughts (of the past or future), or operating on "automatic pilot." | **Dropping Anchor**[6]<br><br>Imagine you are a boat, being tossed around at sea. The weather is rough, representing all the external storms or crises around you that you have no control over, along with the inner emotional storm you may be experiencing. Rather than be tossed around in every direction, by these rough seas, you DROP ANCHOR....<br><br>To steady your boat, steady yourself. Dropping your anchor will hold you steady until the storm passes; it will not stop the storm, the weather is still happening, however you're less affected by it. That's why we drop anchor; to be steady while the storm passes. | **How to Drop Anchor**[6]<br><br>Use the acronym A-C-E:<br><br>A: Acknowledge your thoughts and feelings, and/or bodily sensations. One way to do this is to pause and notice what's showing up inside, in your inner world. Another way to do this is to say to yourself "I'm noticing I'm having the thought...", or "I'm noticing I'm feeling ...," or "I'm noticing a sensation of...." Acknowledging your inner experience at this point in time. While acknowledging your thoughts and feelings, you... move on to "C."<br><br>C: Come back into your body. You can do this in a variety of ways. If you're sitting; you might feel your feet on the floor (press them into the floor, or wiggle your toes), lengthen your spine, move your arms for a stretch, or drop your shoulders with a shrug. You can utilize breathing here as well—take a deep breath in through your nose, and out through your mouth (with or without a sigh), or do a few conscious breaths in a way you might have found helpful before. This step is about expanding your awareness, to be more than the thoughts and feelings that are |

(*Continued*)

Table 1.2 (Continued)

| ACT Process | Description | Metaphor | Exercise |
|---|---|---|---|
| | | | arising; aware that you can also move your body, and control this movement. Now move on to "E." |
| | | | E: Engage in what you're doing; or with where you are. You might look around and notice what you see in the room you're in; or notice what you are doing—if you're in a meeting, consciously engaging in the meeting—listen to what is being said, placing your attention on what you are doing in this moment. |
| | | | You can do this in as little as 30 seconds, or as long as 10 minutes. |
| | | | **Value Clarification Task** (see Table 1.3) |
| Values | Deep in your heart, what do you want your life to be about? What do you want to stand for? What you want to do with your brief time on this planet? What truly matters to you in the big picture? Values are desired qualities of ongoing action. Values describe how we want to behave on an ongoing basis. *Clarifying our values* is an essential step in creating a meaningful life. Much like a compass, values gives us a sense of direction and guides our ongoing journey. | **Trip to Disneyland**[7] Imagine two children on a trip to Disneyland. One child is completely goal focused. The only thing he cares about is arriving at his destination, so he becomes frustrated during the long car ride. The second child, on the other hand, is values focused. During the car ride, he is able to live out his values of being adventurous and curious and notice the world around him. Both | Explore what sort of person you want to be. This task invites reflection on 9 valued areas and asks you to rank the level of importance of each area (from a scale of 1 to 10), as well as the purported effect of this valued area on your life. |
| | | | Which values are most important to you? Which values are least important to you? How did this exercise clarify your values and the sort of person you would like to be? |

| | | |
|---|---|---|
| | children arrive at Disney land and have a wonderful time, but the first child had a frustrating experience up to that point while the second child had a more meaningful experience. In addition, if the car breaks down on the way to Disneyland, even though both children will be upset and frustrated, the value-focused child will have had the satisfaction of living based on his values up to that point in the passengers. | |
| Committed Action | Committed action means taking effective action, guided by our values. It's good to know our values, but it's only via ongoing values-congruent action that life becomes rich, full, and meaningful. In other words, we won't have much of a journey if we simply stare at the compass; our journey only happens when we move our arms and legs in our chosen direction. Committed action means "doing what it takes" to live by our values even if that brings up pain and discomfort. | **Passengers on a Bus**[8]<br>Imagine life is like a journey, and you're the driver of your bus. You want to go places and do what's important for you. Over the course of your life, various passengers have boarded your bus. These passengers reflect your thoughts, feelings, and all kinds of inner states. Some of them you like, such as happy memories or positive thoughts, some you feel neutral about, and then there are passengers that you wish had not boarded the bus; they can be ugly, scary, and nasty.<br><br>**Setting SMART Goals**[9]<br>"SMART" is an acronym. Here's what the acronym means:<br><br>**S = Specific**<br>Do not set a vague, fuzzy, or poorly-defined goal like, "I'll be more loving." Instead, be specific: "I'll give my partner a good, long hug when I get home from work." In other words, *specify what actions you will take.*<br><br>**M = Meaningful**<br>Make sure this goal is aligned with what is important/valuable to you.<br><br>**A = Adaptive**<br>Ask yourself, is this goal likely to improve your life in some way?<br><br>*(Continued)* |

*Self-Care for Behavior Analysts* 41

Table 1.2 (Continued)

| ACT Process | Description | Metaphor | Exercise |
|---|---|---|---|
| | | So, you are driving your bus of life with all sorts of passengers on board. The scary passengers can threaten you and want to be at the front of the bus where you see them. You take this very seriously and stop the bus to struggle and fight with them. You may try to avoid them, distract yourself, or throw them off the bus, but they are your inner states, so you can't get rid of them. However, while the bus is stopped, you're not moving in the direction that's important to you.<br><br>By fighting and struggling with the passengers or giving in to them, you, the driver, are not in control of your journey of life, and it's likely that you are not heading in a direction that is important to you. But what if, even though these passengers look scary, nasty, and threatening, they can't take control unless you allow them | **R = Realistic**<br>Make sure the goal is realistic for the resources you have available. Resources you may need could include: time, money, physical health, social support, knowledge and skills. If these resources are necessary but unavailable, you will need to change your goal to a more realistic one. The new goal might actually be to find the missing resources: to save the money, or develop the skills, or build the social network, or improve health, etc.<br>**T = Time-framed**<br>Put a specific time frame on the goal: specify the day, date and time — as accurately as possible — that you will take the proposed actions.<br><br>Write your **SMART** goal here:<br><br>_____<br>_____<br>_____<br>_____<br>_____<br>_____<br>_____<br>_____ |

to? There can be different ways to respond to the passengers so that you can head in the direction that is important.

**Table 1.1 Citations**
1. Harris, R. (2007). *Accepting Emotions*. https://stoughtonhealth.com/wp-content/uploads/Accepting-Emotions.pdf
2. Harris, R. (2007). *Acceptance and Commitment Therapy (ACT) Introductory Workshop Handout*. https://thehappinesstrap.com/upimages/2007%20Introductory%20ACT%20Workshop%20Handout%20-%20%20Russ%20Harris.pdf
3. ACT for Psychosis Recovery. (n.d.). *A14. Clouds in the Sky Exercise*. http://actforpsychosis.com/pdfs/A14_Clouds_in_the_sky.pdf
4. Masuda, A., Hayes, S. C., Sackett, C. F. and Twohig, M. P. (2004). Cognitive defusion and self-relevant negative thoughts: examining the impact of a ninety year old technique. *Behaviour Research and Therapy, 42,* 477–485.
5. Harris, R. (2016). *Russ Harris: ACT Training Part 1*. www.actmindfully.com.au/upimages/147771_ACT_Workshop_Aug_2016_A4_1up.pdf.
6. Harris, R. (2019). *How To 'Drop Anchor'*. https://survivorsofabuserecovering.ca/wp-content/uploads/2019/10/Dropping-anchor-handout-ACE-formula-Russ-Harris-2019.pdf.
7. Jenkins, J., & Ahles, A. (2019). *When the Going Gets Tough, the Tough Get Mindful: A Toolkit Based on the Principles of Acceptance and Commitment Therapy. Help With ACT*.
8. ACT for Psychosis Recovery. (n.d.). *A3. Passengers on the Bus Metaphor*. http://actforpsychosis.com/pdfs/A3_Passengers_on_the_bus.pdf.
9. Harris, R. (2008). *Goal Setting Worksheet*. www.actmindfully.com.au/upimages/Goal_Setting_Worksheet.pdf.

## References

Bannerman, D. J., Sheldon, J. B., Sherman, J. A., & Harchik, A. E. (1990). Balancing the right to habilitation with the right to personal liberties: The rights of people with developmental disabilities to eat too many doughnuts and take a nap. *Journal of Applied Behavior Analysis, 23*(1), 79–89. https://doi.org/10.1901/jaba.1990.23-79

Behavior Analyst Certification Board. (n.d). *BACB certificant data.* www.bacb.com/BACB-certificant-data

Belisle, J., & Dixon, M. R. (2022). Relational behavior and ACT: A dynamic relationship. *Behavior Analysis in Practice, 15*(1), 71–82. https://doi.org/10.1007/s40617-021-00599-z

Belisle, J., Stanley, C. R., & Dixon, M. R. (2017). The relationship between derived mutually entailed relations and the function of challenging behavior in children with autism: Comparing the PEAK-E-PA and the QABF. *Journal of Contextual Behavioral Science, 6*(3), 298–307. https://doi.org/10.1016/j.jcbs.2017.07.004

Bickel, W. K., Jarmolowicz, D. P., Mueller, E. T., Koffarnus, M. N., & Gatchalian, K. M. (2012). Excessive discounting of delayed reinforcers contributing to addiciton and other disease-related vulnerabilities: Emerging evidence. *Pharmacology & Therapeutics, 134,* 287–297. https://doi.org/10.1016/j.pharmthera.2012.02.004

Billingsley, B., & Cross, L. (1992). Predictors of commitment, job satisfaction, and intent to stay in teaching: A comparison of general and special educators. *The Journal of Special Education, 25*(4), 453–471. https://doi.org/10.1177/002246699202500404

Bloomquist, K. R., Wood, L., Friedmeyer-Trainor, K., & Kim, H. W. (2015). Self-care and professional quality of life: Predictive factors among MSW practitioners. *Advances in Social Work, 16*(2), 292–311. https://doi.org/10.18060/18760

Bottini, S., Wiseman, K., & Gillis, J. (2020). Burnout in providers serving individuals with ASD: The impact of the workplace. *Research in Developmental Disabilities, 100,* 103616. https://doi.org/10.1016/j.ridd.2020.103616

Brown, T. J. (2021). Ethics, burnout, and reported life and job attitudes among board-certified behavior analysts. *Behavior Analysis: Research and Practice, 21*(4), 364–375. https://doi.org/10.1037/bar0000219

Burton, J. (2010). WHO healthy workplace framework and model. *Geneva, Switzerland: World Health Organization.* https://www.who.int/publications/i/item/who-healthy-workplace-framework-and-model

Carroll, L., Gilroy, P. J., & Murra, J. (1999). The moral imperative: Self-care for women psychotherapists. *Women & Therapy, 22*(2), 133–143.

Collins, S. (2021). Social workers and self-care: A promoted yet unexamined concept? *Practice, 33*(2), 87–102. https://doi.org/10.1080/09503153.2019.1709635

Cooper, J. O., Heron, T. E., & Heward, W. L. (2020). *Applied behavior analysis* (3rd ed.). Pearson Education.

Cordes, C., & Dougherty, T. (1993). A review and an integration of research on job burnout. *Academy of Management Review, 18*(4), 621–656. https://doi.org/10.2307/258593

Coster, J. S., & Schwebel, M. (1997). Well-functioning in professional psychologists. *Professional Psychology: Research and Practice, 28*(1), 5.

Dorociak, K. E., Rupert, P. A., Bryant, F. B., & Zahniser, E. (2017). Development of a self-care assessment for psychologists. *Journal of Counseling Psychology*, 64(3), 325.

Dounavi, K., Fennell, B., & Early, E. (2019). Supervision for certification in the field of applied behavior analysis: Characteristics and relationship with job satisfaction, burnout, work demands, and support. *International Journal of Environmental Research and Public Health*, 16(12), 2098. https://doi.org/10.3390/ijerph16122098

Dymond, S., & Rehfeldt, R. A. (2000). Understanding complex behavior: The transformation of stimulus functions. *The Behavior Analyst*, 23(2), 239–254. https://doi.org/10.1007/BF03392013

Emery, D. W., & Vandenberg, B. (2010). Special education teacher burnout and ACT. *International Journal of Special Education*, 25(3), 119–131.

Estanek, S. M. (2006). Redefining spirituality: A new discourse. *College Student Journal*, 40(2), 270–282.

Fiebig, J. H., Gould, E. R., Ming, S., & Watson, R. A. (2020). An invitation to act on the value of self-care: Being a whole person in all that you do. *Behavior Analysis in Practice*, 13(3), 559–567.

Figley, C. R. (2002). Compassion fatigue: Psychotherapists' chronic lack of self-care. *Journal of Clinical Psychology*, 58, 1433–1441. https://doi.org/10.1002/5cip.10090

Garner, J., Peal, A., Klapatch-Totsch, J., & Gamba, J. (2022). Exploitation, freedom, and coercion: The integration of applied behavior analysis in a capitalist system. *Behavior and Social Issues*, 1–16. https://doi.org/10.1007/s42822-022-00100-7

Gibson, J. A., Grey, I. M., & Hastings, R. P. (2009). Supervisor support as a predictor of burnout and therapeutic self-efficacy in therapists working in ABA schools. *Journal of Autism and Developmental Disorders*, 39(7), 1024–1030. https://doi.org/10.1007/s10803-009-0709-4

Gisler, S., Omansky, R., Alenick, P. R., Tumminia, A. M., Eatough, E. M., & Johnson, R. C. (2018). Work-life conflict and employee health: A review. *Journal of Applied Biobehavioral Research*, 23(4), e12157. https://doi.org/10.1111/jabr.12157

Hage, S. M. (2006). Profiles of women survivors: The development of agency in abusive relationships. *Journal of Counseling & Development*, 84(1), 83–94.

Hastings, R. P., Horne, S., & Mitchell, G. (2004). Burnout in direct care staff in intellectual disability services: A factor analytic study of the Maslach Burnout Inventory. *Journal of Intellectual Disability Research*, 48(3), 268–273. https://doi.org/10.1111/j.1365-2788.2003.00523.x

Hayes, S, Luoma, J., Bond, F., Masuda, A., & Lillis, J. (2006). Acceptance and commitment therapy: Model, processes and outcomes. *Behavior Research and Therapy*, 44, 1–25. https://doi.org/10.1016/j.brat.2005.06.006

Henderson, K. A., & Ainsworth, B. E. (2001). Researching leisure and physical activity with women of color: Issues and emerging questions. *Leisure Sciences*, 23(1), 21–34.

Holland, J. G. (1978). Behaviorism: Part of the problem or part of the solution? *Journal of Applied Behavior Analysis, 11*(1), 163–174. https://doi.org/10.1007/s40617-020-00442-x

Hurt, A. A., Grist, C. L., Malesky, L. A., Jr, & McCord, D. M. (2013). Personality traits associated with occupational 'burnout' in ABA therapists. *Journal of Applied Research in Intellectual Disabilities: JARID, 26*(4), 299–308. https://doi.org/10.1111/jar.12043

Jarden, R. J., & Jarden, A. (2022). A systems pathway to self-care in academia. In N. Lemon (Ed.). *Reflections on Valuing Wellbeing in Higher Education: Reforming our Acts of Self-care*. Routledge.

Johnson, S., Cooper, C., Cartwright, S., Donald, I., Taylor, P., & Millet, C. (2005). The experience of work-related stress across occupations. *Journal of Managerial Psychology, 20*, 178–187. https://doi.org/10.1108/02683940510579803

Kissi, A., Harte, C., Hughes, S., De Houwer, J., & Crombez, G. (2020). The rule-based insensitivity effect: A systematic review. *PeerJ, 8*. https://doi.org/10.7717/peerj.9496

Levin, M. E., Pierce, B., & Schoendorff, B. (2017). The acceptance and commitment therapy matrix mobile app: A pilot randomized trial on health behaviors. *Journal of Contextual Behavioral Science, 6*(3), 268–275. https://doi.org/10.1016/j.jcbs.2017.05.003

Lizano, E. L. (2015). Examining the impact of job burnout on the health and well-being of human service workers: A systematic review and synthesis. *Human Service Organizations: Management, Leadership & Governance, 39*, 167–181. https://doi.org/10.1080/23303131.2015.1014122

Martínez, Connelly, C. D., Pérez, A., & Calero, P. (2021). Self-care: A concept analysis. *International Journal of Nursing Sciences, 8*(4), 418–425. https://doi.org/10.1016/j.ijnss.2021.08.007

Maslach, C., & Jackson, S. E. (1981). The measurement of experienced burnout. *Journal of Organizational Behavior, 2*(2), 99–113. https://doi.org/10.1002/job.4030020205

Maslach, C., Schaufeli, W. B., & Leiter, M. P. (2001). Job burnout. *Annual Review of Psychology, 52*, 397–422. https://doi.org/10.1146/annurev.psych.52.1.397

McGowan, S. K., & Behar, E. (2013). A preliminary investigation of stimulus control training for worry: Effects on anxiety and insomnia. *Behavior Modification, 37*(1), 90–112. https://doi.org/10.1177/0145445512455661

Milkman, K. (2021). *How to change: The science of getting from where you are to where you want to be*. Penguin.

Montaner, X., Tárrega, S., Pulgarin, M., & Moix, J. (2022). Effectiveness of acceptance and commitment therapy (ACT) in professional dementia caregivers burnout. *Clinical Gerontologist, 45*(4), 915–926. https://doi.org/10.1080/07317115.2021.1920530

Morrison, K. L., Smith, B. M., Ong, C. W., Lee, E. B., Friedel, J. E., Odum, A., ... & Twohig, M. P. (2020). Effects of acceptance and commitment therapy on impulsive decision-making. *Behavior Modification, 44*(4), 600–623.

Morse, G., Salyers, M. P., Rollins, A. L., Monroe-DeVita, M., & Pfahler, C. (2012). Burnout in mental health services: A review of the problem and its remediation. *Administration and Policy in Mental Health and Mental Health Services Research, 39*(5), 341–352. https://doi.org/10.1007/s10488-011-0352-1

O'Connor, M. F. (2001). On the etiology and effective management of professional distress and impairment among psychologists. *Professional Psychology: Research and Practice, 32*(4), 345.

Oster, E. (2022). *The family firm: A data-driven guide to better decision making in the early school years* (vol. 3). Penguin.

Perrone, K. M., Webb, L. K., Wright, S. L., Jackson, Z. V., & Ksiazak, T. M. (2006). Relationship of spirituality to work and family roles and life satisfaction among gifted adults. *Journal of Mental Health Counseling, 28*(3), 253–268.

Peterson, U., Demerouti, E., Bergström, G., Samuelsson, M., Åsberg, M., & Nygren, Å. (2008). Burnout and physical and mental health among Swedish healthcare workers. *Journal of Advanced Nursing, 62*(1), 84–95. https://doi.org/10.1111/j.1365-2648.2007.04580.x

Pincus, J. (2006). Teaching self-care [Conference session]. Pennsylvania Psychological Association on the Ethics Educators Conference, Harrisburg.

Pistorello, J. (Ed.). (2013). *Mindfulness and acceptance for counseling college students: Theory and practical applications for intervention, prevention, and outreach*. New Harbinger Publications.

Plantiveau, C., Dounavi, K., & Virués-Ortega, J. (2018). High levels of burnout among early-career board-certified behavior analysts with low collegial support in the work environment. *European Journal of Behavior Analysis, 19*(2), 195–207. https://doi.org/10.1080/15021149.2018.1438339

Polk, K. L., & Schoendorff, B. (Eds.). (2014). *The ACT matrix: A new approach to building psychological flexibility across settings and populations*. Context Press/New Harbinger Publications.

Rehfeldt, R. A., & Tyndall, I. (2022). Why we are not acting to save ourselves: ACT, health, and culture. *Behavior Analysis in Practice, 15*(1), 55–70. https://doi.org/10.1007/s40617-021-00592-6

Sandoz, E. K., Gould, E. R., & DuFrene, T. (2022). Ongoing, explicit, and direct functional assessment is a necessary component of ACT as behavior analysis: A response to Tarbox et al. (2020). *Behavior Analysis in Practice, 15*(1), 33–42. https://doi.org/10.1007/s40617-021-00607-2

Shaddock, A. J., Hill, M., & Limbeek, C. A. H. (1998). Factors associated with burnout in workers in residential facilities for people with an intellectual disability. *Journal of Intellectual & Developmental Disability, 23*(4), 309–318. https://doi.org/10.1080/13668259800033379

Skovholt, T. M., & Trotter-Mathison, M. (2014). *The Resilient Practitioner: Burnout prevention and self-care strategies for counselors, therapists, teachers, and health professionals*. Routledge.

Slowiak, J. M., & DeLongchamp, A. C. (2022). Self-care strategies and job-crafting practices among behavior analysts: Do they predict perceptions of work–life balance, work engagement, and burnout?. *Behavior Analysis in Practice, 15*(2), 414–432. https://doi.org/10.1007/s40617-021-00570-y

Stevanovic, P., & Rupert, P. A. (2004). Career-sustaining behaviors, satisfactions, and stresses of professional psychologists. *Psychotherapy: Theory, Research, Practice, Training*, 41(3), 301.

Tims, M., Bakker, A. B., & Derks, D. (2012). Development and validation of the job crafting scale. *Journal of Vocational Behavior*, 80(1), 173–186. https://doi.org/10.1016/j.jvb.2011.05.009

White, C., Holland, E., Marsland, D., & Oakes, P. (2003). The identification of environments and cultures that promote the abuse of people with intellectual disabilities: A review of the literature. *Journal of Applied Research in Intellectual Disabilities*, 16(1), 1–9. https://doi.org/10.1046/j.1468-3148.2003.00147.x

# 2
# Interpersonal Skills

*Ana Luisa Santo and Kimberley Taylor*

# 2.1 20 Key Behaviors for Rapport Building

*Ana Luisa Santo and Kimberley Taylor*

Introduction

Although Applied Behavior Analysis (ABA) has been viewed by some as a harsh, robotic, and cold science, there has been a more recent focus on ABA as a compassionate science, with research and practice focusing more on building effective therapeutic relationships through the use of compassionate care skills and values-based approaches such as Acceptance and Commitment Training (ACT). Just like the fields of social work and psychology, ABA is committed to addressing socially significant behaviors that improve the human condition (Baer et al., 1968, 1987; National Association of Social Workers, 2018).

ABA is prosocial, in that it involves communicating and engaging with clients and families in order to create meaningful changes in their lives. Therapeutic relationship variables may be integral to understanding the success of behavioral interventions (Hayes et al., 2003; Hayes et al., 2004; Robins & Chapman, 2004).

Lejuez and colleagues (2005) make the case that ignoring the role of the therapeutic relationship in behavior analytic services may be inconsistent with the basic principles of behavior analysis. Making efforts toward understanding the client and mediator's unique environment and experiences not only increases the likelihood of identifying functional reinforcers or punishers within particular contexts, but also demonstrates empathy and understanding that can enhance that therapeutic relationship (Lejuez et al., 2005).

A recent survey evaluating caregiver satisfaction (Taylor et al., 2019) on various components of behavior analytic service revealed some areas which were rated relatively low. While the study sample was not large, it can shed light on how Behavioral Clinicians are perceived and serve as a starting point to supplementing some of the gaps in current behavioral education. In the study, Behavioral Clinicians were perceived as displaying weaknesses in the following areas:

DOI: 10.4324/9781003300465-7

1. Compromising during a disagreement
2. Inquiring about satisfaction
3. Role clarification
4. Caring about the entire family
5. Acknowledging mistakes/treatment failures
6. Being patient and reassuring
7. Underestimating the client's ability
8. Clinician having their own agenda
9. Having an authoritarian demeanor

A list of behaviors for increasing rapport was created by targeting these nine perceived areas of weakness. They were broken down further into 20 distinct behaviors that have been clearly defined so that they are more easily teachable.

*Key Behaviors*

We outline 20 behaviors for increasing rapport, divided into 3 main categories: Active Listening and Communication, Empathy, and Compassion (categories based on Taylor et al., 2019).

Active Listening and Communication is the first category of behaviors broken down in this chapter. The specific skills in this category are perhaps some of the easiest to provide operational definitions for. The chapter will then delve further into the more elusive-to-define categories of Empathy and Compassion.

Leaf and colleagues (2016) write that "Unfortunately, much of the general public believes that ABA procedures are rigid and cannot be used to teach complex behaviors and that practitioners are merely technicians, lacking clinical sensitivity and prowess" (p. 722). The intent is to make these rapport-building skills, which behavior analysts have traditionally relegated to the category of un-teachable (or even worse, mentalistic), accessible to anyone wishing to teach them in an evidence-based manner.

*Category 1: Active Listening and Communication*

List of teachable behaviors:

1. Eye contact
2. Body language
3. Use verbal and non-verbal continuers
4. Paraphrase
5. Ask open-ended questions
6. Allow for silence; allow space and time for client to respond

7 Summarize
8 Use appropriate language (avoid jargon)

"Active Listening" is a term that is frequently used in social work and counseling psychology practice (Rogers & Farson, 1957). This term is defined in a variety of different ways depending on the field of study, the theoretical underpinnings being used to inform the practice, or the individual definitions of the practicing clinician. "Communication" is similarly subject to a plurality of definitions.

The attempt to break down these terms into eight operationally defined skills is not meant to supersede or replace any existing definitions in other fields, but rather to approach this category of skills from a behavior analytic framework, in order to facilitate the explicit teaching of these skills.

One popularly referred to list of active listening skills is a figure developed by the Center for Creative Leadership (see Figure 2.1). Several of the eight skills broken down in this section are taken directly from this list of skills, while others were modified or taken from reflections by Taylor et al. (2019). In their article, they clarify the importance of active listening and explain that "by engaging the parent in conversation and *actively listening* to the parents' concerns, the behavior analyst may proactively identify potential barriers to adherence" (Taylor et al., 2019, p. 655). By understanding and appreciating parental concerns, both parent and provider can work together to create a plan that increases the likelihood of effective treatment outcomes. (Taylor et al., 2019).

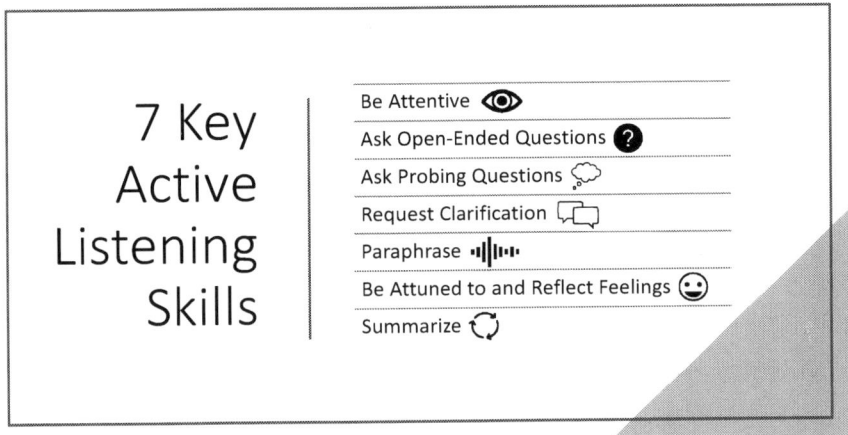

*Figure 2.1* 7 Key active listening skills.

### Eye Contact

One of the items identified by Taylor et al. (2019) that may contribute to problems in the therapeutic relationship is that "The behavior analyst often seems distracted during meetings" (p. 658). Since active listening requires one to direct and focus one's attention (Lown et al., 2014), making eye contact is an important skill to display in order to express presence to what is being said during a conversation between a behavior analyst and their client, and to increase the likelihood that clients do not feel as if the behavior analyst seems distracted during their meeting.

"Eye contact" can be operationally defined as the behavior analyst looking directly at the eyes of the person to whom they are speaking. The likelihood of this behavior being exhibited by the behavior analyst can be increased by setting aside distractions (e.g., phone, computer, notebook). Making eye contact with another is of course a more complex skill to teach than simply instructing someone to gaze indefinitely into the eyes of another person. There are culturally appropriate ways of initiating and maintaining eye contact, as well as cultural and physical barriers to exhibiting this behavior (Lown et al., 2014).

Research that has specifically studied the preferred length of eye contact that makes individuals feel comfortable (in North American culture, at least) found that a gaze of 3.3 seconds (on average) was the most preferred length of time eye contact was made, with brief moments of looking away in between each gaze (Binetti et al., 2016). Differentiations from this gaze pattern tends to make individuals feel uncomfortable. Therefore, any training of eye contact behavior in behavior analysts should aim to teach them not to deviate from this optimal length of gaze and intermittent pattern. For behavior analysts practicing outside the North American culture or with individuals who primarily have a different culture, care should be taken to learn more about the specifics of eye contact that apply, because what is comfortable for North American speakers in terms of eye contact may differ across cultures (Akechi et al., 2013).

### Body Language

Body language is another component of active listening and communication that can be used in a therapeutic relationship to increase the likelihood that mediators or clients feel the provider is attending to the meeting and not distracted. These non-verbal cues are an essential component of communication.

As "body language" is not just one skill but encompasses a variety of different behaviors, they can each be identified and further broken down into definitions that can be used in training. While not an exhaustive list, some of the main important behaviors involved in body language are: smiling, sitting up straight, orienting one's body to face the speaker (as opposed to a 45 or 90 degree angle), leaning slightly forward, and maintaining uncrossed arms and legs. These are relatively simple behaviors that can facilitate the building and maintenance of rapport. They help communicate interest in what the client is saying.

Another important point to include in training of body language skills is to remember that each individual culture, family, and individual may have differences in what they find comfortable. It is important for the provider to be aware of the client's own body language, and match accordingly. Observing the client's facial expressions and bodily expressions of emotion will help the provider understand what the client is experiencing during the conversation.

For example, if a client is expressing distress via facial expressions (but not in the content of their verbal behavior), it is important that the provider respond to this body language by matching their own facial expressions and tone of voice. This helps communicate to the client that the provider is aware of the client's underlying emotions or distress, and increases the feeling of rapport between both people. This is known as "mirroring" in the literature, and there is substantial evidence to support its use to increase feelings of rapport and empathy (Pfeifer et al., 2008).

It is also important for the provider to monitor their own emotional reactions to the clients' communication, and remain mindful of any changes in body language that may be occurring and adjust accordingly. For example, if a provider feels uncomfortable because of something that is being said, this expression of discomfort may show in their body language (e.g., facial expression, turning slightly away from the client), and decrease rapport with the client.

*Use Verbal and Non-Verbal Continuers*

Two other items identified by Taylor et al. (2019) that may contribute to problems in the therapeutic relationship are that "The behavior analyst is too busy to discuss things about my child's program that are important to me" (p. 658); and low scores for "When I have concerns about my child's program, the behavior analyst actively listens to my concerns without being defensive" (p. 657).

One effective method of communicating interest in what the client is saying and decreasing the likelihood that the behavior analyst comes across as too busy or uninterested is to use verbal and non-verbal continuers while the client is speaking. Non-verbal continuers are aspects of speech that indicate ongoing interest, and may or may not be words. Some examples are "mm-hm," "uh huh," "yes," "right," "go on," "mmm," "ah!" and so forth. When interjected at appropriate pauses during the speech of a communication partner, these continuers can be used to demonstrate active listening (Lown et al., 2014), and thereby increase rapport.

Nodding one's head is also an example of a non-verbal continuer that can indicate active listening (Taylor et al., 2019). While nodding of one's head is also an example of body language, it is included in this section due to its utility in expressing interest in a conversation when use of verbal or non-verbal continuers are not possible (e.g., when a parent is speaking quickly and appropriate pauses in the conversation are not present).

*Paraphrase*

Paraphrasing is a commonly used tool in Social Work and Counseling practice (Rogers & Farson, 1957). Stating what the other person has said, in one's own words, can be useful in ensuring the content of what is being said is understood correctly. It gives the client/caregiver the opportunity to correct the behavior analyst if something has been misunderstood. This skill is helpful in increasing rapport by addressing some of the issues with behavior analyst communication identified by Taylor et al. (2019) that are barriers to developing rapport.

Paraphrasing can also be thought of as a loose form of tacting. By acknowledging (tacting) the client's true concerns, the behavior analyst is communicating that they have heard correctly what has been expressed to them and, if not correct, checking to see if what they have understood is correct. This facilitates feelings of rapport because the client is able to hear their concerns reflected back to them, and therefore know that the behavior analyst has truly heard and understood what they have expressed.

It should be noted what paraphrasing is *not*—it is not simply repeating back the client's sentences word for word. This can appear to the client to be disingenuous, as well as become challenging to do during lengthy and complex conversations. Paraphrasing is also not "interpretation" of what the client is saying. It focuses on the content of the conversation as much as possible, rather than on assumptions about underlying meanings. However, paraphrasing can definitely include tacting of the client's emotional or body-language responses (e.g., "You said you aren't getting enough sleep at night, is that correct? I also noticed your voice seemed sad and teary when you mentioned that"). By tacting the observation of emotions expressed through body language, the behavior analyst is communicating that they are not only listening to the client's words, but also remaining attentive to the entirety of a client's experience. This allows them to confirm clients' emotional responses in a non-judgmental way. The section below on Empathy provides further context on the rapport that can be built through paraphrasing.

*Ask Open-Ended Questions*

As mentioned above, Taylor et al. (2019) identified that clients perceive when behavior analysts seem being too busy to discuss things that are of importance to them. Clients feel more of a sense of rapport when "the behavior analyst regularly asks [them] if [they are] happy with how things are going" (p. 657).

Behaviors identified by the clients in Taylor and colleagues' (2019) article that may increase their feelings of rapport include the behavior analyst regularly asking the caregiver if they are happy with how things are progressing for their child, asking the parent how they themselves are doing, and seeking input and feedback from the parents. These types of questions are all open-ended questions—questions that allow the caregiver to respond with what matters and is relevant to them.

Other examples of open-ended questions could be asking about the family's general functioning, asking how individual members of the family are doing regardless of whether they are the recipients of behavioral service, and asking questions to elicit the caregiver's values and preferences (Lown et al., 2014). Seeking input and feedback from parents via open-ended questions also gives caregivers the opportunity to express thoughts and feelings they may not otherwise feel comfortable bringing up when presented with closed-ended questions.

Above all else, open-ended questions asked of the family should remain contextually specific and attuned to the individual family's experiences and responses. Once responses are provided to open-ended questions, the behavior analyst can seek further clarification through paraphrasing and use of verbal and nonverbal continuers.

*Allow for Silence; Allow Space and Time for Client to Process and Respond*

Another item in Taylor and colleagues' (2019) article that may contribute to problems in the therapeutic relationship is, "The behavior analyst interrupts me during my meetings about my child."

The opposite behavior to interrupting a caregiver would be to remain silent, even during the caregiver's silences. Pauses are a natural part of conversation, and don't always indicate that the speaker has completed their thought. If the behavior analyst refrains from jumping in too quickly during pauses, and instead allows for silence, it may allow the caregiver to feel that their thoughts and feelings are not being interrupted, which can increase feelings of rapport. Indeed, Taylor and colleagues (2019) state that "A clinician rushing to fix the problem may undermine or invalidate the parent's expression of sadness or frustration, potentially jeopardizing investment in treatment" (p. 656). Behavior analysts allow for silence in a conversation to express that they are maintaining presence and focus on the caregiver (Lown et al., 2014).

Following a particularly difficult expression of emotion, a caregiver may also go silent. As mentioned above, they may be readying themselves to go deeper and share very personal experiences or feelings or may be considering what the behavior analyst has said (Cochran & Cochran, 2015). The information subsequently shared may otherwise not have been disclosed if the behavior analyst had not paused for silence.

*Summarize*

While similar to paraphrasing, summarizing what a caregiver or client has expressed during the course of a conversation can be helpful in creating rapport (Miville et al., 2011). This differs from paraphrasing in that the summary should be done at the end of a lengthy and complex conversation (or periodically throughout a longer conversation). It allows the caregiver to feel that the behavior analyst has taken away the important points of what has been discussed. It also gives the caregiver the opportunity to correct anything that the behavior analyst may have misunderstood (Lown et al., 2014).

Summarizing not only includes a summary of the content of the conversation and the treatment plan or next steps, but can also be done to summarize the caregiver's overall feelings expressed during the conversation (i.e., "You're feeling very overwhelmed by the treatment plan we discussed, is that right?"). Feelings of rapport occur when the behavior analyst expresses what is truly important to the client.

*Use of Appropriate Language (Avoid Jargon)*

It is well known that the use of technical jargon, especially when used frequently, can be off-putting, and "may lead the behavior analysts to be perceived as authoritarian or 'expert' rather than collaborative and flexible" (Critchfield et al., 2017, p. 659). Taylor and colleagues (2019) found this in their research as well, as parents indicated that "The behavior analyst uses too much technical language that I don't understand," one of the more frequent behaviors that contributed to problems in the therapeutic relationship. Another related item was "The behavior analyst has an authoritarian demeanor rather than a collaborative one when discussing decisions about my child's program" (Taylor et al., 2019, p. 658).

The Ethics Code for Behavior Analysts (Behavior Analyst Certification Board, 2020) also identifies the importance of the relationship of the behavior analyst and caregiver. Guideline 2.08, Communicating About

Services, states that "Behavior analysts use understandable language in, and ensure comprehension of, all communications with clients" (p. 11).

What this looks like in practice is that the behavior analyst should describe procedures and concepts in layman's terms; in other words, precise, everyday language that describes the concepts being discussed. The language used should also be accessible to and understandable by the caregiver and their family. Special care should be given to ensure that the vocabulary used is matched to the individual's repertoire. For example, a caregiver whose first language is not English may need concepts explained with vocabulary they are familiar with.

If it is necessary to use behavior analytic terms or jargon, these terms can be defined and explained. Use of words that have negative connotations in everyday language, such as "extinction," "discrimination," or "punishment," should be avoided where possible, as caregivers may already have an extensive learning history with the common meaning of these words, and it may be difficult for them to associate a new, neutral meaning to them even if carefully explained by the behavior analyst.

*Category 2: Empathy*

List of teachable behaviors:

9 Novel responding (stay present)
10 Tact the client's feelings, emotions, and perspective prior to moving on
11 Tact your own feelings, emotions, and experiences that may be similar to the client's
12 Match facial expression and tone of voice to the client's
13 Be open and vulnerable
14 Use a non-judgmental approach

In most other health disciplines, both empathy and compassion are regarded as important characteristics of clinical care and are associated with positive client outcomes, increased treatment adherence, client satisfaction, and clinician well-being (Taylor et al., 2019). Empathy, in particular, is one of several factors crucial to a strong therapeutic alliance (American Psychological Association, 2019).

The common definition of empathy usually involves imagining oneself in another's shoes, but empathy can also be understood "as a sequence of reciprocal turns-of talk, starting with the patient's expression of emotion, followed by (…) empathic response by the clinician. These patterns of reciprocity may also include the patient's experience of and response to the clinician's emotions" (Finset & Ørnes, 2017, p. 64). Moreover, Taylor and colleagues (2019) describe empathy as "listening attentively to the parent, taking the parent's perspective, acknowledging and accepting the parent's feelings, and allowing [him or herself] to feel what the parent is feeling in the moment" (p. 655). It is about the clinician seeing the client's world as they see it; to respectfully perceive what the client is bringing from their frame of reference and to communicate that back in a way that makes the client feel they've been understood.

In a review by Feller and Cottone (2003), the authors emphasized the importance of empathy to the therapeutic relationship, which has been long-studied and confirmed within the field of counselling psychology. This dates back to the original work of Carl Rogers (1957). Further, a meta-analysis by Neinhuis and colleagues (2018) revealed a strong relationship between therapeutic alliance, empathy and genuineness.

It is important to also recognize that empathy takes energy, time, and work (Riess, 2017). It requires the clinician to become more open to their own and the client's feelings and to make them visible in the relationship with the client (Cochran & Cochran, 2015; Greenberg, 2011). Empathy is needed for the client to feel understood, empowered to make choices, and to take on an active role in the assessment and treatment process. It requires more than providing solutions. Focusing too early on what the behavior analyst can do to address the client's goals may get in the way of sharing experiences and showing empathy. When searching for a solution when talking, it may be a sign of straying too far from empathizing and listening (Cochran & Cochran, 2015). This section will focus on six skills related to empathy that will help the clinician to increase rapport with clients and mediators, which will in turn help to increase adherence, fidelity and overall treatment outcome.

*Novel Responding (Stay Present)*

One of the most important behaviors that a clinician can exhibit is "being present." As mentioned previously in Acceptance and Commitment Training (ACTr), contact with the *present moment* involves shifting attention to what is happening here and now (Fletcher & Hayes, 2005). This means "contacting both internal stimuli, such as bodily sensations, thoughts, and feelings, and external stimuli, such as sounds, sights, smells, and touch" (Fletcher & Hayes, 2005, p. 320–321). Contact with the present moment allows for observation of relevant thoughts and feelings as they arise and allows for greater overall psychological flexibility (Fletcher & Hayes, 2005). By responding to clients with spontaneous reflections and feedback, behavior analysts can demonstrate to clients that they genuinely care about their well-being and values, which leads to increased collaboration and meaningful goal selection.

## Tact the Client's Feelings, Emotions and Perspective Prior to Moving on

Skinner (1957) defined *tact* as a response "evoked by a particular object or event or property of an object or event" (p. 82) and considered it to be one of the most important verbal operants. Tacts are maintained by generalized social reinforcement (praise) and, thus, they are central to many social interactions (Marchese et al., 2012).

From an evolutionary perspective, emotions serve several functions for us as human beings. These include: motivating us to take action; helping us to survive, strive and avoid danger; helping us to make important decisions; helping other people to understand us and relate to us and, finally, helping us to understand others (All Relationship Matters, 2022). The latter two points are especially important in developing positive rapport and understanding what is of meaning or value to those around us.

Emotions have long been studied in the field of psychology and counselling; however, far less research has occurred within the field of behavior analysis. Although it would be inaccurate to suggest that no work has been done in the field of behavior analysis in regard to emotions, as many behavior analysts have addressed emotions and private events in both experimental studies and from theoretical perspectives, it would be reasonable to say that we, as a field, have yet to come to a consensus on the ways that private events should be defined and studied (Catagnus, 2021). In 1998, Friman et al. (1998) asserted that anxiety and emotions are in fact suitable subjects for behavior–analytic study. They concluded that the most interesting aspects of anxiety disorders may occur as a function of derived rather than direct relations between public events and overt and private responses with avoidance functions (Friman et al., 1998).

In a more recent survey by Catagnus (2021), behavior analysts were asked five questions to gain an understanding of participants' perception of the role that emotions play in their day-to-day work in ABA. Specifically, participants were asked

> how important it is for them to help their clients feel better emotionally, how likely they were to consider emotions in assessment and intervention development, and how their intervention development or revision might be impacted by emotional reactions from clients.
> (Catagnus, 2021, p. 3)

They were also asked "how empathetic they felt their colleagues were and how well the field of behavior analysis responds to emotion and considers emotions in the analysis of and intervention on behavior" (Catagnus, 2021, p. 3). Data from these survey questions suggest that most participants believed emotions play an important role in their work within the field of ABA (Catagnus, 2021). The majority reported that they agreed or strongly agreed that their job was to help clients feel better emotionally and that they were likely or somewhat likely to consider the emotions of caregivers, clients, or coworkers during assessment and intervention development and revision (Catagnus, 2021). By taking the time during each meeting to check in with the client, and surface feelings or emotions related to progress, positive rapport can be established relatively early on in the client–clinician relationship and maintained more easily over time. As stated above, empathy takes time and effort and it is our duty as behavior analysts to provide the opportunity for clients to feel understood and heard prior to moving on to more technical aspects of service in a discussion.

*Tact Your Own Feelings, Emotions, and Experiences that May Be Similar to the Client's*

Words are powerful tools and can easily be used to name the feelings and emotions a client is experiencing, demonstrating empathic listening (Cochran & Cochran, 2015). When a behavior analyst can tact the feelings and emotions of the client, this allows them to show genuine care and concern for their well-being, as the client feels "understood" and that their feelings are an important part of the clinical relationship. This can be done through paraphrasing or summarizing (see Active Listening section above) and also by staying present as the client speaks. In addition, tacting the feelings or worries of clients can help to surface barriers to implementation and prevent avoidance of participating in service.

Just as one can tact another's feelings or emotions, one can also tact one's own emotions. By voicing and sharing our emotions during clinical meetings, clients are better able to see the behavior analyst as a human being who experiences struggle just as they do, and clients are able to relate to the clinician on a more equal level. This helps to minimize unhealthy dynamics in the therapeutic relationship due to power imbalances as the clinician appears less like an "expert" and more as an equal. This skill relates closely to the ability to be open and vulnerable, as explained in more depth below.

Tacting or sharing personal emotions has been defined in psychological literature as "self-disclosure." A balanced approach to self-disclosure is a valuable empathy skill. Indeed, sharing experiences with the client can help "normalize" their feelings while reminding the clinician they may have walked similar paths (Nelson-Jones, 2014). Henretty and Levitt (2010) suggest that clinicians should consider self-disclosing information such as: demographics, feelings and thoughts, mistakes, relevant past struggles, similarities between the client and clinician, and values. When

*Table 2.1* Reasons to Self-Disclose as a Practitioner

| **Reasons to Self-Disclose as a Practitioner** | |
|---|---|
| • Ethical obligation | • To show similarities |
| • Encourage client disclosure | • To provide reassurance |
| • To foster the therapeutic alliance | • To build client self-esteem |
| • To model disclosure for clients | • To reinforce and/or shape desirable client behavior |
| • To encourage clients' autonomy | |
| • To equalize power | • To offer alternative ways to think or act |
| • To repair an impasse or rupture in rapport | |
| | • To help clients recognize boundaries |
| • To correct misconceptions, such as tendencies to perceive clinicians as experts | |
| | • To create a safe environment |
| | • To provide clients with authentic, human-to-human communication |
| • To assist clients in identifying and labeling their emotions | |

Henretty & Levitt, 2010.

clinicians choose to self-disclose, it is essential that they do so with a clear rationale (Henretty & Levitt, 2010). One rationale is the belief that it is an ethical obligation. Other appropriate reasons are: (a) to encourage client disclosure; (b) to foster the therapeutic alliance; (c) to model disclosure for clients; (d) to encourage clients' autonomy; (e) to equalize power; (f) to repair an impasse or rupture in rapport; (g) to correct misconceptions, such as tendencies to perceive clinicians as experts; (h) to assist clients in identifying and labeling their emotions; (i) to show similarities; (j) to provide reassurance; (k) to build client self-esteem; (l) to reinforce and/or shape desirable client behavior; (m) to offer alternative ways to think or act; (n) to help clients recognize boundaries; (o) to create a safe environment; and (p) to provide clients with authentic, human-to-human communication (Henretty & Levitt, 2010).

Sharing one or more similar feelings or experiences with a client can lead to increased positive rapport by making the client feel understood and that they are not alone. It can also be a helpful way to surface barriers to effective treatment and to suggest changes in approach or goals that are more meaningful to the family (Table 2.1).

*Match Facial Expression and Tone of Voice to the Client's*

Listening with empathy is a way of *being*. When doing so, it is apparent in body language among other things (see previous section on Active Listening and Communication). For example, leaning in with arms and legs uncrossed, communicates to the client interest in what they are saying and empathizes with what they are feeling (Cochran & Cochran, 2015).

In addition to body language, empathy can be expressed by matching the client's tone of voice and facial expression (Cochran & Cochran, 2015). If a client expresses that they feel hurt and appear sad, then the clinician can match that hurt and sadness with their own tone of voice when responding. It is not recommended to maintain a cold or neutral expression or tone. Let the client see that what they have shared has affected the clinician on a deeper level by indicating it with one's face and tone, as well as hand gestures or how the body is held (see Body Language section above). Again, this increases the likelihood that the client feels heard and they feel the behavior analyst is authentic, empathetic, and genuine.

*Be Open and Vulnerable*

Vulnerability is the core of all emotions and feelings (Brown, 2012). It is about sharing feelings and experiences with people. Being vulnerable and open is mutual and an integral part in building trust (Brown, 2012). Vulnerability includes letting people see that one is not perfect, that everyone make mistakes, and that no one person knows everything. Being vulnerable is a step to show people that the clinician is worthy of trust, as clients are trusting them with information that could be potentially embarrassing (Brown, 2012). A simple way to demonstrate vulnerability is to acknowledge one's humanness. Everyone has preferences and may not agree with others depending on the topic and personal experiences. Although each person has ways they like to do things, it is fair to be open to discussing other opinions and approaches, engaging in open conversation, and exchanging different ideas. Willingness to listen to others is essential because no one will have the perfect answer (Bloom, 2020).

Sharing feelings, emotions, and experiences with clients therefore increases the likelihood that we, as behavior analysts, can develop a strong sense of rapport through mutual understanding and vulnerability, leading to more socially significant conversations, goals, and interventions.

*Use a Non-Judgmental Approach*

Being non-judgmental means not criticizing, shaming, or rejecting others based on their perspectives or beliefs. It is an important part of learning to experience the present moment "as it is" (Gilbert, 2009).

Non-judgmental listening is about genuinely trying to understand the other person. It is about going beyond just hearing the words spoken and involves understanding exactly what the other person is saying (Oakes, 2020). As the listener, behavior analysts try to put their own views aside and avoid being distracted by personal thoughts and feelings. In addition, it is not appropriate in every situation to react to what the other person is saying but, instead, listening should be continuous, regardless. The goal as clinicians is not to judge or criticize, but simply to listen to and receive what the person is saying at face value (Oakes, 2020). Listening without judgment allows the client (the speaker) to feel heard and valued. It is important to acknowledge that views and opinions are a product of our upbringing and experiences, and that it is a person's individual right to hold their views and opinions (Oakes, 2020).

Three conditions are necessary to form a safe environment in which the client feels comfortable enough to speak openly without fear of being judged (Oakes, 2020):

a Acceptance: Even if a client's views are different from our own, it is crucial that we respect and accept them. Behavior analysts should be understanding of the client's experiences, regardless of their own.
b Genuineness: Being mindful of one's body language to ensure it matches what is said demonstrates that the message is genuine and honest.
c Empathy: As clinicians, try to understand and hear what the client is saying. Imagining oneself in their position truly allows one to feel what they are feeling.

By truly listening to clients' feelings and concerns from a non-judgmental stance, behavior analysts can foster a safe environment in which they feel comfortable sharing honest thoughts and experiences. Such an environment can lead to increased positive rapport and better overall outcomes for the behavioral service.

*Category 3: Compassion*

List of teachable behaviors:

15 Respond to and/or include clients' ideas in behavioral goals
16 Offer to discuss or troubleshoot concerns, even if not behavioral
17 Express interest in/and include relevant information related to family and culture
18 Model calm behavior and language when client is upset/defensive
19 Recognize client and client successes
20 Acknowledge your own mistakes when something is not working

Compassion and empathy are often used with overlapping meanings. One commonly cited definition of compassion is a "feeling that arises in witnessing another's suffering and that motivates a subsequent desire to help" (Goetz et al. 2010, p. 352).

Taylor et al. (2019) similarly defined compassion as the conversion of empathy into "an act aimed at the alleviation of suffering," and Lown et al. (2014) also highlights that compassion goes a step further than empathy and writes that compassion is "the recognition, empathetic understanding of and emotional resonance with the concerns, pain, distress, or suffering of others coupled with motivation and relational action to ameliorate these conditions" (p. 3).

Considered from a behavior-analytic perspective, the function of the six behaviors described in the empathy category is to *understand the client's feelings*, while the function of the behaviors in this final section on compassion is to use this understanding to *better the situation of the client*. Therefore, for the purposes of the framework presented here, compassion behaviors cannot exist without first mastering empathy behaviors.

*Respond to and/or Include Clients' Ideas in Behavioral Goals*

Including client ideas in their own treatment is well-known to increase feelings of empowerment (e.g., Bombard et al., 2018), which can then lead to feelings of rapport. These feelings of rapport, in turn, are known to increase the likelihood of treatment adherence, which improves client outcomes (Stewart, 1995). While this research on client outcomes has not been conducted specifically with the clients of behavior-analytic treatments, this evidence from other healthcare fields can be reasonably extrapolated to also apply to the field of ABA.

Where possible, behavior analysts should aim to include client ideas in their treatment plans. For example, clients may request that materials be created a certain way, that particular actions or activities should be included as part of the treatment plan, or have ideas about which days or times the treatment plan should be implemented. Even if the inclusion of some of these ideas does not directly increase the likelihood of positive treatment outcomes directly, as long as they are not harmful to the client or to the treatment goals, it is preferable to include these ideas as much as is reasonable, given the circumstances. This aligns with Section 2 of the Ethics Code for Behavior Analysts, where "behavior analysts prioritize clients' rights and needs in service delivery" and "make appropriate efforts to involve clients and relevant stakeholders throughout the service relationship" (2.01, 2.08; BCBA, 2020, p. 10). This will allow clients to feel included, which will not only increase their feelings of rapport with the behavior analyst, but hopefully increase the likelihood that the outcomes of the treatment will be successful.

Of course, it is not always possible to include client ideas. Perhaps there are not enough resources, the inclusion of certain ideas would be detrimental to the client or treatment outcome, or the ideas are not evidence-based. In these circumstances, it is important for the behavior analyst to respond to and acknowledge the ideas that the client is proposing, rather than ignoring them. Even responding to ideas that are not included has the benefit of the client feeling heard and can still increase feelings of rapport. More about responding to or including ideas that are not behavioral in nature are discussed in the next section.

*Offer to Discuss or Troubleshoot Concerns, Even if Not Behavioral*

Feeling safe enough to speak what is on one's mind is vital amid times of uncertainty, a state that is unfortunately all too familiar for the caregivers we work with (Kerrissey et al., 2022). "Feeling heard" may help mitigate burnout and enable adaptation during times of uncertainty or crisis, and expressing empathy and compassion through problem-solving is one way to let mediators know they are "heard" (Kerrissey et al., 2022). Kerrissey et al. (2022) go on to clarify that feeling heard is defined as "the belief that the content of one's voiced ideas or questions will be recognized and responded to" (p. 308).

When behavior analysts are confronted with concerns from a client that are not behavioral in nature, the tendency is to ignore them in favor of remaining focused on behavioral goals. However, in order to built rapport with clients, allowing their concerns to feel heard is an important component that cannot be overlooked. The compassionate component of this skill, however, goes one step further by reminding the behavior analyst to consider how non-behavioral goals can be addressed, and to provide resources or assistance to that effect. For example, a family could express something as simple as not knowing how to access an application for social assistance as they don't know how to use a computer; the behavior analyst could offer to print it for the client and bring it to the following session. This is an example of a very straightforward non-behavioral concern (to provide assistance) that takes minimal effort on the part of the behavior analyst.

Of course, there are often more complex non-behavioral concerns brought up by a client. Frequently, these concerns require consultation with interdisciplinary professionals, who may not always be readily available or accessible, depending on the individual situation and whether clinical services are being accessed through out-of-pocket funding from the family or social service organizations with long waitlists. Every effort should be made on the part of the behavior analyst to access interdisciplinary consultation where appropriate. Even if ultimately unsuccessful, rapport can be built and compassion expressed through this behavior.

Brodhead (2015) developed a decision-making model for assessing non-behavioral treatments with the goal of increasing collaboration with clients and other professionals, and maximizing treatment success, safety, social validity, and overall resources for clients and families (see Figure 2.2). If non-behavioral treatments could interfere with goals, it is important to listen and to understand the mediator's concerns, discuss alternatives and share knowledge and experiences related to the concerns raised.

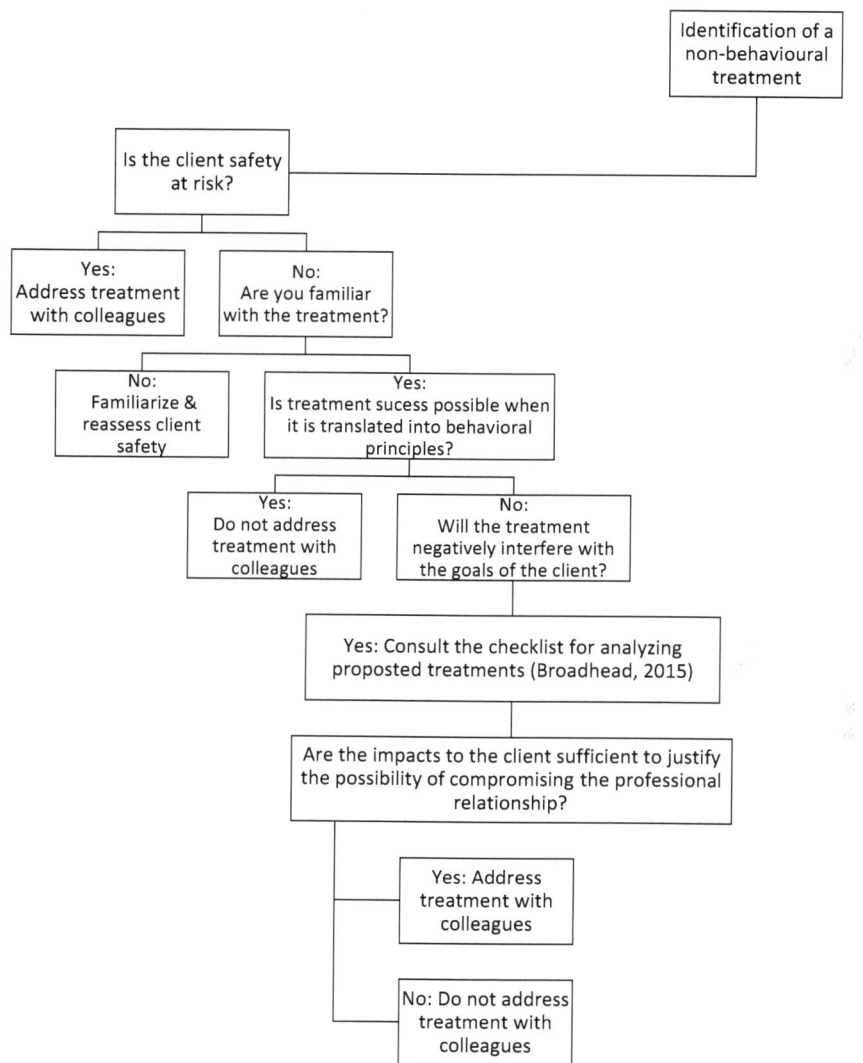

*Figure 2.2* A decision-making model for assessing non-behavioral treatments to determine if clinicians should address the implementation of the treatment with the colleague (Broadhead, 2015).

*Express Interest in/and Include Relevant Information Related to Family and Culture*

Skinner (1971) loosely defines culture as the various contingencies of reinforcement present in our local environment, and that we experience throughout our lives. All individuals are a part of at least one culture. These contingencies shape our behavior, and these behaviors are not always familiar or acceptable if the culture of the behavior analyst and the client are not the same (Fong et al., 2016). As culture is therefore an important influencer of behavior, it would be remiss for a behavior analyst to miss the opportunity to express interest in the cultural contingencies affecting their clients.

Beaulieu et al. (2019) assert that culture intersects with behavior analysis in several ways. All of behavior is affected by our culture in some way. They go on to write that, "because culture underlies much of our behavior, understanding the differences in cultures and how to work effectively with individuals from diverse backgrounds is critical when working in applied settings" (Beaulieu et al., 2019, p. 557) and that, furthermore, "ignored or unidentified cultural variables may become barriers to the delivery of effective treatment" (Betancourt et al., 2016 & Kodjo, 2009, as cited in Beaulieu et al., 2019, p. 558), including by damaging rapport between the behavior therapist and client (Parette & Huer, 2002). It is vital to the relationship and to rapport building to be aware of potential barriers as much as possible.

One way that being attentive to a client's culture can help build rapport is that it allows the behavior analyst to be aware of the client's personal values, preferences, and reinforcers. Being aware of these contingencies can also help build rapport by assisting the behavior analyst in coming up with methods to reduce factors that may make access to services more difficult, and improving the quality of these services, especially for populations where access and quality to behavioral services have historically been paltry due to cultural and racial differences (Fong et al., 2016). Being attentive to these factors also helps the behavior analyst express that they care about the entire family, not just the client, which was one of the areas of weakness identified by Taylor and colleagues (2019).

This type of cultural awareness requires a significant level of humility, and a commitment to a lifelong exploration of one's own internal biases (Wright, 2019). These internal biases may create inadvertent power imbalances and also prevent behavior analysts from developing rapport-filled relationships with clients on an equal footing (Tervalon & Murray-Garcia, 1998).

### Model Calm Behavior and Language When Client Is Upset/Defensive

The importance of modeling has long been studied within the field of behavior analysis. Lovaas (1966) showed that relatively complex repertoires can be established through a combination of modeling and reinforcement procedures.

O'Connor (1969) studied the effects of symbolic modeling on preschool children and found significant increases in social behavior among children with relatively severe social deficits after they viewed videos of positive social interactions between children. Similarly, it is possible that the modeling of calm behaviors and language by the behavior analyst during an emotional conversation with clients can increase the likelihood that clients will also engage in calm behaviors. This, in turn, can create more effective conversations that are not derailed by strong emotional behaviors, and can also create an overall increase in feelings of rapport.

Modeling calm language can take many forms. One non-verbal way of expressing calm with one's voice is to use a volume that is just below normal speaking voice. Previous studies have shown that the human ear is very attuned to empathy and compassion that is expressed through the sound of a calm voice (Kraus, 2017). Changing one's voice volume is a simple way of increasing the likelihood that the client feels a sense of rapport when the behavior analyst is speaking, especially during difficult conversations. Using a low, calm-sounding volume also facilitates co-regulation, a term used in the field of psychology that refers to the effect that modeling calm behaviors have on the listener—namely, a co-regulating that occurs (Murray et al., 2015). This can be an effective way to provide support and reassurance to the client and increase therapeutic rapport.

Word choice also matters. Tan and colleagues (2016) found that selecting words that had calm or positive associations were more likely to persuade another person when having arguments online. While persuasion should not necessarily be the goal of all conversations between the behavior analyst and client, this finding highlights that word choice does matter when attempting to build rapport with another person.

### Recognize Client and Mediator Successes

In their 2019 article, Leblanc and Nosik provide recommendations and tools for planning and leading effective meetings. Meetings can have many purposes, but one of the most important aspects is "establishing and maintaining rapport and interpersonal relationships" (Leblanc & Nosik, 2019, p. 697).

The authors suggest that providing social reinforcement for the efforts and participation of others is a vital part of leading effective meetings. Client contributions and successes can be acknowledged during conversation with nonverbal behaviors like smiles and nods (see section on "continuers"), as well as short praise statements (e.g., "I hadn't thought of that!" or "That's a great idea."). This type of positive feedback is also an important aspect of Behavioral Skills Training (BST) and that is frequently used when training mediators, and has been stated as one of the most effective components of BST (Ward-Horner & Sturmey, 2012). Similar to the importance Taylor et al. (2019) placed on the need for behavior analysts to avoid underestimating client abilities, it is equally important to recognize the contributions and successes that a client and/or their caregiver may brainstorm or achieve as an aspect of rapport building that cannot be overlooked.

Including successes in a written format, where appropriate (e.g., in a client's assessment report), or bringing up the success with others in the client's interdisciplinary team, can be powerful recognition. Since this recognition is intended to function as a reinforcer of feelings of rapport in the client, ensuring the recognition is provided in a timely manner (shortly after it is expressed) is important. Any praise or acknowledgement should also be as specific as possible, so that the client feels the behavior analyst is expressing genuine thoughts and feelings. Behavior-specific praise is a commonly used strategy to increase behaviors in a variety of contexts. Ennis et al. (2018) define behavior-specific praise as adding a specific description of behavior to clarify what is being acknowledged. An example of this would be, "Your idea to include music time on his visual schedule was really great!" This is in contrast to general praise, which simply uses statements like "Great idea!" or "I like that." Using behavior-specific praise can increase the likelihood that clients share more of their thoughts in future conversations, which in turn provides more opportunities for rapport-building.

*Acknowledge Your Own Mistakes When Something Is Not Working*

Taylor et al. (2019) found families perceived behavior analysts to have a weakness in acknowledging mistakes and treatment failure. Clinician mistakes and failures in treatment are an inevitable part of providing behavior analytic services, and families prefer honesty from their clinician when these things happen.

This is not a new idea. In 1991, Neuringer wrote, "If behaviorists were more humble, their effectiveness as scientists would increase" (p. 1). Humility is difficult to operationally define, but he goes on to express that behavior analysts should keep at the forefront of their minds that "all knowledge is provisional and that one's most deeply held positions must be continually reconsidered" (Neuringer, 1991, p. 1). When working with families from a variety of cultures and life experiences, it is especially important to keep this in mind. Strategies that may work in a family that is part of their society's main culture or racial group may not work in a family with similar resources but which is part of a minority group. Behavior plans that worked for one client may not work for another client with similar problem behaviors, especially in situations where clients or families have experienced some form of trauma (Rajaraman et al., 2022). And so on.

Being willing to acknowledge when a plan is not working, or even that a mistake has been made in selecting the plan (due to a mismatch with the family's culture or resources, a misunderstanding of the function of the client's behavior, mistakes made in the assessment, not effectively training the mediators, etc.) is something that should be at the forefront for behavior analysts.

Being open to criticism and being willing to consider alternative views expressed by the family are other ways of acknowledging when something is not working as intended (Neuringer, 1991). It may be useful for a behavior analyst to practice different ways of admitting mistakes. Writing a list of different phrases and referring to them frequently may help this behavior be more readily expressed when confronted with one's own mistakes. Some examples of these phrases could be:

> I did not take <factor x> into account, so my program is not working. I am sorry for that.
> I did not realize how important <factor x> was to your family, and I think that is why we are not seeing results. Can I get your input about how we make changes?
> I was pressed for time and forgot to read through <client's> previous reports, I apologize for missing this.

Even though this may be one of the most uncomfortable changes for a behavior analyst to make, the immediate benefits to the therapeutic relationship and increased feelings of rapport between both parties will hopefully reinforce this behavior and encourage clinicians to make it a regular part of their interactions with clients.

## Conclusion

"People skills" within the domains of empathy and compassion have long been viewed as essential parts of clinical service in other health disciplines. Such skills are related to positive outcomes for clients, increased adherence to treatment, and an overall increase in client satisfaction.

Terms like active listening, empathy, and compassion are frequently used to refer to skills that clinicians should have when working with families, but these terms are not often defined, leaving behavior analysts struggling to know exactly how to implement these behaviors with clients. Behavior analysts who supervise other behavioral clinicians are also left wondering how to train their staff to increase their skill repertoires in these areas. The difficulty in operationally defining these skills to teach them in a behavior analytic context could be argued as holding back the science of behavior analysis in tackling them despite the plentitude of research from other health disciplines. More recently, the behavior analytic research itself has shown that behavior analysts who are adept at demonstrating empathy and active listening are more likely to be successful in overcoming barriers to treatment and identifying factors for success (Taylor et al., 2019).

The hope is that by breaking down skills that were previously considered abstract constructs into 20 observable and measurable behaviors, the discipline of behavior analysis can begin to consider how to provide training to behavior analysts on such skills.

There is also a larger goal to strive for. Jon Bailey (1991) highlighted that "We are never going to sell our approach to our society as long as we stress the control of human behavior. We must instead begin focusing on important values held by all behavior analysts" (1991). Behavior analysts all over the world, regardless of environment or culture, have this in common with the clients we encounter: the desire for compassion towards ourselves and the instinctive inclination to express compassion towards others. These are values that are universally shared by all human beings, behavior analyst or not.

**The authors would like to acknowledge Danielle Flood for her research contributions to this chapter.

## References

Akechi, H., Senju, A., Uibo, H., Kikuchi, Y., Hasegawa, T., Hietanen, J. K. (2013). Attention to eye contact in the West and East: autonomic responses and evaluative ratings. *PLoS One, 8(3)*, e59312. https://doi.org/10.1371/journal.pone.0059312

American Psychological Association. (2019) *What the evidence shows.* www.apa.org/monitor/2019/11/ce-corner-sidebar.html

Baer, D. M., Wolf, M. M., & Risley, T. R. (1968). Some current dimensions of applied behavior analysis. *Journal of Applied Behavior Analysis, 1*, 91–97. https://doi.org/10.1901/jaba.1968.1-91

Baer, D. M., Wolf, M. M., & Risley, T. R. (1987). Some still-current dimensions of applied behavior analysis. *Journal of Applied Behavior Analysis, 3(20)*, 13–327. https://doi.org/10.1901/jaba.1987.20-313

Bailey, J. S. (1991). Marketing behavior analysis requires different talk. *Journal of Applied Behavior Analysis, 24*, 445–448.

Beaulieu, L., Addington, J. & Almeida, D. (2019). Behavior analysts' training and practices regarding cultural diversity: The case for culturally competent care. *Behavior Analysis in Practice, 8*, 557–575. https://doi.org/10.1007/s40617-018-00313-6

Behavior Analyst Certification Board. (2020). *Ethics code for behavior analysts.* https://bacb.com/wp-content/ethics-code-for-behavior-analysts/

Betancourt, J. R., Green, A. R., Carrillo, J. E., & Owusu Ananeh-Firempong, I. I. (2016). Defining cultural competence: A practical framework for addressing racial/ethnic disparities in health and health care. *Public Health Reports, 118*, 293–302. https://doi.org/10.1093/phr/118.4.293

Binetti, N., Harrison, C., Coutrot, A., Johnston, A., Mareschal, I. (2016). Pupil dilation as an index of preferred mutual gaze duration. *Royal Society Open Science, 3*: 160086. https://doi.org/10.1098/rsos.160086

Bloom, T. J. (2020). The importance of vulnerability in pharmacy educators. *American Journal of Pharmaceutical Education, 84*(7), 884–885. https://doi.org/10.5688/ajpe7939

Bombard, Y., Baker, G. R., Orlando, E., Fancott, C., Bhatia, P., Casalino, S., Onate, K., Denis, J. L. & Pomey, M. P. (2018). Engaging patients to improve quality of care: a systematic review. *Implementation Science, 13*(1), 1–22. https://doi.org/10.1186/s13012-018-0784-z

Brodhead, M. T. (2015). Maintaining professional relationships in an interdisciplinary setting: Strategies for navigating nonbehavioral treatment recommendations for individuals with autism. *Behavior Analysis in Practice, 8*, 70–78. https://doi.org/10.1007/s40617-015-0042-7

Brown, B. (2012). *Daring greatly: How the courage to be vulnerable transforms the way we live, love, parent, and lead.* Penguin.

Catagnus, R. M., Griffith, A. K., & Umphrey, B. J. (2021). Anger, fear, and sadness: how emotions could help us end a pandemic of racism. *Behavior Analysis in Practice*, 1–12. https://doi.org/10.1007/s40617-021-00581-9

Cochran, J. L., & Cochran, N. H. (2015). *The heart of counseling: Counseling skills through therapeutic relationships.* Taylor & Francis Group.

Critchfield, T. S., Doepke, K. J., Epting, K. L., Becirevic, A., Reed, D. D., Fienup, D. M., Kremsreiter, J. L., & Ecott, C. L. (2017). Normative emotional responses to behavior analysis jargon or how not to use words to win friends and influence people. *Behavior Analysis in Practice, 10*, 97–106. https://doi.org/10.1007/s40617-016-0161-9

Ennis, R. P., Royer, D. J., Lane, K. L., Menzies, H. M., Oakes, W. P., & Schellman, L. E. (2018). Behavior-specific praise: An effective, efficient, low-intensity strategy to support student success. *Beyond Behavior, 27*(3), 134–139. https://doi.org/10.1177/1074295618798858

Feller, C. P. & Cottone, R. R. (2003). The importance of empathy in the therapeutic alliance. *The Journal of Humanistic Counseling, Education and Development, 42*(1), 53–61. https://doi.org/10.1002/j.2164-490X.2003.tb00168.x

Finset, A. & Ørnes, K. (2017). Empathy in the clinician-patient relationship: The role of reciprocal adjustments and processes of synchrony. *Journal of Patient Experience, 4*(2), 64–68. https://doi.org/10.1177/2374373517699271

Fletcher, L. & Hayes, S. (2005). Relational frame theory, acceptance and commitment therapy, and a functional analytic definition of mindfulness. *Journal of Rational-Emotive & Cognitive-Behavior Therapy, 23*(4), 315–336.

Fong, E. H., Catagnus, R. M., Brodhead, M. T., Quigley, S. & Field, S. (2016). Developing the cultural awareness skills of behavior analysts. *Behavior Analysis in Practice, 9*, 84–94. https://doi.org/10.1007/s40617-016-0111-6

Friman, P. C., Hayes, S. C. & Wilson, K. G. (1998). Why behavior analysts should study emotion: The example of anxiety. *Journal of Applied Behavior Analysis, 31*, 137–156. https://doi.org/10.1901/jaba.1998.31-137

Goetz, J., Keltner, D. & Simon-Thomas, E. (2010). Compassion: An evolutionary analysis and empirical review. *Psychological Bulletin, 136*, 351–374. https://doi.org/10.1037/a0018807

Gilbert, P. (2009). Introducing compassion-focused therapy. *Advances in Psychiatric Treatment, 15*(3), 199–208.

Greenberg, L. S. (2011). *Emotion-focused therapy.* American Psychological Association.

Hayes, S. C., Masuda, A., Bissett, R., Luoma, J., & Guerrero, L. F. (2004). DBT, FAP, and ACT: How empirically oriented are the new behavior therapy technologies? *Behavior Therapy, 35*(1), 35–54.

Hayes, S. C., Masuda, A., & Demey, H. (2003). Acceptance and commitment therapy and the third wave of behavior therapy. *Gedragstherapie, 36*, 69–96. https://doi.org/10.1016/S0005-7894(04)80013-3

Henretty & Levitt (2010). The role of therapist self-disclosure in psychotherapy: A qualitative review. *Clinical Psychology Review, 30*, 63–77. https://doi.org/10.1016/j.cpr.2009.09.004

Kerrissey, M. J., Hayirli, T. C., Bhanja, A., Stark, N., Hardy, J., Peabody, C. R. (2022). How psychological safety and feeling heard relate to burnout and adaptation amid uncertainty. *Health Care Management Review, 47*(4), 308–316. https://doi.org/10.1097/HMR.0000000000000338

Kraus, M. W. (2017). Voice-only communication enhances empathic accuracy. *American Psychologist, 72*(7), 644–654. http://dx.doi.org/10.1037/amp0000147

Kodjo, C. (2009). Cultural competence in clinician communication. *Pediatrics in Review/American Academy of Pediatrics, 30*(2), 57. https://doi.org/10.1542/pir.30-2-57

Leaf, J. B., Leaf, R., McEachin, J., Taubman, M., Ala'i-Rosales, S., Ross, R. K., Smith, T., & Weiss, M. J. (2016). Applied behavior analysis is a science and, therefore, progressive. *Journal of Autism and Developmental Disorders, 46*(2), 720–31. https://doi.org/10.1007/s10803-015-2591-6

Leblanc, L.A. & Nosik, M.R. (2019). Planning and leading effective meetings. *Behavior Analysis in Practice, 12*, 696–708. https://doi.org/10.1007/s40617-019-00330-z

Lejuez, C., Hopko, D., Levine, S., Gholkar, R., & Collins, L. (2005). The therapeutic alliance in behavior therapy. *Psychotherapy: Theory, Research, Practice, Training.* 42, 456–468. https://doi.org/10.1037/0033-3204.42.4.456

Lovaas, I., Berberich, J. P., Perlof, B. F., & Schaeffer, B. (1966). Acquisition of imitative speech by schizophrenic children. *Science, 151*, 705–707. https://doi.org/10.1126/science.151.3711.705

Lown, B. A., McIntosh, S., McGuinn, K., Aschenbrener, C., DeWitt, B. B., Chou, C., Durrah, H., Irons, M., King, A., & Schwartzberg, J. (2014). *Triple C conference framework tables.* www.theschwartzcenter.org/media/Triple-C-Conference-Framework-Tables_FINAL.pdf

Marchese, N., Carr, J., Leblanc, L., Rosati, T. & Conroy, S. (2012). The effects of the question "What is this?" On tact-training outcomes of children with autism. *Journal of Applied Behavior Analysis, 45*, 539–547. https://doi.org/10.1901/jaba.2012.45-539

Miville, M. L., Redway, J. A. K., & Hernandez, E. (2011). Microskills, trainee competence, and therapy outcomes: Learning to work in circles. *The Counseling Psychologist, 39*(6), 897–907. https://doi.org/10.1177/001100001140443

Murray, D. W., Rosanbalm, K., Chrisopoulos, C. & Hamoudi, A. (2015). *Self-Regulation and Toxic Stress: Foundations for understanding self-regulation from an applied developmental perspective.* OPRE Report #2015-21. Washington, DC: Office of Planning, Research and Evaluation, Administration for Children and Families, US Department of Health and Human Services.

National Association of Social Workers. (2018). *Careers.* www.socialworkers.org/Careers/CareerCenter/Explore-Social-Work/WhyChoose-the-Social-Work-Profession

Neinhuis, J. B., Owen, J., Valentine, J. C., Winkeljohn Black, S., Halford, T. C., Parazak, S. E., Budge, S. & Hilsenroth, M. (2018). Therapeutic alliance, empathy, and genuineness in individual adult psychotherapy: A meta-analytic review. *Psychotherapy Research, 28*(4), 593–605. https://doi.org/10.1080/10503307.2016.1204023

Nelson-Jones, R. (2014). *Practical counselling and helping skills.* Sage.

Neuringer A. (1991). Humble behaviorism. *The Behavior Analyst, 14*(1), 1–13. https://doi.org/10.1007/BF03392543

O'Connor, R. D. (1969). Modification of social withdrawal through symbolic modeling. *Journal of Applied Behavior Analysis, 2*, 15–22. https://doi.org/10.1901/jaba.1969.2-15

Oakes, L. (January 6, 2020). *Non-judgemental listening: How and why? Happy, safe, ready psychology.* https://hsrpsychology.co.uk/blog/non-judgemental-listening-how-and-why/

Parette, P., & Huer, M. B. (2002). Working with Asian American families whose children have augmentative and alternative communication needs. *Journal of Special Education Technology, 17*(4), 5–13. https://doi.org/10.1177/016264340201700401

Pfeifer, J. H., Iacoboni, M., Mazziotta, J. C., Dapretto, M. (2008). Mirroring others' emotions relates to empathy and interpersonal competence in children. *Neuroimage, 39*(4), 2076–2085. https://doi.org/10.1016/j.neuroimage.2007.10.032

Rajaraman, A., Autins, J. L., Gover, H. C., Cammilleri, A. P., Donelly, D. R., & Hanley, G. P. (2022). Toward trauma-informed applications of behavior analysis. *Journal of Applied Behavior Analysis, 55*(1), 40–61. https://doi.org/10.1002/jaba.881

Riess, H. (2017). The science of empathy. *Journal of Patient Experience, 4*(2), 74–77. https://doi.org/10.1177/2374373517699267

Robins, C. J., & Chapman, A. L. (2004). Dialectical behavior therapy: Current status, recent developments, and future directions. *Journal of Personality Disorders, 18*, 73–89. https://doi.org/10.1521/pedi.18.1.73.32771

Rogers, C. & Farson, R. E. (1957). *Active listening*. Guilford Press.

Skinner B. F. (1957). *Verbal behavior*. Appleton-Century-Crofts.

Skinner, B. F. (1971). *Beyond freedom and dignity*. Knopf.

Stewart, M.A. (1995). Effective physician-patient communication and health outcomes: A review. *Canadian Medical Association Journal, 152*(9), 1423–1433.

Tan, C., Niculae, V., Danescu-Niculescu-Mizil, C., & Lee, L. (2016). Winning arguments: Interaction dynamics and persuasion strategies in good-faith online discussions. *Proceedings of the 25th International Conference on World Wide Web*, 613–624. https://doi.org/10.1145/2872427.2883081

Taylor, B. A., LeBlanc, L. A., & Nosik, M. R. (2019). Compassionate care in behavior analytic treatment: Can outcomes be enhanced by attending to relationships with caregivers? *Behavior Analysis in Practice, 12*(3), 654–666. https://doi.org/10.1007/s40617-018-00289-3

Tervalon, M., & Murray-Garcia, J. (1998). Cultural humility versus cultural competence: A critical distinction in defining physician training outcomes in multicultural education. *Journal of Health Care for the Poor and Underserved, 9*, 117–125. https://doi.org/10.1353/hpu.2010.0233

Ward-Horner, J., & Sturmey, P. (2012). Component analysis of behavior skills training in functional analysis. *Behavioral Interventions, 27*(2), 75–92. https://doi.org/10.1002/bin.1339

Why Talk About Feelings? (2022). *All relationship matters*. www.allrelationshipmatters.com.au/insights-healthy-relationships/why-talk-about-feelings.

Wright, P. I. (2019). Cultural humility in the practice of applied behavior analysis. *Behavior Analysis in Practice, 12*, 805–809. https://doi.org/10.1007/s40617-019-00343-8

# 2.2 Cultural Competence in Communication

*Tasmina Khan*

### What Is Culture?

Hofstede defined culture as "the collective programming of the mind which distinguishes the members of one group or category of people from another" (1997, p. 226). Many scholars from the cross-cultural school of thought maintain that culture and communication are intertwined.

Skinner (1953) defined culture as variables "arranged by other people" (p. 419). Sugai et al. (2011) further elaborated the definition as "the extent to which a group of individuals engage in overt and verbal behavior reflecting shared behavioral learning histories, serving to differentiate the group from other groups, and predicting how individuals within the group act in specific setting conditions" (p. 200). These include the distinguishable stimuli (e.g., race, socioeconomic class, age, religion, sexual orientation, ethnicity, disability, nationality, and geographic context) that vary across cultures.

An individual's cultural identity is defined as a unique set of distinguishable stimuli and response classes collectively (Sugai et al., 2011). People grow up with this unique set of distinguishable stimuli and response classes that influence one's behaviors in many different environments. Being aware of one's own and other individuals' cultural identities allows one to see and understand the various behavioral responses in each situation. This unique blend of behavioral responses to various situations (i.e., social communication at a party) is the root of cultural diversity.

In the past fifty years, cultural competence has been a key aspect of psychological thinking and practice (DeAngelis, 2015). When behavior analysts are culturally competent, their attitudes or beliefs will reflect a greater understanding of cultural conditioning that affects personal beliefs, values, and attitudes. This understanding of the differences of each person's behavior through past learning histories will assist in competence to understand the worldviews of individuals and groups with cultures

DOI: 10.4324/9781003300465-8

different from one's own. In terms of skills, cultural competency assists in utilizing culturally appropriate intervention and communication skills.

Consider the first case study, where a cultural practice may be seen as "crossing boundaries" or even "inappropriate."

---

**Case Study 1**

Robert, a businessman, was sent to the airport to receive a British lady on behalf of their business team and was trained on how he should welcome the lady by kissing her hand.

Cheek or hand kissing is a common culture of greeting in many places. Robert successfully received and welcomed the lady client and brought her to their office for the meeting. Robert later shared his experience with his friend, Daniel, who is a behavior analyst. Robert also stated that it was very important to greet in that way for the person to feel welcomed and respected. However, from Daniel's perspective, kissing a client, whether it was the hand or cheek could be seen as acting in compliance with Guideline 1.14—Romantic and Sexual Relationships (BACB, 2020). However, in some cultures and situations, kissing is often paired with romantic relations.

Daniel remembered starting his career and hearing stories of a colleague who met with a Muslim family. After a handshake, the husband, an Arab man, slapped Daniel on his face for being disrespectful to his wife by attempting to touch her hand. While kissing on the cheek or hand is appreciated in some cultures, it is prohibited in other cultures such as Islam. In Islam, kissing anyone from the opposite sex, who is not an immediate family member, is a crime and a great sin.

---

Although this may be seen from many different perspectives, the case study highlights how a common, polite behavior in one culture can be seen as offensive and crossing boundaries in another, highlighting the need to analyze and not jump to conclusions as to the function of each behavior.

**What Is Cultural Competence?**

Flaskerud (2007) defined *cultural competence* as "a set of congruent behaviors, attitudes, policies, and structures that come together in a system or agency or among professionals that enables the system, agency, or professionals to work effectively in cross-cultural situations" (p. 121).

In everyday life, cultural competence is crucial when people from multidisciplinary backgrounds work together. In communication, people need to know about other parties' likes, dislikes, beliefs, races, religions, and cultures to develop and maintain trust while enabling them to work successfully as a team.

Today's world is competitive and is a collective world. North America is a place where people of a variety of cultures congregate and thrive together. For behavior analysts to be successful in their professional lives, they need to pay close attention to the details of the diverse cultures around them. While some practices are prohibited in some cultures, those could be extremely important in other cultures.

## A Working Definition for Cultural Competence

In past literature, the descriptors of different disciplines (e.g., public health, gerontology, and nursing) spelled out skills, attitudes, and values in three areas: cultural knowledge, cultural sensitivity, and collaboration with the community to be served. Flaskerud (2007) further explained cultural knowledge as actively learning about the community—its ethnicities, languages, origins, immigration or migration history, acculturation level, economy, sources of income, family and social structures and roles, value systems and beliefs, education levels and literacy, geography, and ecologic environment. Nair (2022) states that "the things you do and the practices you were taught inform whom you become." This learning of knowledge allows the behavior analyst to understand some commonly held cultural practices and histories behind them (see Figure 2.3).

However, this knowledge does not always allow one to act appropriately. Similarly, cultural sensitivity is an ethic or a moral imperative to value and respect the beliefs, norms, and practices of the people to be served (Flaskerud, 2007). This underlying belief assists the behavior analyst to be open when new situations or perspectives arise that they may or may not know about—the acceptance. During communication, staying sensitive to the cultural differences of both parties respects the diverse cultural populations clients live in.

In terms of collaboration with the community to be served, Flaskerud (2007) highlights the people served by a certain agency, program, or a group of people in a geographic boundary who have shared identity and experiences, similar beliefs, values, and norms. By working together with this community, the behavior analyst will be able to integrate their practice continuing to gain cultural knowledge and cultural sensitivity. When working in a multicultural collaborative team, one needs to be inquisitive, sympathetic, and knowledgeable about the cultures of every stakeholder of the team.

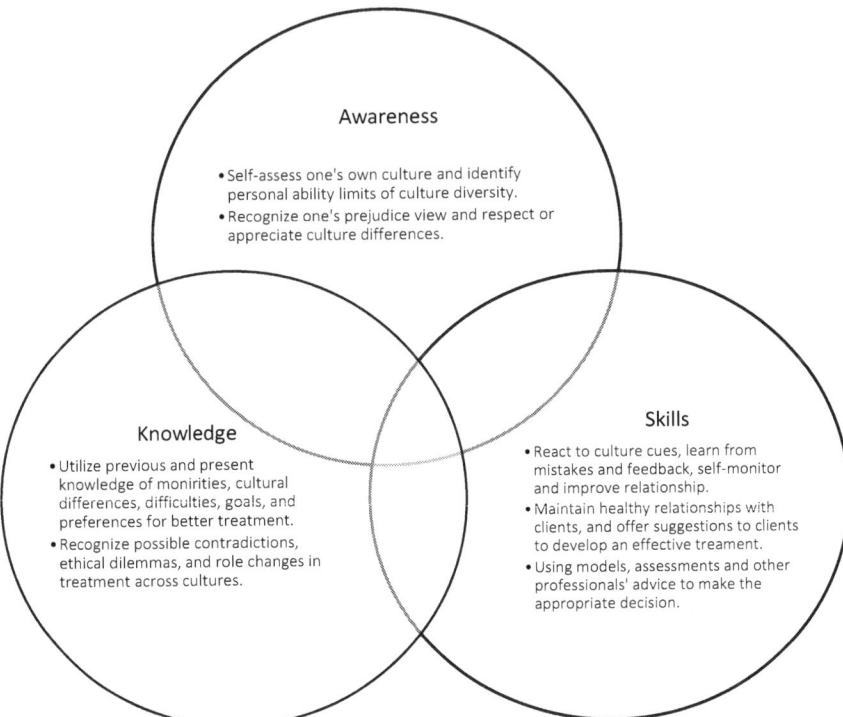

*Figure 2.3* A working definition of cultural competence. Adapted from Sue et al. (1982, p. 49).

## What BACB Says About Having Cultural Competency as a Behavior Analyst (BA)? What Are the Ethical Obligations to Maintain Those?

The Ethics Code for Behavior Analysts (BACB, 2020) reflects the urgency of having cultural competence in their core principles section. They suggested that behavior analysts ensure competency by "Working to continually increase their knowledge and skills related to cultural responsiveness and service delivery to diverse groups" (p. 4). Jimenez-Gomez & Beaulieu (2022) claim that the term "culturally responsive" is aligned with the terminology used in the Ethics Code for Behavior Analysts (BACB, 2020), explaining how cultural responsiveness and ABA overlap in several ways. For example, both avoid gross generalizations (e.g., stereotypes) and stress the importance of individualization in terms of individuals and their individual behaviors that are impacted by individual cultures. Code 1.07 (Cultural Responsiveness and Diversity) highlights behavior analysts' responsibilities to engage in regular professional development to address and acquire skills in diversity and cultural competency as well as identifying personal/supervisee biases to serve each client's diverse needs (BACB, 2020). The American Psychological Association's (APA; 2003) multicultural guidelines recognize that each clinician will have attitudes and behaviors as cultural human beings that could negatively influence their interactions with and perceptions of those with different cultural backgrounds. Staying aware of one's personal beliefs, biases, and norms is needed to be culturally competent. Fong et al. (2017) suggest that developing self-awareness may prevent our biases from impeding how we serve culturally diverse clients.

---

**Case Study 2**

Jackie, a fifteen-year-old girl with an intellectual disability, prefers to identify himself as a male. He is taking medication to help his physical transformation with the help of his physician. He is in the process of changing himself internally and externally from a female to a male and is very engaged with the entire process. When his new 1:1 support staff, Nickie comes, she uses the pronouns "she" and "her" to talk about Jackie and offers him clothing based on how he still looks based on Nickie's judgment. It makes Jackie very upset and frustrated. Jackie does not like the new staff and keeps yelling at her. He also slams his door and locks it as soon as the new staff shows up for the shift.

---

From the above scenario, Nickie could first learn about her client's preferences, values, likes, and dislikes before starting to work with him and could be successful in making a good rapport with her client. Sugai et al. (2011) discussed by acknowledging the importance of culture, how behavior analysts can help achieve socially meaningful goals such as reducing disparities in access to services and improving the quality of services for diverse populations in behavioral health systems (U.S. Department of Health and Human Services, n.d.). The social significance of goals is different for every individual. Choosing the most appropriate and socially significant goal that will increase the quality of a person's life can be possible when the intervention planner is aware of the person's social and moral values, cultures, and preferences. In this regard, Spring (2007) suggests that evidence-based services require a combination of clinical expertise and knowledge of the client's preferences and learning histories. Behaviorally, cultural awareness may be defined as the discriminated operant of tacting contingencies of reinforcement and punishment administered by a group of individuals. Before implementing a contingency based on an individual's behavior, it is necessary to discriminate whether the contingency is culturally appropriate for the individual. Cultural awareness is very important because behavioral patterns that are viewed as problematic in one culture may be the norm in other cultures (Goldiamond, 2002; Vandenberghe, 2008). Fong, et al. (2017) has said that culturally aware interventions, which seek an understanding of client values, characteristics, preferences, and circumstances would honor the client's culture and allow the client to be successful in each environment. Everybody's needs are different and, while providing services, the service provider needs to know what is more important and appropriate for a specific person to live an independent and successful life.

> **Case Study 3**
>
> Sherron has just started working with Abid last month. She goes to Abid's home three days a week. She carries a huge transparent toy bag loaded with different kinds of Abid's age-appropriate toys. She paired very well with Abid, and he waits at the window for her on the days when she is scheduled to provide services. Abid's mom, Lina, brings him snacks in the middle of the session and checks with Sherron how everything is going. On a certain day, Lina observes that Sherron passes the glass of water to Abid using her left hand and Abid also receives it using his left hand. Later, when the mom offers the snack, Abid again receives it using his left hand. Mom keeps wondering as she taught Abid to use his right hand only specifically for food, books, and all the things that he respects. After the therapist leaves, mom asks Abid to open his religious scripture (The Holly Quran) and read a few pages. Abid picks up the Holly Quran from the shelf using his left hand and attempts to open it using his left hand. The mom got furious since it was something extremely big for her as she was a practicing Muslim. In their religious book Sunnah, Al-Bukhaari (5376) and Muslim (2022) narrate: "O young boy, say the name of Allaah and eat with your right hand, and eat from what is nearest to you." Mom calls the service provider and cancels Sherron's service and asks for someone who is familiar with Islamic etiquette.

Fong et al. (2017) communicated that it is possible that without information about cultural preferences and norms, behavior analysts may unintentionally provide less than optimal service delivery. The above case projects that, for not knowing the culture of the client's family, Sherron had to lose her client. In Islamic culture, it is very important to exchange things only using the right hand. Using the left hand to give or take items is disrespectful. Since Sherron was not aware of this fact of Abid's culture and taught Abid to utilize both hands (as we would usually do when one hand is busy), she misled the child and caused unnecessary confusion for the client's family.

Cultural diversity is within our differences, and it benefits everyone. There are diverse people from different corners of the world in today's congregated multicultural universe. In North America, people from different cultures, races, and ethnicity are common. Rassool (2015) mentions the Muslim community is experiencing Islamophobia, microaggression, prejudices, hate crimes, and social exclusion related to their cultural and religious identity, which correlates with mental health problems in the Muslim community. To work efficiently with this population as a professional, we need to know about those variables. Also, cultural variables impact access to healthcare services as noted by Jimenez-Gomez and Beaulieu (2022, p. 650). The lack of cultural competency may influence the assessment process and subsequent quality of care and interventions (p. 323). It's almost impossible to know about every culture, but practitioners at least need to know about the key factors of their clients' cultures. The case studies (below) will explain the scenarios further.

Jafari (1993) mentioned that Muslim patients report fear that their values will be undermined by secular counselling. This fear might have been caused by a culture shock from any previous counselling session with someone who was not aware of the client's culture and was disrespectful. The author also explained that "the most effective therapeutic treatment is based on an understanding of who is providing the treatment to whom, under what conditions, and for what specific problem." (p. 329). For example, Jafari (1993) claims that "Unlike western counseling, Islamic counseling is based upon commonly held ideological beliefs and value systems and is therefore universal" (p. 329). Therefore, to get the best therapeutic result, the counsellor needs to be aware of Islamic culture to provide counselling to a Muslim.

Cultural diversity is extremely important in the field of Behavior Analysis since individuals participating in behavior change programs and those who provide significant support for them should determine what is important to them, to their society, and their culture (Fong et al., 2017). Without knowing cultural diversity, it is not possible to design appropriate behavior change programs to meet the needs of clients from diverse backgrounds. For example, in some cultures, people traditionally eat using their hands instead of using utensils. Without knowing about the significance and importance of this practice, implementing a program to teach the child to use utensils can cause distress for the learner and the family.

Fong et al. (2017) also describe strategies to understand a client's cultural values and contingencies, as well as those of the behavior analyst. They also clarify that behavior analysts need to be aware of their cultural values and beliefs or the lack of understanding of their consumers' cultures can negatively impact treatment and service delivery (p. 103).

One strategy to enhance cultural self-awareness is talking about our diverse client interactions with a professional community in group discussions, written forums, journals, mentorship meetings, verbal feedback sessions, or self-reflective exercises (Tervalon and Murray-Garcia, 1998). For example, we can share the cultural values of a particular client with the other members of the client's team who are not aware of that specific culture. I remember how, while working with a Jewish client for the first time, I learned how strictly they maintain the use of Kosher products in their meals. I alerted the client's team so that unknowingly they do not cause any cultural shock to the client or the family.

Nair (2022) described cultural diversity by interpreting some actions such as (1) recognizing that many cultures exist, (2) respecting each other's differences, (3) acknowledging that all cultural expressions are valid, (4) valuing what cultures must bring to the table, (5) empowering diverse groups to contribute, and (6) celebrating differences, not just tolerating them.

For example, when a new client with a disability uses their hands to eat rice, based on most Asian cultures, and is bullied by peers at the group session, it is not appropriate. Eating without utensils is allowed by their culture. Second, the teacher can educate the clients to recognize and respect the cultures of their peers.

Similarly, while some children are in their swimming costumes in a swimming class, some religious beliefs can hinder other children from wearing clothing that uncovers their skin (e.g., the females). In such situations, the teachers, educators, therapists, and other stakeholders need to be aware of these cultural differences and educate the associated people to acknowledge, respect, value, empower, and celebrate those differences. By performing these actions, we can move the world forward. If people are not judged by their races, genders, religions, languages, cultures, or other diverse characteristics, they can significantly grow and prove that every single person is different, but if they are not discriminated against, they can also bring significant changes to society.

## How Should We Know about Them and Pay Respect to Them?

To know about cultures, we need to do our homework first. We need to be open-minded, empathetic, and good receivers to understand how other people from different cultures see, say, and hear about things. We need to know about the perspectives of the collaborative team members or the people around us before dealing with them so that we do not hurt their feelings unknowingly. It is okay to ask questions instead of making mistakes. We can ask people what they value more in their life and what their priorities are. After knowing their cultures, we need to respect those in our communication and interaction with them.

> **Case Study 4**
>
> Ralph is a group home manager, and he manages twelve adults with developmental disabilities. For the Christmas party, he orders all food items that contain either ham, pork, or beef. There were people from other cultures for whom, having beef is prohibited and for three others, having pork is prohibited. Ralph arranges the party with lots of fun activities and food items. However, during dinner time, Bijoy sees beef on the table, and he refuses to eat anything from that table. Similarly, seeing pork, Ali, Bahauddin, and Jashim left the table without eating anything from there. They remained hungry and started aggression toward others. The fun party turned into an aversive one for all participants.

## Why Are There So Many Cultural Diversities? Why Do We Care?

During the 1970s, professionals were concerned about the access to and quality of care given to ethnic minority clients who sought psychotherapy (Sue, 2003, p. 964). Sue (2003) highlighted that some interpreted multiculturalism as political correctness rather than understanding the lack of research completed in cultural competency. Many did not understand that the inclusion of individuals from different cultures was only a component of cultural competency, overall. In history, people from different cultures achieved success, still being in their individual cultures. It's high time, we combine all those merits to make a better world. To accomplish that, we need to be aware of diverse cultures. Individuals fought against their difficulties and diverse background to win success but still maintained their cultures. For example, Lindsay (2022) listed five incredibly successful entrepreneurs, all women of color (Madam C.J. Walker, Farah Mohammed, Weili Dai, Vickie Wessel, and Nely Galan) who obtained great success despite being from diverse cultures, ethnicity, educational, and economic status. Cultural diversity increases creativity and promotes innovation. It also increases our flexibility when we deal with multiple people of diverse cultures and work respecting their cultures that are different from ours.

*Example of the Utilities of Diverse Cultures*

Cultural diversity is a blessing in almost every sector of life. It is beneficial in several sectors, such as in the classroom. Among several other ways, it promotes deep learning when a student learns by discussing the concept with peers from diverse backgrounds. It also boosts the students' confidence and growth when they come out of their comfort zone to deal with diverse concepts and thus eventually strengthens their understanding, character, pride, and confidence.

By promoting the knowledge, creativity, and success of people of diverse cultures, we can enhance the quality and quantity of our progress. To ensure everyone is heard and understood, teachers and other professionals need to be culturally competent and aware of cultural diversities.

# References

American Psychological Association. (2003). Guidelines on multicultural education training, research, practice, and organizational change for psychologists. *American Psychologist, 58,* 377–402. https://doi.org/10.1037/0003-066X.58.5.377

Behavior Analyst Certification Board. (2020). *Ethics code for behavior analysts.* https://bacb.com/wp-content/ethics-code-for-behavior-analysts/

DeAngelis, T. (2015, March). *In search of cultural competence.* Monitor on Psychology. www.apa.org/monitor/2015/03/cultural-competence

Flaskerud. (2007). Cultural competence: What is it? *Issues in Mental Health Nursing, 28*(1), 121–123. https://doi.org/10.1080/01612840600998154

Fong, E. H., Ficklin, S., & Lee, H. Y. (2017). Increasing cultural understanding and diversity in applied behavior analysis. *Behavior Analysis: Research and Practice, 17*(2), 103–113. https://doi.org/10.1037/bar0000076

Goldiamond, I. (2002). Toward a constructional approach to social problems: Ethical and constitutional issues raised by applied behavior analysis. *Behavior and Social Issues, 11*(2), 108–197. https://doi.org/10.5210/bsi.v11i2.92

Jafari, M. F. (1993). Counseling values and objectives. *American Journal of Islam and Society, 10*(3), 326–339. https://doi.org/10.35632/ajis.v10i3.2490

Jimenez-Gomez, C., & Beaulieu, L. (2022). Cultural responsiveness in applied behavior analysis: Research and practice. *Journal of Applied Behavior Analysis, 55*(3), 650–673. https://doi.org/10.1002/jaba.920

Lindsay, L. (2022, July 25). *How these 5 incredibly successful women of color entrepreneurs got started*. University of the People. www.uopeople.edu/blog/how-these-5-incredibly-successful-women-of-color-entrepreneurs-got-started/

Nair, M. (2022, July 12). *What is cultural diversity and why is it important?* University of the People. www.uopeople.edu/blog/what-is-cultural-diversity/

Rassool, G. H. (2015). Cultural competence in counseling the Muslim patient: Implications for mental health. *Archives of Psychiatric Nursing, 29*(5), 321–325. https://doi.org/10.1016/j.apnu.2015.05.009

Skinner, B. F. (1953). *Science and human behavior.* Collier-Macmillan.

Spring, B. (2007). Evidence-based practice in clinical psychology: What it is, why it matters; what you need to know. *Journal of Clinical Psychology, 63*(7), 611–631. https://doi.org/10.1002/jclp.20373

Sue, S. (2003). In defense of cultural competency in psychotherapy and treatment. *American Psychologist, 58*(11), 964–970. https://doi.org/10.1037/0003-066x.58.11.964

Sugai, G., O'Keeffe, B. V., & Fallon, L. M. (2011). A contextual consideration of culture and school-wide positive behavior support. *Journal of Positive Behavior Interventions, 14*(4), 197–208. https://doi.org/10.1177/1098300711426334

Tervalon, M., & Murray-García, J. (1998). Cultural humility versus cultural competence: A critical distinction in defining physician training outcomes in multicultural education. *Journal of Health Care for the Poor and Underserved, 9*(2), 117–125. https://doi.org/10.1353/hpu.2010.0233

U.S. Department of Health and Human Services. (n.d.). *U.S. department of health and human services agency details*. USAGov. www.usa.gov/federal-agencies/u-s-department-of-health-and-human-services

Vandenberghe, L. (2008). Culture-sensitive functional analytic psychotherapy. *The Behavior Analyst, 31*(1), 67. https://doi.org/10.1007/BF03392162

# 2.3 A Critical Look at Autism

*Christopher Peters*

Autism is a disability. There are two common models of disability: a medical model and a social model. The medical model of disability views autism as a disorder to be fixed, focusing on an autistic person's supposed deficits and trying to eliminate or reduce them through behavioral therapy, medication, and other methods. Goodley et al. (2019) discussed how the public education system pathologizes difference by adopting the medical model and employing a range of professionals to address disabilities, such as social workers, behavior analysts, and psychologists. The social model locates disability in social environments rather than within individual people (McGuire, 2012). In other words, the social model defines disability as "a relationship between people with impairment and a disabling society" (Shakespeare, 2017, p. 197). An impairment can be thought of as belonging to the biological domain and can include physical, mental, or intellectual conditions (Barnes & Mercer, 2010). In contrast, disability belongs to the social domain, and refers to social barriers that prevent people with impairments from participating fully in mainstream economic and social life. The social model has been effective at improving the self-esteem of disabled people, as it provides them with an explanation for the difficulties they have experienced in life that shifts the blame from individuals to society. If the social model of disability is applied to the school system, it offers an opportunity to reimagine how disabled individuals are treated (Goodley et al., 2019).

Throughout this chapter, identity-first rather than person-first language is used when referring to autistic people. The American Psychological Association (2020) discussed how identity-first language allows disabled people to claim their disability as an integral part of their identity and states that the preferences of the community being written about should be honored. The autistic community overwhelmingly prefers identity-first language. For example, in a survey of over three thousand autistic people,

DOI: 10.4324/9781003300465-9

parents of autistic children, and professionals working with autistic people in the United Kingdom, Kenny et al. (2016) found that identity-first language was the clear preference for autistic people and parents. Many autistic people believe that person-first language perpetuates the notion that there is something wrong or shameful about being autistic (Kenny et al., 2015). Sinclair (2013) discussed how person-first language is misleading because it implies that autism can be removed from a person, or that it does not define an autistic person to a significant degree. Bottema-Beutel et al. (2021) stated that person-first language is potentially ableist and could contribute to the further marginalization and stigmatization of autistic people. However, it should be noted that language preferences evolve over time, and that person-first language was an important component of the early history of the disability rights movement (Fletcher-Watson & Happé, 2019).

The critical disability studies framework is useful for understanding how race, class, feminism, capitalist economics, and other societal influences and aspects of identity intersect with the disability experience (Goodley et al., 2019; O'Dell et al., 2016). The intent of the framework is to provide a wide variety of voices and perspectives on disability (Goodley et al., 2019). Critical disability studies theory is committed to political organization and improving the everyday lives of disabled people. Critical autism studies are a subfield of critical disability studies in which researchers focus specifically on the oppression of autistic people within social institutions. The term critical autism studies was coined in 2010 at a workshop in Ottawa by Joyce Davidson and Michael Orsini, two Canadian scholars. Davidson and Orsini (2013) identified three important elements to the critical autism studies framework:

1 Careful attention to the ways in which power relations shape the field of autism
2 Concern to advance new, enabling narratives of autism that challenge the predominant (deficit-focused and degrading) constructions that influence public opinion, policy, and popular culture
3 Commitment to develop new analytical frameworks using inclusive and nonreductive methodological and theoretical approaches to study the nature *and* culture of autism. The interdisciplinary research required (particularly in the social sciences and humanities) demands sensitivity to the kaleidoscopic complexity of this highly individualized, relational (dis)order.

(p. 12)

Critical autism studies is a powerful framework for understanding how autistic people are marginalized in society, but there have not been many attempts in the literature to use it as a lens to understand how the public-school environment impacts autistic students.

The neurodiversity paradigm is closely aligned with critical disability studies. The term neurodiversity refers to the biological reality that there exists variation (diversity) in brains and minds amongst human beings, and the neurodiversity paradigm is a philosophical framework that states there is no such thing as normal neurocognitive functioning, but that there exist social power imbalances between groups with different types of brains and minds, such as autistic and allistic (non-autistic) people (Walker, 2021). Social institutions (e.g., schools) are designed to meet the needs of allistic people, thereby marginalizing autistic individuals socially, emotionally, academically, and environmentally. Without frameworks like critical disability studies and the neurodiversity paradigm, this marginalization would be more difficult to identify and understand, especially because autism is often conceptualized as a *hidden* disability. Hidden disabilities can be difficult to recognize because they usually do not have visible, physical presentations (Couzens et al., 2015). Using the above-mentioned theoretical frameworks could lead to school environments where the power imbalance between allistic and autistic people is eliminated and all students have their needs met (Wood, 2019).

The criteria for an autism diagnosis in the *Diagnostic and Statistical Manual of Mental Disorders* (5th ed., text rev.; DSM-5-TR; American Psychiatric Association, 2022) include deficits in social communication and interaction and restricted, repetitive patterns of behavior, interests, or activities. Alternatively, autism can be described in less pathologizing language as being characterized by differences in social communication and interaction, as well as focused interests and differences in sensitivity to sensory input (Perry et al., 2022). Since autism is a developmental disability, an autistic person will show the above-mentioned characteristics early in their life, but these symptoms could become increasingly noticeable as children mature and social expectations become more complex.

In the past few decades, the understanding of autism has continued to grow and shift, and a substantial amount of important and influential work has been done by advocates and researchers who are autistic themselves. Dr. Jim Sinclair is an important figure in the early history of the neurodiversity movement, and perhaps more than anyone else was crucial in beginning the shift from a medical model understanding of autism to a social model understanding of autism. Sinclair founded the autistic advocacy and community group Autism Network International in 1992, and in 1993 described the following view of autism (Fletcher-Watson & Happé, 2019):

> Autism is a way of being. It is pervasive; it colors every experience, every sensation, perception, thought, emotion, and encounter, every

aspect of existence. It is not possible to separate the autism from the person–and if it were possible, the person you'd have left would not be the same person you started with.

(p. 22)

Sinclair is also responsible for establishing and running Autreat, an annual autistic-run event designed to allow autistic people to meet each other and spend time in an environment that was built with autistic sensory and social needs in mind (Buckle, 2020). The success and popularity of events like Autreat (and Autscape in the United Kingdom) demonstrate that it is possible to create environments in which autistic people feel comfortable and have their needs met.

One of the most important autistic researchers in the past decade, associated with the neurodiversity movement, is Damian Milton, a lecturer at the University of Kent in England. Possibly his most influential theory is known as the double empathy problem. The double empathy problem puts forth the notion that, since two people are responsible for a successful social interaction, if a social interaction fails, both people should take mutual responsibility for the failure, rather than the autistic person always taking the blame (Fletcher-Watson & Happé, 2019). Milton (2017) pointed out that allistic people often claim that autistic people lack empathy, and that allistic people can be very cruel toward autistic people, demonstrating a lack of empathy themselves. There has been a great deal of research into this theory over the past ten years, and much evidence to support the idea that allistic people are just as responsible for communications breaking down with autistic people as vice versa.

The DSM-5-TR lists hypersensitivity or hyposensitivity to sensory input among the diagnostic criteria for autism. Autistic people commonly have differences from allistic people in their experiences of noise, light, smell, temperature, and other forms of sensory input. Fulton et al. (2020) proposed that sensory differences are just as integral to understanding autism as social differences, as they wrote, "autistic people have a range of social competences and capacities but the interactive expression of these may be hampered by the ongoing sensory experience of their body in the world" (p. 12). In other words, the supposed deficits in social interaction that have historically characterized the clinical view of autism may partly be understood as autistic people feeling overwhelmed by sensory input from their environments. In many circumstances, autistic people may be unable to devote the same mental attention and focus to social participation and communication as allistic people, despite having a genuine desire for friendship and social belonging. Fulton et al. (2020) pointed out that ordinary activities of daily life, such as attending school, can be traumatic for autistic children if their sensory environments are overwhelming.

Boroson (2016) wrote that "since sensory assaults often spring up without warning, students are often on guard against them. Living with that relentless tension contributes significantly to the anxiety that is so common among students on the spectrum" (p. 79). To calm themselves, many autistic people self-regulate by stimming. Boroson (2016) stated that *stims*, or self-stimulatory behaviors, can be used to stimulate sensory input and help students focus, block out overwhelming input, or release excess energy. Some examples of common stims autistic children may use at school include humming, flapping their hands, scratching, and stomping. Some professionals have tried to extinguish these stims through behavior modification techniques, but Boroson (2016) explained that stimming is usually a response to a deeper problem with the sensory environment that must be addressed.

In addition to overwhelming and traumatic sensory experiences, there are other problems that autistic children in public schools face at higher rates than allistic students, including bullying. There are different categories of bullying, such as physical, verbal, relational, and cyber (Kloosterman et al., 2013). Kloosterman et al. (2013) found that autistic students were significantly more likely than allistic students to have been bullying victims, including physical bullying (e.g., being hit, kicked, or pushed) and social exclusion. In another study, 90 percent of autistic students reported being victims of bullying, compared to an average of about 30 percent of allistic students as determined in earlier research studies (Ung et al., 2016). Other studies found similar results about the prevalence of bullying victimization for autistic children in non-specialized public schools (e.g., Cook et al., 2016; Zeedyk et al., 2014). A study by Haruvi-Lamdan et al. (2020) found that 32 percent of autistic adults met the criteria for posttraumatic stress disorder (PTSD) compared to 4 percent of allistic adults. Other mental health problems related to frequent bullying in autistic students include significantly higher rates of anxiety, depression, loneliness, and suicidal ideation compared to students who are not bullied (Ung et al., 2016). A study by Crane et al. (2019) involving autistic young people between the ages of 16 and 25 found that about 80 percent of the participants had a mental health problem (most commonly anxiety and depression), and the majority felt unhappy, worthless, and lacked confidence.

One strategy often used by autistics to try to reduce bullying victimization and other issues is known as camouflaging (Mandy, 2019). Halsall et al. (2021) described three types of camouflaging: assimilation, when autistic people try to follow social norms; masking, when they hide their autistic traits; and compensation, which can involve more complex strategies like using the Internet to research what it is like to be allistic. Some examples of camouflaging strategies include forcing eye contact, using conversational scripts, and suppressing repetitive hand movements (Cook et al., 2021). Mandy (2019) explained that "for many autistic people, camouflaging is experienced as an obligation, rather than a choice" (p. 1879). Although allistic people can also camouflage during their day-to-day lives, the pressure felt by autistic individuals to hide their natural ways of being is especially strong. The social pressure to camouflage at school, for example, could pose a barrier for autistic children who are in the process of working out their authentic identity (Cook et al., 2021; Perry et al., 2022).

Society must adopt a perspective on autism that is founded on the social model of disability and informed by the neurodiversity paradigm in order to become truly accessible and inclusive for autistic people. There is nothing inherently wrong with preferring not to make eye contact, feeling more relaxed when working individually rather than in a group, and remaining silent instead of making small talk, even if Western cultural norms attach negative interpretations to these social behaviors. As an autistic child, I was totally content with myself until I entered the education system and many of my allistic peers made sure I knew daily that my social differences were unacceptable. Fighting against one's own neurology and attempting to conform to allistic social standards for many hours a day, every day, is enormously stressful and ultimately traumatic for most autistic children and adults. My advice to allistics is to recognize their allistic privilege by understanding that they are members of the dominant neurocognitive group in Western society and remain vigilant about how their words, attitudes, and actions can potentially contribute to the marginalization of autistic people.

## References

American Psychiatric Association. (2022). *Diagnostic and statistical manual of mental disorders* (5th ed.). https://doi.org/10.1176/appi.books.9780890425787

American Psychological Association. (2020). *Publication manual of the American psychological association* (7th ed.). https://doi.org/10.1037/0000165-000.

Barnes, C., & Mercer, G. (2010). *Exploring disability* (2nd ed.). Polity Press.

Boroson, B. (2016). *Autism spectrum disorder in the inclusive classroom* (2nd ed.). Scholastic.

Bottoma-Beutel, K., Kapp, S. K., Lester, J. N., Sasson, N. J., & Hand, B. N. (2021). Avoiding ableist language: Suggestions for autism researchers. *Autism in Adulthood*, 3(1), 18–29. https://doi.org/10.1089/aut.2020.0014

Buckle, K. L. (2020). *Autistic community and the neurodiversity movement* (1st ed). Palgrave Macmillan.

Cook, A., Ogden, J., & Winstone, N. (2016). The experiences of learning, friendship and bullying of boys with autism in mainstream and special settings: A qualitative study. *British Journal of Special Education*, 43(3), 250–271. https://doi.org/10.1111/1467-8578.12143

Cook, J., Hull, L., Crane, L., & Mandy, W. (2021). Camouflaging in autism: A systematic review. *Clinical Psychology Review, 89*, 102080. https://doi.org/10.1016/j.cpr.2021.102080

Couzens, D., Poed, S., Kataoka, M., Brandon, A., Hartley, J., & Keen, D. (2015). Support for students with hidden disabilities in universities: A case study. *International Journal of Disability, Development and Education, 62*(1), 24–41.

Crane, L., Adams, F., Harper, G., Welch, J., & Pellicano, E. (2019). "Something needs to change": Mental health experiences of young autistic adults in England. *Autism, 23*(2), 477–493. https://doi.org/10.1177/1362361318757048

Davidson, J., & Orsini, M. (2013). Critical autism studies: Notes on an emerging field. In J. Davidson & M. Orsini (Eds.), *Worlds of Autism: Across the Spectrum of Neurological Difference* (1st ed., pp. 1–28). University of Minnesota Press.

Davis, R., den Houting, J., Nordahl-Hansen, A., & Fletcher-Watson, S. (2022). Helping autistic children. In P. K. Smith & C. H. Hart (Eds.), *The Wiley-Blackwell handbook of childhood social development* (3rd ed., pp. 729–746). John Wiley.

Fletcher-Watson, S., & Happé, F. (2019). *Autism: A new introduction to psychological theory and current debate*. Routledge.

Fulton, R., Reardon, E., Richardson, K., & Jones, R. (2020). *Sensory trauma: Autism, sensory difference and the daily experience of fear*. Autism Wellbeing Press.

Goodley, D., Lawthorn, R., Liddiard, K., & Runswick-Cole, K. (2019). Provocations for critical disability studies. *Disability & Society, 34*(6), 972–997. https://doi.org/10.1080/09687599.2019.1566889

Halsall, J., Clarke, C., & Crane, L. (2021). "Camouflaging" by adolescent autistic girls who attend both mainstream and specialist resource classes: Perspectives of girls, their mothers and their educators. *Autism, 25*(7), 2074–2086. https://doi.org/10.1177/13623613211012819

Haruvi-Lamdan, N., Horesh, D., Zohar, S., Kraus, M., & Golan, O. (2020). Autism spectrum disorder and post-traumatic stress disorder: An unexplored co-occurrence of conditions. *Autism, 24*(4), 884–898. https://doi.org/10.1177/1362361320912143

Kenny, L., Hattersley, C., Molins, B., Buckley, C., Povey, C., & Pellicano, E. (2016). Which terms should be used to describe autism? Perspectives from the UK autism community. *Autism, 20*(4), 442–462. https://doi.org/10.1177/1362361315588200

Kloosterman, P. H., Kelley, E. A., Craig, W. M., Parker, J. D. A., & Javier, C. (2013). Types and experiences of bullying in adolescents with an autism spectrum disorder. *Research in Autism Spectrum Disorders, 7*, 824–832. https://doi.org/10.1016/j.rasd.2013.02.013

Mandy, W. (2019). Social camouflaging in autism: Is it time to lose the mask? [Editorial]. *Autism, 23*(8), 1879–1881. https://doi.org/10.1177/1362361319878559

McGuire, A. (2012). Representing autism: A sociological examination of autism advocacy. *Atlantis, 35*(2), 62–71. https://journals.msvu.ca/index.php/atlantis/article/view/918

O'Dell, L., Bertilsdotter Rosqvist, H., Ortega, F., Brownlow, C., & Orsini, M. (2016). Critical autism studies: Exploring epistemic dialogues and intersections, challenging dominant understandings of autism. *Disability & Society*, *31*(2), 166–179. https://doi.org/10.1080/09687599.2016.1164026

Perry, E., Mandy, W., Hull, L., & Cage, E. (2022). Understanding camouflaging as a response to autism-related stigma: A social identity theory approach. *Journal of Autism and Developmental Disorders*, *52*(2), 800–810. https://doi.org/10.1007/s10803-021-04987-w

Shakespeare, T. (2017). The social model of disability. In L. J. Davis (Ed.), *The Disability Studies Reader* (5th ed., pp. 195–203). Routledge.

Sinclair, J. (2013). Why I dislike "person first" language. *Autonomy*, *1*(2), 1.

Ung, D., McBride, N., Collier, A., Selles, R., Small, B., Phares, V., & Storch, E. (2016). The relationship between peer victimization and the psychological characteristics of youth with autism spectrum disorder. *Research in Autism Spectrum Disorders*, *32*, 70–79. https://doi.org/10.1016/j.rasd.2016.09.002

Walker, N. (2021). *Neuroqueer heresies: Notes on the neurodiversity paradigm, autistic empowerment, and postnormal possibilities*. Autonomous Press.

Wood, R. (2019). *Inclusive education for autistic children: Helping children and young people to learn and flourish in the classroom*. Jessica Kingsley Publishers.

Zeedyk, S. M., Rodriguez, G., Tipton, L. A., Baker, B. L., & Blacher, J. (2014). Bullying of youth with autism spectrum disorder, intellectual disability, or typical development: Victim and parent perspectives. *Research in Autism Spectrum Disorders*, *8*, 1173–1183. https://doi.org/10.1016/j.rasd.2014.06.001

# 2.4 Accepting Feedback Makes the World Go 'Round
## From Student to Practitioner

*Don Togade and Marina Jiujias*

Introduction

Behavior analysts are skilled in arranging conditions for meaningful behavior change to occur within individuals and among groups. As supervisors in clinical and educational settings responsible for training future behavior analysts, this skill necessarily extends to shaping the behavior of students and supervisees. A critical component of this work is the delivery of feedback and, subsequently, the recipient's acceptance and application of said feedback.

Feedback, when delivered effectively and efficiently, can promote lasting and meaningful change in an individual's behavior. Whether one is aware of it or not, almost every behavior that we engage with produces some type of effect on our environment. Whether we are learning a new skill, working with others, or conversing with loved ones, the effects associated with our activities determine whether we will (or will not) engage in the same behaviors in the future.

Acceptance and application of feedback is an essential component in the feedback process to ensure that significant behavior change occurs. Sellers and colleagues (2016a) identified the inability or unwillingness of supervisees to accept and apply supervisor feedback as a significant barrier to a successful supervisory experience. To overcome this barrier, some literature recommendations include preparing supervisees on how to accept delivered feedback, including the direct teaching of this behavior (Sellers et al., 2016b). Although there is a wide variety of research within the behavior analytic community focused on how, when, and where to deliver feedback, there remains a paucity of practical suggestions for how to specifically ensure that feedback is accepted and applied. In a recent survey asking behavior analysts providing supervision to evaluate their supervisory practices, a noted area requiring improvement was setting expectations for, and teaching supervisees how, to respond to the delivered feedback (Sellers et al., 2019). For the field of behavior analysis to work towards continuous improvement, there is a great need to provide effective training in not only implementing evidence-based procedures, but in the development of effective interpersonal skills such as the acceptance of performance-based feedback.

**What Is Feedback?**

To understand suggestions for how to accept feedback, and to teach others how to accept feedback, it is first important to review some components of feedback itself. Feedback occurs when, upon observation of someone's behavior, information related to that performance is relayed with the intent to maintain certain aspects of the performance and/or change some aspect of the future occurrence of that performance (Mangiapanello & Hemmes, 2015). In a review of articles published in four behavior analytic/organizational behavior management journals between 1998 and 2018, Sleiman et al. (2020) summarized key evidence-based qualities of effective feedback. In general, feedback is most effective when provided immediately following someone's performance, on a daily or weekly basis, and provided via multiple modes such as verbally, graphically, and/or in written form (Sleiman et al., 2020).

In the delivery of feedback, both supportive and corrective feedback are important components. Supportive feedback constitutes providing the performer with information around which aspects of the behavior(s) were completed correctly, thus providing a maintenance function for the future. Conversely, corrective feedback highlights which aspects of the performance were done incorrectly, functioning to shape future behavior towards the target performance conditions (Parsons & Rollyson, 2012).

As experts in selecting, defining, and teaching new skills, the field of behavior analysis can treat following through with feedback as a learned behavior. Though scarce, some literature is beginning to explore how to specifically teach this behavior. Ehrlich et al. (2020) demonstrated how to teach accepting feedback behaviors to employees working in an administrative role using behavioral skills training. Through specifically identifying and teaching behaviors related to maximally absorbing and applying delivered feedback, such as asking follow-up questions, engaging in active listening, and committing to behavior change, the researchers observed a significant increase in accepting-feedback behaviors and subsequent improvement in job performance across three participants (Ehrlich et al., 2020). More recent recommendations for effective supervisor–supervisee relationships echo these results; for example, it is suggested that supervisors evaluate supervisee demonstrations of accepting feedback, including both immediate (active listening and acknowledgement) and future behaviors (changing behavior related to feedback) following receiving feedback (Helvey et al., 2022).

## Accepting Feedback: From Student to Practitioner

Pursuing certification to become a behavior analyst starts by attending formal education, and further refining one's analytic skills with real world application of behavior change tactics. Given that accepting feedback can be defined and taught as an operant response, there are several stages within the professional "life cycle" of a behavior analyst in which the acquisition of this skill occurs. The beginning of the acquisition phase occurs when students learn to accept feedback in formal education, with receiving grades on papers and exams. Following graduation, as students transition to become front-line therapists, they begin to build on the prerequisite skills for accepting feedback learned as a student. At this stage, individuals are required to contact relevant motivating operations to learn the skills that are relevant to eventually obtaining the Board Certified Behavior Analyst (BCBA) designation. Finally, as a practicing behavior analysts, a supervisor, and/or a professor of behavior analysis, we can assume mastery of this skill, thus closing the acquisition loop. At this point, individuals then become responsible for teaching the "accepting feedback" repertoire to new students and therapists.

This chapter describes the specific process of learning how to accept feedback at each stage of the cycle in the skill-acquisition journey described above. Specific recommendations and discussion around qualities of the teaching environment(s), the teaching itself, and the individual characteristics of learners going through the cycle are presented. Anecdotal experiences around learning how to accept feedback from individuals currently experiencing these distinct phases of becoming a behavior analyst, are presented to highlight the idiosyncrasies in various accepting feedback learning histories (see Figure 2.4).

## Accepting Feedback as a Student in ABA

Whether pursuing an undergraduate or graduate degree, students who have chosen Behavior Analysis as their field of education and training generally have an intrinsic aspiration for continuous self-improvement, complemented with their desire to positively impact the lives of others in meaningful ways. To support students in these endeavors, it is critical to teach them how to accept feedback, both positive and corrective, early on in their behavior-analytic education and training. Systematically teaching students how to accept feedback will allow them to have an enriched learning experience, thus facilitating both their personal growth and professional development.

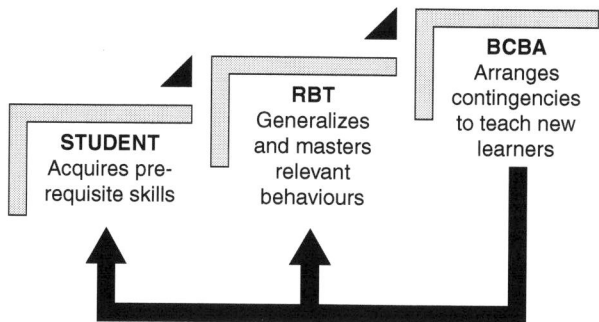

*Figure 2.4* Life cycle of learning to accept feedback, from learning as a student to arranging contingencies as a supervisor.

In academia, the primary task of an educator is to skillfully arrange meaningful and thriving learning conditions for students. Designing context-specific learning opportunities is largely influenced by the professor's ability to establish rapport and maintain positive relations with students. These relationship qualities will further allow the professor to incorporate teaching the important life skill of developing a more positive attitude in accepting feedback.

Positive or corrective, the goal of delivering feedback is to provide support for students to become more self-aware and accountable for their performance. Preparing post-secondary students to become more accepting of feedback is an essential skill to learn in their journey of becoming a well-rounded behavior analyst. In the context of their academic performance and character development, arguably, most students are more accepting of positive than corrective feedback. Most students would typically relate to corrective feedback as a personal attack, shaming, and/or punitive. Negatively relating to corrective feedback may be a result of their past experiences, which may explain their avoidance of receiving meaningful supports from their professor.

In recognition of the benefits in facilitating meaningful change in one's behavior, how do we help prepare students to become more open to feedback? What are some ways we can further develop the student's greater self-awareness and personal accountability?

With professors having the capacity to arrange meaningful teaching opportunities for their students, below are some anecdotal excerpts from a former ABA undergraduate student's experience of learning to become more accepting of feedback. The following sections will identify and discuss some practical antecedent steps, including (1) Professor–Student

*Table 2.2* Practical Techniques Aligning to the BACB 5th Edition Task List and BACB Ethics Code

| Practical Suggestions | BACB 5th Edition Task List | BACB Ethics Code |
| --- | --- | --- |
| Professor-Student Pairing | I-2. Establish clear performance expectations for the supervisor and supervisee | 4.02 Supervisory Competence<br>4.04 Accountability and Supervision |
| Assessing History of Receiving Feedback | I-8. Evaluate the effects of Supervision. | 4.07 Incorporating and Addressing Diversity |
| On Designing Multiple and Individualized Opportunities for Feedback | I-5. Use performance monitoring, feedback, and reinforcement systems. | 4.07 Incorporating and Addressing Diversity<br>4.08 Performance Monitoring and Feedback |
| Tact, The Function of Feedback & Value of Feedback | I-2. Establish clear performance expectations for the supervisor and supervisee. | 4.02 Supervisory Competence<br>4.10 Evaluating Effects of Supervision and Training |

Pairing, (2) Assessing History of Receiving Feedback, and (3) Imbedding multiple opportunities for differential feedback that was helpful in shaping students' positive attitude in being receptive to feedback (see Table 2.2).

**On Professor–Student Pairing**

An important first step is to make it known to students that your primary role is to support their learning and facilitate their best interests.

> I believe the reason why my professor had a positive impact was due to how he built a strong rapport with me. When I had concerns about his marking system, I emailed him my concerns, that same day he emailed me back, asking if I wanted to meet during his office hours to discuss it. I was very pleased with how quick he responded to my email, I remember being in another university and waiting for almost a week to get a reply.

In this scenario, having a rapport with the student might have been a function of the professor's timeliness in addressing the student's needs. Responding in a timely manner and offering alternative supports allowed the student to feel that their concerns were heard and valued.

The immediacy of support the student received positively enhanced the student's confidence and trust with their professor.

> During our first office hour meeting, he started off explaining how strong of a student I was and how he saw a lot of potential. As he mentioned my strengths, he goes on to explain how I needed to use behavioral techniques in my writing—that was my weakness. What stuck with me was how he mentioned to not let the grade affect me as a student nor a person. I liked how he was willing to use his time to help become a better student. That meeting truly shows how much compassion he has for his students and how he wants all his students to succeed.

In providing in-person feedback, it is helpful to start the feedback process by first acknowledging the student's efforts. Often, most students may not be aware of their strengths and capacities until someone they trust makes it known to them. Highlighting specific areas of improvement increases the likelihood that the student will follow through with feedback. Equally important is to also explain that feedback is "performance-based not personal." Helping the student understand this concept early on will help them relate to feedback as a tool to improve performance and not as a criticism to them as an individual. As they continue to experience meaningful supports, their professor can continue to provide relevant teaching opportunities where the student's performance can thrive and their character development is facilitated.

**Assessing History of Receiving Feedback**

Often, a student's capacity to accept feedback is largely influenced by their past experiences; therefore, it is a wise tactic to understand the way students have received supports in the past. By doing so, the professor gains necessary context for determining the most effective way to facilitate a student's greater capacity in receiving feedback.

> Throughout my undergraduate journey in another university, I was never prepared on how to receive feedback. The feedback given seemed to be based on the individual rather than on the behavior or product we have produced. Going into second undergraduate program, ... I was feeling uneasy on the type of feedback I would receive and how I would internalize the feedback. Based on my learning history, I did not feel prepared receiving and interpreting feedback. It was not until my second year of George Brown, where I had to engage in behaviors to understand the feedback that was received.

By conducting some type of an assessment (e.g., asking the student's preference between vocal and written feedback), the professor can maximize the likelihood of accepting feedback by capitalizing on the student's strengths. Consequently, the professor can also skillfully expand a student's feedback repertoire by teaching alternative forms of receiving feedback. By exposing them to variety of modalities, students will eventually develop a more generalized and flexible repertoire in their capacity in accepting and following through with feedback.

> I have learned it is better to receive feedback vocally, rather than written because written feedback can appear vague, whereas with vocal feedback, the individual is able to go into detail and you are able to ask questions during the meeting.

By taking the student's past experiences into consideration, an individualized approach in giving feedback is maximized. Being intentional of determining student's preferences not only communicates care but, most importantly, it makes students feel that their experiences are acknowledged, and that their progress and success are their professor's priority.

### Imbedding Multiple Opportunities for Feedback

Just like any learned behavior, teaching someone to be more receptive to feedback also requires multiple training opportunities. For instance, a strategy that can done is to divide projects into multiple parts with different due dates.

> During my second year, I had a professor who had us write a skill acquisition plan. The paper was broken down into four parts, with four different due dates. When I handed in the first part of the paper, I lost two marks from the total, the feedback given were more comments rather than suggestions for improvements, therefore I ignored it. It was not until I received the same marks deducted with similar feedback given from part one for part two and three of the assignment. After feedback from the third assignment, I decided to meet with my professor. He was able to explain his reasoning for the marks deducted a lot better than what was written on paper and that boosted my confidence not only within my undergraduate studies but my current position in this field.

Since the project is divided into parts, such instructional design will allow the professor to gradually shape the student's performance. Not only are students exposed to various feedback modalities, but this style

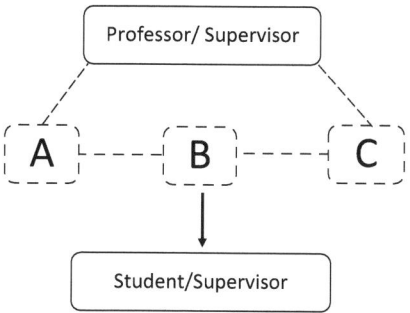

*Figure 2.5* Shaping student/supervisee receptiveness to feedback.

also encourages student assertiveness while also promoting instructional integrity and accountability from their professor. When students are afforded multiple opportunities to work collaboratively with their professor, the frequency of interaction will also allow the professor to model behaviors indicative of being receptive to meaningful supports. Students' periodic access to differential feedback complemented with their professor's care and commitment will eventually enable them to become more appreciative of the intended benefits of feedback. While these skills are initially introduced and practiced in the classroom, continued opportunities to master and generalize accepting feedback are present when students transition to being direct service implementers (see Figure 2.5).

### Accepting Feedback as a Registered Behavior Technician (RBT)

After students graduate from their respective behavior science programs, if motivated to do so, they pursue Registered Behavior Technician certification and obtain experience working directly in the field of behavior analysis. This work often involves working directly with individuals of varying needs and engaging in direct intervention to teach socially significant behaviors and reduce challenging behaviors. There are numerous tasks involved in the delivery of behavioral interventions that require a precise skill set. This may include collecting data, graphing, and precisely implementing components of programming such as prompting, error correction, reinforcement delivery, and contingency management.

The role of the supervising behavior analyst is to ensure that the RBTs are continually being provided with the necessary oversight to ensure that the clinical goals set for their clients are being reached with proper program fidelity. The delivery of feedback based on RBT performance is an important component of ensuring this success, but where does the feedback go once delivered?

In accordance with the "life cycle of a behavior analyst" framework that has been presented, one can reasonably assume that an RBT already has some level of prerequisite skills with accepting and applying feedback. There are significant differences in engaging in this behavior during and after post-secondary education, differences that impact how the skill acquisition process will continue in this next phase of learning.

**On Promoting Relevant Motivating Operations**

The driving motivating operation that impacts student performance and the subsequent likelihood of accepting feedback is frequently external and tied to grades. While students may learn to accept and apply feedback specifically pertaining to learning material for assignments and tests, it cannot be assumed that this behavior will generalize to the clinical setting following graduation. As an RBT, there are significant shifts in the variables that impact how frequently or accurately the accepting feedback response will occur: ensuring good employee performance and maintaining good client outcomes. Similarly, the frequency and rate of feedback is on a more variable schedule in the workplace when regular scheduled assignments and grades dictated by the academic cycle are no longer present.

The behavior of the supervising behavior analyst is a significant factor in contriving the relevant motivating operations for an RBTs acceptance and application of performance feedback. When a behavior analyst can specifically relay the importance of accepting feedback to a supervisee in terms that are individualized and specific to that supervisee's needs, it can have a tremendous impact. One RBT with several years' experience noted the relevance of a supervisor relating practicing the skill of accepting feedback to practicing their sport:

> I had a good manager who promoted me as a team leader. During that time, she would give feedback and I wouldn't accept it. She sat me down and asked how I expected to be better if I didn't accept feedback? .... Once she put that sport analogy into it, I was more accepting.

Conversely, the actions of a supervising behavior analyst themselves play a role in creating either an establishing or an abolishing operation for accepting feedback.

> The feedback was never verbal; it was written down and sent in an email. One piece of feedback once was, "You didn't record data immediately after a probe." The next week, I applied the feedback—on the

next supervision form, it said I didn't go with the child during reinforcement because I was immediately entering data. I stopped applying their feedback because it was contradictory.

It is important for supervising behavior analysts to recognize the impact of their feedback on not only the behaviors of their supervisees, but the motivation to engage (or not engage) in certain behaviors on the job. Similarly, receiving feedback can serve a reinforcing or punishing function; therefore, an awareness of providing individualized feedback that meets the needs of the therapist and the client is an important component of ensuring that feedback is going to be accepted.

**Arranging Meaningful Teaching Conditions**

As a student, individuals rarely receive feedback face to face or immediately after the expected behavior (e.g., completing an assignment or writing a test). This is contradictory to the consensus amongst researchers and practitioners that feedback should be delivered immediately if it is to be accepted and applied (Cooper et al., 2020). In a clinical setting, the conditions should be much more favorable due to availability of immediate feedback, direct and immediate application, and feedback provision in multiple formats (face to face, written, etc.). Therefore, it is the supervisor's job to construct an optimal learning environment for RBTs to accept and apply feedback in a generalization setting differing from their original instructional setting (a.k.a., in school). Occasionally, this may pertain to how feedback is delivered. One RBT recalled having a supervisor instruct them to record themselves implementing intensive trial instruction to review the recording and critique their own performance:

> Feedback for me is most effective when done in a unique fashion and paired with specific instructions to self-monitor (e.g., recording yourself and watching it back).

In this instance, the exercise provided by the supervisor allowed the RBT to engage in self-monitoring behavior that allowed feedback from the supervisor to be fully accepted. When supervisors create unique learning opportunities that allow RBTs to become more involved in the feedback process, the acceptance of that feedback in that moment and in the future is more likely.

The construction of the environment is especially important in consideration of the influence that a teaching environment has on the acquisition of learning new behaviors. Ehrlich et al. (2020) note in their study on teaching employees how to accept feedback, that an essential component of their intervention was to provide feedback in a systematic way. Behavior analysts are skilled in arranging conditions towards promoting the acquisition of target behaviors; as such, in the pursuit of teaching others to accept feedback, recognizing the impact of these variables should be a primary consideration.

## On Acknowledging and Identifying Value of Feedback

RBTs are required to learn how to accept feedback when the consequences of applying feedback may be delayed. There may occasionally be immediate and tangible contingencies in the clinical environment that directly relate to the application of feedback during client sessions, but sometimes this requires ongoing monitoring of program progress to fully recognize. It is not enough for supervisors to create the necessary environmental contingencies to foster learning how to accept feedback; RBTs must be able to identify the value in feedback and how it pertains to their own growth, both clinically and personally, as a behavioral professional. When asked if RBTs believed they accepted feedback well:

> I think I've learned over time that it's not usually out of a place of negativity. When it's done correctly, it is a gift. It's an opportunity to learn.
>
> The first thing is not just accepting but applying it [feedback]. You can hear it and accept it, but not necessarily know how to apply it ... accepting in the moment, being able to hear exactly what your supervisor is saying and knowing that it has nothing to do with you. Yes, it's about your skills, but more about how you can structure your session and the way you present things for an overall better experience and learning environment for your client.
>
> Without feedback, you're stagnant and you won't grow ... it's important to improve yourself. Especially in this field, it's changing so much and [there are] so many ways to do things—being open to different ways is important.

Some therapists may not always have the skills to recognize the importance of accepting feedback immediately, especially if there is a history of avoidance or punishment with feedback. Tarbox et al. (2022) highlight the relevance of integrating components of acceptance and commitment therapy (ACT) into behavior analytic practice to better manage professional and personal barriers that can impact work in this field. Importantly, the researchers define values as "rules that function as verbal motivating operations that increase or decrease the effectiveness of stimuli as reinforcers or punishers, thereby supporting overt behaviors that produce those stimuli" (Tarbox et al., 2022). Teaching RBTs to identify the values that brought them to this work in the first place may function effectively to assist in the acquisition of accepting feedback, despite potentially aversive learning histories.

## Summary

Every individual, regardless of their background, learning history, or current circumstances, is equally capable of changing their behavior in meaningful ways.

Accepting feedback can be operationally defined as a behavior under the effective control of relevant antecedent instructional conditions. Therefore, the onus on developing a proficient "accepting feedback repertoire" is both on the learner and on the teacher, as has been outlined in this chapter. As students become supervisors or instructors, they learn from their previous experiences, which in turn they will translate what they have learned in practice. Knowing this, the life cycle of this important interpersonal skill has important ramifications for the growth of behavior analytic practice. If supervisors do not undergo effective learning experiences in accepting feedback themselves, it is difficult to expect that they will be able to impart this same knowledge to future students, RBTs, and behavior analysts.

The recommendations presented in this chapter are an important step towards ensuring that professors and supervisors understand their roles in teaching accepting feedback. Applying these practical suggestions helps facilitate the cycle of accepting feedback that is necessary for continued professional development. By skillfully arranging learning opportunities for their students and supervisees to practice and master this professional skill, supervising behavior analysts can meaningfully contribute towards the betterment of professional behavior repertoires amongst behavior analysts.

## References

Cooper, J. O., Heron, T. E., & Heward, W. L. (2020). *Applied behavior analysis* (3rd edn.). Pearson.

Ehrlich, R. J., Nosik, M. R., Carr, J. E., & Wine, B. (2020). Teaching employees how to receive feedback: A preliminary investigation. *Journal of Organizational Behavior Management,* 40(1–2), 19–29. https://doi.org/10.1080/01608061.2020.1746470

Helvey, C. I., Thuman, E., & Cariveau, T. (2022). Recommended practices for individual supervision: Considerations for the behavior-analytic trainee. *Behavior Analysis in Practice,* 15(1), 370–381. https://doi.org/10.1007/s40617-021-00557-9

Mangiapanello, K. A., & Hemmes, N. S. (2015). An analysis of feedback from a behavior analytic perspective. *The Behavior Analyst, 38*(1), 51–75. http://doi.org/10.1007/s40614-014-0026-x

Parsons, M. B., Rollyson, J. H., & Reid, D. H. (2012). Evidence-based staff training: A guide for practitioners. *Behavior Analysis in Practice, 5*(2), 2–11. http://doi.org/10.1007/BF03391819

Sellers, T. P., LeBlanc, L. A., & Valentino, A. L. (2016a). Recommendations for detecting and addressing barriers to successful supervision. *Behavior Analysis in Practice, 9*(4), 309–319. http://doi.org/10.1007/s40617-016-0142-z

Sellers, T. P., Valentino, A. L., Landon, T. J., & Aiello, S. (2019). Board certified behavior analysts' supervisory practices of trainees: Survey results and recommendations. *Behavior Analysis in Practice, 12*(3), 536–546. https://doi.org/10.1007/s40617-019-00367-0

Sellers, T. P., Valentino, A. L., & LeBlanc, L. A. (2016b). Recommended practices for individual supervision of aspiring behavior analysts. *Behavior Analysis in Practice, 9*(4), 274–286. http://doi.org/10.1007/s40617-016-0110-7

Sleiman, A. A., Sigurjonsdottir, S., Elnes, A., Gage, N. A., & Gravina, N. E. (2020). A quantitative review of performance feedback in organizational settings (1998–2018). *Journal of Organizational and Behavior Management, 40*(3–4), 1–30. https://doi.org/10.1080/01608061.2020.1823300

Tarbox, J., Szabo, T. G., & Aclan, M. (2022). Acceptance and commitment training within the scope of practice of applied behavior analysis. *Behavior Analysis in Practice, 15*(1), 11–32. http://doi.org/10.1007/s40617-020-00466-3

# 2.5 How to Teach Rapport Building Skills to Behavior Analysts

*Ana Luisa Santo and Kimberley Taylor*

## Introduction

Historically, post-secondary programs in the Behavioral Sciences have not explicitly taught frameworks for conceptualizing or building therapeutic relationships with clients and mediators. Targeted training in developing such relationship-building skills can increase mediator adherence and treatment fidelity, improve satisfaction and client outcomes, and would be consistent with the Behavior Analyst Certification Board's (BACB) Ethics Code for Behavior Analysts and training requirements (BACB, 2020; Martin et al., 2000).

Current research in the medical field suggests that it is possible to teach compassion through education and training (Lown, 2016). Therefore, teaching compassion to behavior analysts should be possible as well. A multi-year project was embarked upon at Surrey Place, a social services agency in Toronto, to develop a method of training behavior analysts in rapport-building behaviors. This workshop was developed with the intention of addressing several concerns that were identified in an open-ended survey given to behavior therapists working with the adult population, who primarily use a mediator model of training caregivers and support staff when working with clients who have challenging behaviors.

The main concerns identified by the clinicians surveyed included mediators/clients leaving service early, not adhering to recommended strategies, and being unable to maintain procedural fidelity. In the comments received from clinicians, it appeared that both fidelity and adherence could be affected by poor therapeutic rapport. Fidelity (also referred to as treatment integrity) is defined as how well the treatment is implemented; in other words, the degree to which an intervention is implemented as designed by the behavior analyst. Adherence is defined as whether the treatment is being implemented at all, or the degree to which a mediator implements the intervention components without the behavior analyst present. Mediator "rapport" can thus potentially be defined as

DOI: 10.4324/9781003300465-11

"the mitigation of reasons for drop-out, and improvement in adherence and fidelity with program procedures due to behaviors of the clinician" (Biran et al., 2021a).

The success of evidence-based behavioral interventions depends on both fidelity and adherence (Allen & Warzak, 2000). Other competing contingencies and unique family contexts also affect the success of interventions; even the most well-designed treatments will be unsuccessful at changing client behaviors if they are too complex for the mediators to follow, if the mediators lack the financial or social resources to implement the strategies, or if they experience other environmental challenges such as family stress or isolation (Allen & Warzak, 2000). However, simply commenting on treatment integrity (or lack thereof) may not elicit the true reasons for mediators' nonadherence (Fallon et al., 2016).

Some ways to increase fidelity and adherence have been identified in the literature. These include helping families identify barriers and modifying the interventions accordingly, incorporating the intervention strategies and ensuring compatibility with everyday life and family routines, identifying the resources required to remove burdens, and discussing solutions to anticipated barriers (Fallon et al., 2016; Hock et al., 2015). While these strategies to increase fidelity and adherence have been shown to have some success on their own, evidence shows that the training of compassionate care skills (such as empathy, compassion and active listening) in a variety of other clinical and health professions (e.g., medicine, nursing, psychology, social work) results in increased therapeutic rapport, client satisfaction, and better client outcomes (i.e., it ameliorates the problems mentioned above) (Hojat et al., 2011; Kelley et al., 2014).

Formal training to increase rapport in behavior analysts has only very recently been a subject of scientific investigation. Roher & Weiss (2022) taught eleven compassionate care skills using a Behavioral Skills Training (BST) model to four students of behavior analysis, and Canon & Gould (2021) used clicker training and verbal instruction to teach three relationship-building skills in behavioral clinicians. Both studies showed promising results in that participants were able to acquire and maintain these skills upon completion of the training.

Similarly operationalizing and teaching rapport-building skills via a BST model, a training curriculum that uses the principles of ABA to increase these rapport building skills among behavior analysts was developed at the aforementioned agency as part of a project to improve clinical outcomes, including increasing mediator adherence to interventions, and decreasing drop-out rates (Biran et al., 2021a). The workshop includes a brief overview of Anti-Oppressive Practice (AOP) and theoretical underpinnings from Social Work and Counseling fields, the use of a Mediator Check-in Tool and follow-up questions, and explicit skills training in the behaviors of active listening, compassion, and empathy. Adaptations to programming and individualized support are also reviewed.

The pilot workshop was provided on a single day for approximately six hours; the subsequent two trained groups were provided with an eight-hour workshop split over two separate days—after feedback from the initial pilot group. This monograph will elaborate on some of the different elements of the workshop and present some preliminary results on its effectiveness at increasing rapport behaviors in the clinicians trained.

**Workshop Design: Behavioral Skills Training**

In order to ensure maximum efficacy, the training manual and subsequent workshop were developed using evidence-based methods—namely, Behavioral Skills Training (Miltenberger, 2004). This training model is widely accepted by behavior analysts as the gold standard for providing effective training with lasting behavior changes, and includes four major components: Instruction, Modeling, Rehearsal (i.e., role plays) and Feedback (immediate and individualized).

The workshop includes a Task List comprised of three main sections that break down compassionate care skills and anti-oppressive practices into 20 observable and measurable skills geared directly toward behavior analysts. These categories take counselling and social work literature into account and were based largely on Taylor and colleagues' 2019 article on compassionate care in the field of behavior analysis: (1) Active Listening and Communication, (2) Empathy, and (3) Compassion (see Monograph 2.1 for a full overview).

Two versions of the Task List were developed that outline and operationalize key clinician behaviors related to mediator rapport building. The "In Person" version lists the behaviors relevant when delivering face-to-face clinical services. The "Telehealth" version contains additional behaviors that are specific to delivering services via videoconferencing platforms. Such considerations include: looking at the camera at least once per response; following telehealth guidelines for camera set up and appropriate background/environment; and checking in with the client after each main point to see if they understood everything that was said or if they have clarifying questions.

Instruction was provided by using a combination of written information (in the manual) and verbal expansion of this information by the trainers. The instruction covered the theoretical underpinnings of the content (e.g., Anti-Oppressive Practice) and the rapport-building skills from a target list. Modeling was provided with a variety of pre-recorded video models designed to show each of the specific rapport-building behaviors being trained. In certain cases, non-examples were also modeled.

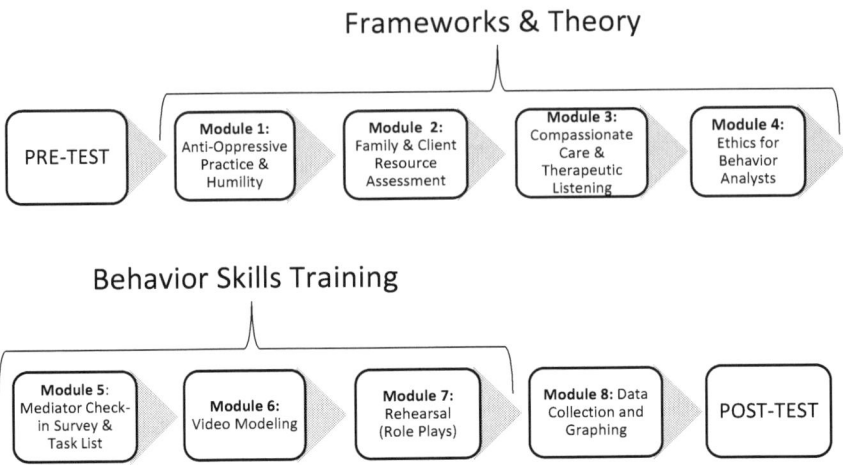

*Figure 2.6* Breakdown of Modules from Mediator Rapport Building Workshop (Biran et al., 2021b).

Participants were given multiple opportunities to practice the skills being taught. Carefully constructed clinical scenarios were provided for each of the practice opportunities. Each scenario was designed to elicit multiple rapport-building skills from a list of targets. While participants role-played the given scenarios, trainers scored how many of the rapport-building skills were exhibited, and immediately at the end of the rehearsals provided the participants with feedback that was specific to their performance. An overview of the different workshop components is summarized in Figure 2.6.

## Mediator Check-In Survey: A Unique Tool to Elicit Opportunities for Rapport-Building

An accompanying component to the workshop manual that was developed was a succinct check-in survey, with follow up questions, designed to facilitate conversation with mediators and increase the likelihood of mediators sharing barriers to program adherence and fidelity. It may not be valid to ask mediators to rate the level of rapport they feel with the clinician directly because they may not be comfortable to answer truthfully, they may worry about hurt feelings, or they may worry their comments will affect the service received. Behaviors that attempt to avoid or terminate aversive private events are referred to as experiential avoidance in the ACT literature (Hayes et al., 2003). In an effort to escape contact with aversive thoughts and feelings, a mediator may avoid doing things that are important to them and to the person to whom they provide care. For example, to avoid feeling anxious, a mediator may increasingly avoid meeting with the client's clinician. Mediator involvement or compliance issues might thus be related to avoidance of uncomfortable emotions that tend to accompany engagement in treatment. Previous research suggests that this sort of experiential avoidance is associated with high levels of mental health distress in parents of children with autism (Hastings et al., 2005).

A group of Behavioral Clinicians were poled and a comprehensive list of reasons why services or interventions have been unsuccessful were gathered. This included reasons why clients and families ended behavioral services prematurely. The list of reasons was then grouped into four main categories based on common themes:

1 Complexity of strategies and data-collection procedures
2 Comfort and understanding of the intervention strategies
3 Relevance of goals to current circumstances
4 Long-term relevance of strategies and goals

These themes were further refined into four questions that make up the "Mediator Check-In Survey" (see Figure 2.7).

The four survey questions (and their follow-up questions) were designed to measure and address the mediator's main concerns with service provision, while phrasing the question in such a way as to encourage them to feel comfortable enough to answer honestly. The discussion held following the survey response also provides space and opportunity for the behavior analyst to apply the rapport-building skills presented in the workshop.

It was recommended that the behavior analyst begin every session or meeting with the mediator by having them complete the Mediator Check-in Survey, and then lead a discussion based on their responses, both positive and negative. Follow-up questions are provided for any scores that elicited concern in one or more of the four areas. Some examples of follow-up questions are: "Ask what the biggest challenges are in implementing the program (listen to what the challenges are in implementing protocols)"; "What are your concerns with the current direction relative to the future?"; "Ask if there are other family members or support systems that can assist in program implementation, respite, or other aspects of the mediator's life" (Biran et al., 2021a).

These questions were designed help to surface barriers, concerns, and guide solution-focused discussion with the mediator, while also providing ample opportunities for the behavior analyst to exhibit rapport-building skills. The hope is that by engaging in this type of conversation at the beginning of each meeting, it will help to build and maintain rapport, support

Date: _____

Respondent: _____     Client: _____

| Question | Rating (please circle) 0 = Not at all 1 = A little 2 = Mostly 3 = Very Much |
|---|---|
| 1. I can handle doing the strategies and data collection as required. | 0   1   2   3 |
| 2. I understand the reason why I am doing these strategies and am comfortable doing them. | 0   1   2   3 |
| 3. I still think this is the right goal to work on. | 0   1   2   3 |
| 4. I think these strategies will be effective and appropriate in the long-term. | 0   1   2   3 |
| **Total Score:** | _____ / 12 |

*Figure 2.7* Mediator Check-in Survey (Biran et al., 2021a).

inclusion and buy-in, and ensure that the intervention is fitting the client and mediator's needs. Scores given by mediators on the four questions can be recorded and graphed on an ongoing basis as a way of measuring and monitoring treatment acceptability. While this is not a direct measure of rapport, it may function as an indirect way to measure rapport with the clinician. Individual questions can be monitored in order to determine the most prominent areas of dissatisfaction, and once total scores are relatively high the total scores can be monitored for maintenance purposes.

Then, the rapport-building skills from the Task List can be used to express empathy and compassion toward the mediator's concerns and to build rapport during the conversation. These conversations can be brief or can encompass the entirety of the allotted time the clinician and mediator have scheduled together, depending on how complex the situation is.

If appropriate, the behavior analyst can adjust the intervention plan based on the results of the conversation. Adjustments can vary from small changes to changing the treatment strategies entirely, can be related to environmental factors such as seeking resources to provide staffing or family supports to assist with the intervention plan, making changes to data collection, taking family preferences into account when modifying the intervention, or a variety of other adjustments depending on the dynamic needs of the family.

This methodology is aligned with the principles of AOP as the mediator is given time to evaluate the service and able to voice their feedback and experiences each session, thus reducing the typical power imbalance and clinician-driven paradigm significantly. Changes are applied each session based on the mediator's feedback. By working alongside the family and readily seeking to recognize forms of oppression or social barriers, the behavior analyst is using their position of power to help the client navigate services that have inherent inequalities for marginalized groups (Burke & Harrison, 1998).

### Monitoring "Rapport" Over Time

Using the scores from the Mediator Check-in Survey, clinicians are able to graph the individual survey responses and monitor responses to each of the four questions over time. By graphing this measure of "rapport" (or "service acceptability") on an ongoing basis, clinicians can monitor this dimension of the relationship with their mediators over time and use the information to make adjustments in service provision when needed.

In order to effectively track progress related to rapport building, clinicians should collect client responses to the Mediator Check-in Survey at the beginning of each meeting or session when possible. Aside from using the data in the moment as described earlier, it is important to track and monitor this rapport data on an ongoing basis to detect any concerns that may inadvertently affect the client's progress in treatment due to the mediator's lack of rapport with the clinician.

Figures 2.8 and 2.9 show an example of what such a graph could look like. In Figure 2.8, each of the four questions is graphed separately, while in Figure 2.9, the total of the same scores is graphed as a single data point. In these examples, we can see that treatment acceptability/rapport, as measured by the survey, is increasing steadily over time.

### Monitoring Adherence and Fidelity Over Time

Additional data related to mediator adherence and fidelity can also be recorded in order to detect correlations between these measures and

116  *Ana Luisa Santo and Kimberley Taylor*

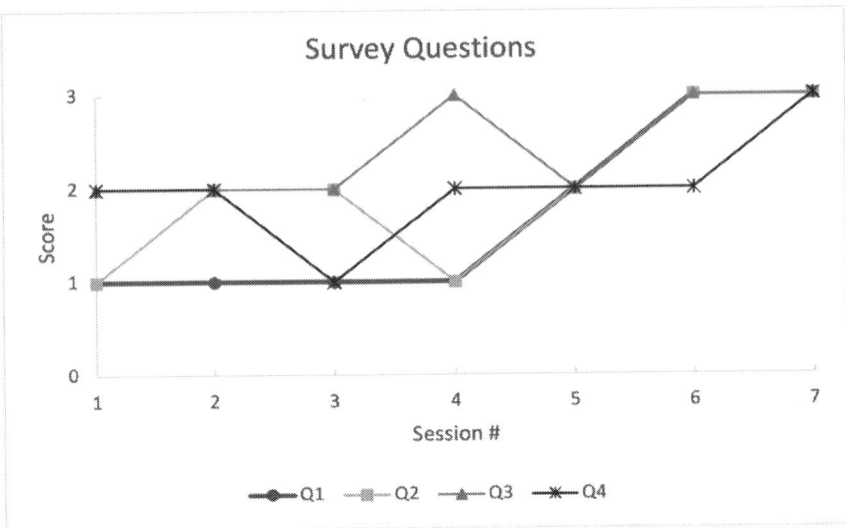

*Figure 2.8* Example of graphed responses to a Mediator Check-in Survey over seven days.

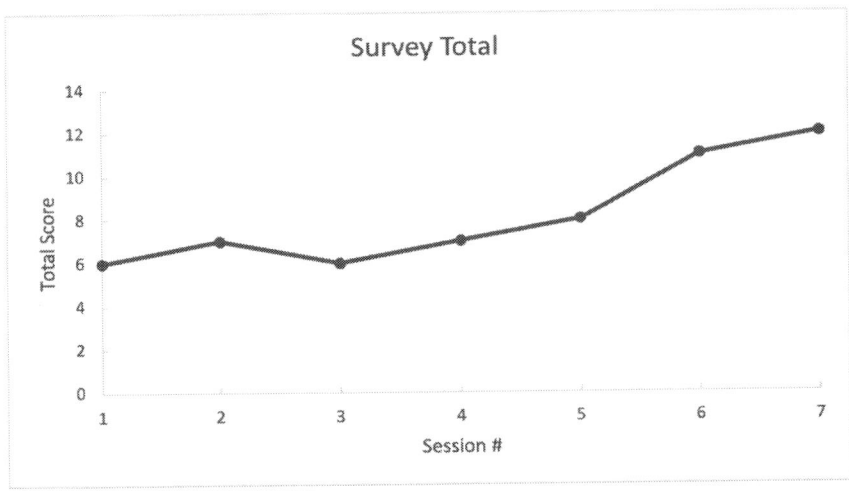

*Figure 2.9* Example of graphed total responses to a Mediator Check-in Survey over seven sessions.

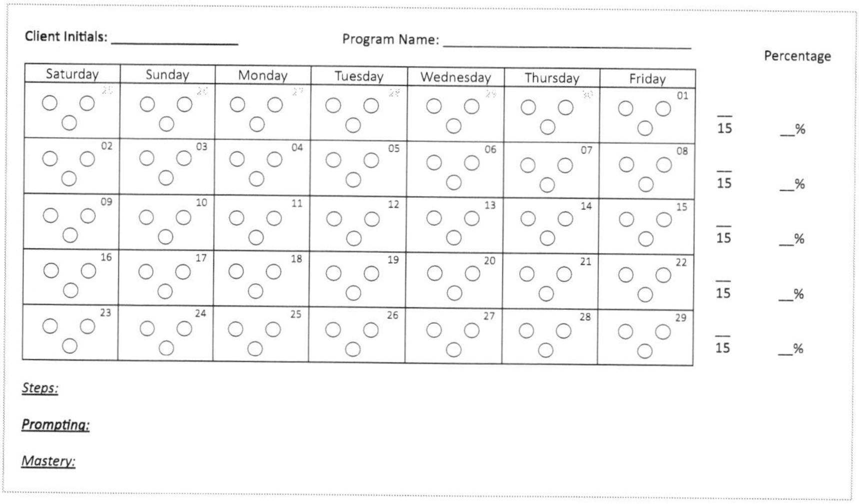

Client Initials: ______________ Program Name: ______________________________

| Saturday | Sunday | Monday | Tuesday | Wednesday | Thursday | Friday | | Percentage |
|---|---|---|---|---|---|---|---|---|
| | | | | | | 01 | __ 15 | __% |
| 02 | 03 | 04 | 05 | 06 | 07 | 08 | __ 15 | __% |
| 09 | 10 | 11 | 12 | 13 | 14 | 15 | __ 15 | __% |
| 16 | 17 | 18 | 19 | 20 | 21 | 22 | __ 15 | __% |
| 23 | 24 | 25 | 26 | 27 | 28 | 29 | __ 15 | __% |

Steps:

Prompting:

Mastery:

*Figure 2.10* Example datasheet for tracking daily adherence to treatment (Biran et al., 2021a).

measures of rapport/service acceptability. In the workshop, clinicians were provided with examples of datasheets that could be used to track adherence to treatment. One example is shown in Figure 2.10. In this example, mediators have been instructed to implement an intervention three times per day, represented by the three circles in each calendar day. Mediators can simply check off how many times each day they have implemented the strategies. Clinicians can then use this data to graph adherence as a percentage. Figure 2.11 shows an example of what this data could look like. Depending on the type of intervention, it may not be appropriate to provide an extra datasheet for mediators to fill out to track adherence. For example, if a clinician wanted to track a mediator's adherence with running a communication program that has its own datasheet, the clinician could simply look at this datasheet and count how many days the program was run since the date the program was implemented and calculate percentage that way.

Tracking treatment fidelity is common practice in the field of behavior analysis, especially when teaching individuals to implement discrete trial-training with the developmental disability population (Fingerhut & Moeyaert, 2022). Fidelity can therefore also be tracked with a variety of datasheet formats that are individualized to the steps from the intervention in question. An example of one such

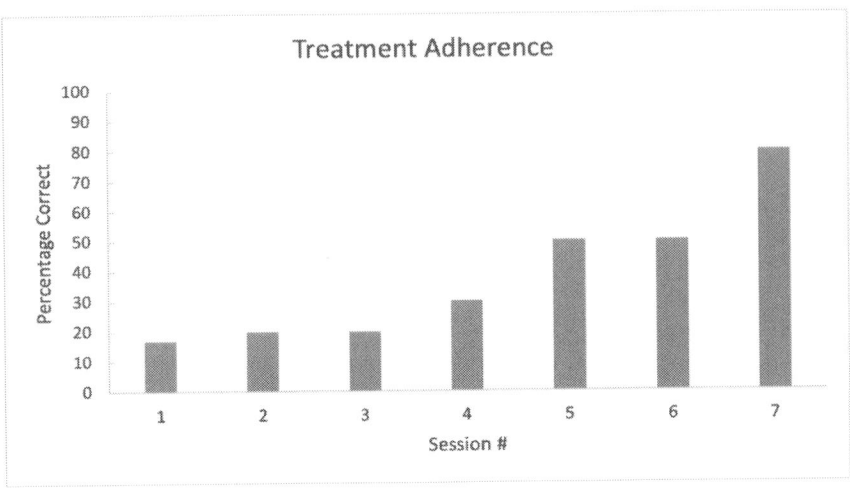

*Figure 2.11* Example of adherence percentage graphed over seven sessions.

datasheet is shown in Figure 2.12. In the workshop, clinicians are encouraged to track fidelity of treatment interventions when possible and monitor their progress over time (see Figure 2.13). The main reason for doing so, of course, is to ensure that the procedural steps are being implemented as intended. This helps with quick identification of errors in implementations that may lead to treatment failure, and alerts the behavior analyst as to which steps in the treatment need to be re-trained or modified.

The secondary reason for tracking both adherence and fidelity over time is to enable the clinician to compare the percentage of these two treatment variables with the measure of rapport/treatment acceptability. As adherence and fidelity with treatments in many healthcare fields are known to be affected by the rapport a mediator or client has with their clinician (Kelley et al., 2014), it is hoped that tracking these measures alongside each other will enable the clinician to observe some correlations amongst the data. An example of what this information would look like together on the same graph is shown in Figure 2.14. This information can then help the clinician modify subsequent conversations with the mediator in order to increase rapport, which will in turn hopefully increase treatment adherence and fidelity.

*How to Teach Rapport Building Skills to Behavior Analysts* 119

**Fidelity Checklist**

Program: _____
Mediator: _____ Client: _____
Date: _____

**How to use this datasheet:**
1. Fill in the blanks with each of the steps in your intervention.
2. When observing a mediator, mark whether each step was completed correctly or not.
3. Calculate the percentage of correctly completed steps divided by total steps involved.

| Step | Response Dependent Step | Correct |
|---|---|---|
| 1. | | Y / N |
| 2. | | Y / N |
| 3. | | Y / N |
| 4. | | Y / N |
| 5. | | Y / N |
| 6. | | Y / N |

**% Fidelity**
a. Total steps completed correctly (Y): _____
b. Divided by: Total # of steps circled: _____
= _____ %

*Figure 2.12* Example of a datasheet to track fidelity (Biran et al., 2021a).

## Workshop Results and Feedback from Trainees

### Pre- and Post-Evaluation

At the time of writing, the workshop was run with three different groups of clinicians. Individual participant pre- and post-scores are shown in Figure 2.15 for rapport-building skills from the task list. In order to score the participants, trainers counted the total number of skills from the task list that was exhibited during a three-minute role-play. Data was collected and graphed in this manner rather than as a percentage of correct responses, because it was too difficult to determine what a "missed opportunity" was during a dynamic role-playing conversation.

120  *Ana Luisa Santo and Kimberley Taylor*

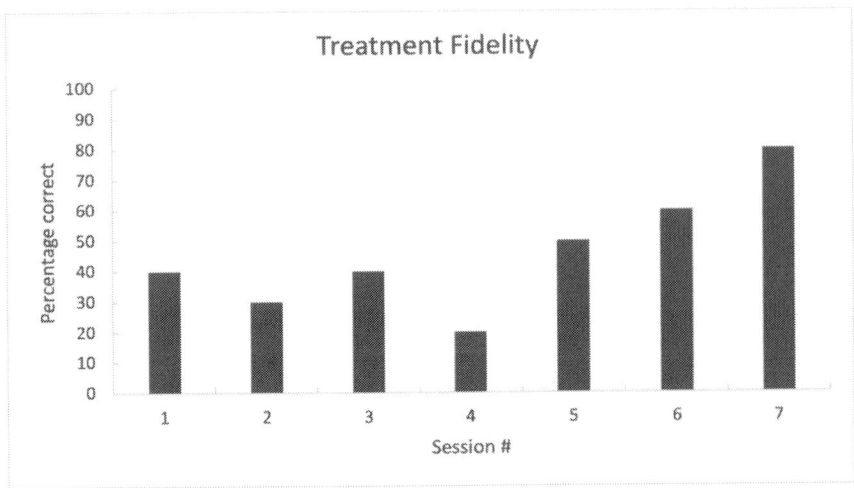

*Figure 2.13* Example of treatment fidelity percentage graphed over seven sessions.

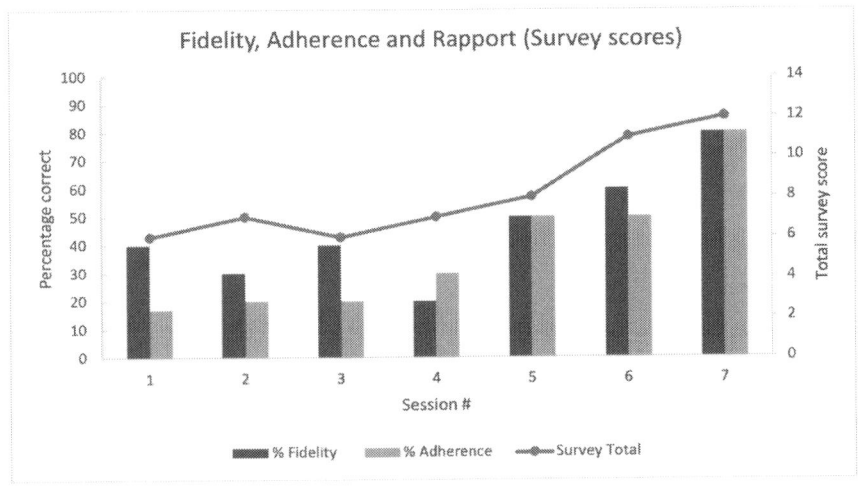

*Figure 2.14* Example graph of adherence, fidelity, and Mediator Check-in Survey scores graphed over seven sessions.

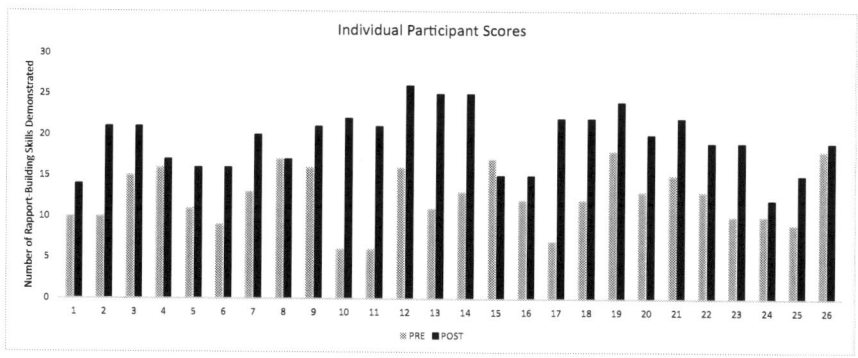

*Figure 2.15* Individual participant scores on rapport-building skills from the task list.

Figure 2.16 shows the increase in number of skills displayed (from an average of 12 in the pre-test to an average of 19 in the post-test), and a modest increase over the two practice sessions. Only one participant showed a slight decrease in scores, and four participants demonstrated minor changes from pre- to post-test. This could potentially be because these individuals already had significant rapport-building skills prior to attending the workshop.

It is important to note that the number of behaviors exhibited by each participant is not to be compared to each other—that is, Participant 12 did not acquire "more" skills than Participant 24 (see Figure 2.15). This is because conversations with mediators (whether role-played or actual) are dynamic, and there is not always the opportunity to exhibit a large number of skills. For example, in some conversations it may be appropriate to exhibit the skill of remaining silent and listening to what is being said for a significant period of time. This would count as one single behavior, but would not mean that the behavior analyst had acquired less skills than an individual who exhibited eye contact, paraphrasing, expressing interest about other family members, and tacting the mediator's feelings in the same amount of time.

### Participant Feedback

Participants provided anecdotal, anonymous feedback upon completion of the workshop. Feedback was overwhelmingly positive and appreciative. The clinicians trained commented on factors they had never considered before:

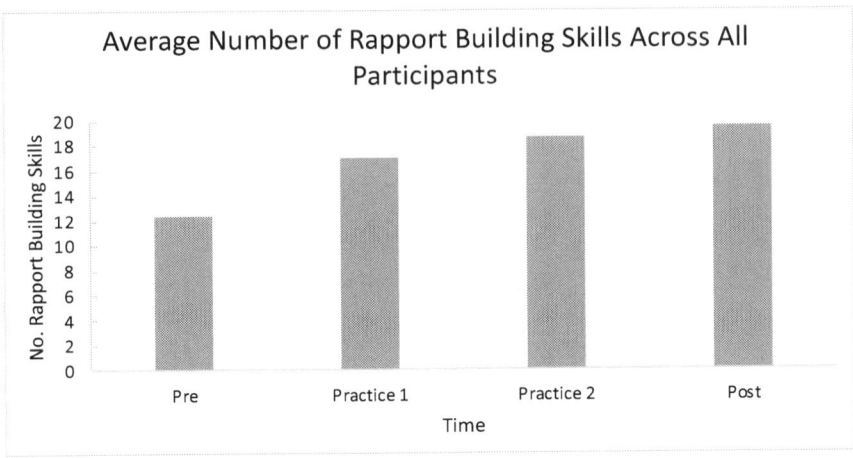

*Figure 2.16* Average scores on rapport-building skills during workshop from all participants.

I liked learning about how to manage difficult conversations with caregivers. [The] big take away was the importance of just listening, and I think that's really important; ... It's made me re-think how I'll approach my mediator training sessions going forward.

Participants also gave examples of how they had used specific skills in conversations with mediators soon after the workshop:

Letting [the client's mother] know that I had also had the experience of a family member with mental health challenges made her feel more comfortable with sharing things with me, knowing I understood [what she was going through].

Being able to give examples from my own personal experience and sharing or empathizing with parents in workshops made it seem like I was a real person and not just a clinician.

[During an] intake conversation and discussion about non-behavioral goals, parent felt they were heard and appreciated it a lot.

Participants expressed they enjoyed the BST and preferred this type of dynamic training over those that are didactic only:

Great balance of presenting materials with use of visuals, time for discussion, and practice. Feeling more confident going into future potentially challenging conversations with caregivers. Will be keeping in mind that sometimes they just want a listening ear and not a solution right off the bat!.

I liked the opportunity to practice the skills and get immediate feedback.

**Ethics**

The workshop was developed and presented with all four core principles within the Behavior BACB Ethics Code for Behavior Analysts in mind. Training behavior analysts to increase their skills in rapport-building behaviors directly benefits others (mediators and clients), allows for increased demonstration of compassion, dignity, and respect, encourages clinicians to behave with integrity, and increases their competence as practicing behavior analysts (Behavior Analyst Certification Board, 2020).

Eight specific Guidelines of the BACB's ethical code were targeted in the workshop and are briefly summarized below:

1.01: **Being Truthful.** One of the targeted skills for increasing rapport-building is to acknowledge your own mistakes when something is not working. This includes being truthful about when the behavior change program is not being effective, or the clinician has not appropriately considered all environmental factors.

1.07: **Cultural Responsiveness and Diversity.** Specific skills like eye contact, being non-judgmental, and expressing interest in the family and their culture are discussed within the framework of AOP in order to respect the diversity of each family and to remain culturally sensitive and responsive.

1.08: **Nondiscrimination.** Similar to Guideline 1.07, the skills of being non-judgmental and using an AOP lens to respond to family concerns are a focal point of the workshop.

1.10: **Awareness of Personal Biases and Challenges.** Personal biases are inevitable for all humans, and AOP helps individuals become aware of these biases and engage in behaviors to overcome them. There is an entire module in the workshop dedicated to exploring what AOP is. While it is by no means exhaustive, this model does provide a starting point for clinicians to begin to examine how they interact with the mediators and clients to whom they provide service.

2.01: **Providing Effective Treatment.** As mentioned previously, research from other healthcare fields indicates that treatment outcomes are more effective when the mediator, client or patient feels a sense of rapport with the clinician providing the treatment (Kelley et al., 2014). It is therefore in alignment with this guideline of providing effective treatment for behavior analysts to ensure they are providing the most effective treatment possible, which includes acquiring the needed skills to increase rapport with families.

**2.09: Involving Clients and Stakeholders.** Two of the rapport-building skills in the workshop target this guideline specifically: responding to and including client ideas in behavioral goals and acknowledging the mediator and client successes. These are key ways to involve clients and stakeholders (i.e., mediators) in their own behavioral treatments.

**2.19: Addressing Conditions Interfering with Service Delivery.** Lack of a sense of rapport with clinicians is one main condition that interferes with effective behavioral service delivery (e.g., Taylor et al., 2019). Addressing this concern by receiving training in how to increase rapport skills is one way that behavior analysts can ensure they are following this Guideline.

The skills presented in the workshop directly and indirectly address many other conditions that could interfere with service delivery. These include identifying a lack of resources (whether social, financial, emotional, etc.), and staying present to what is being discussed if the conversation veers away from the behavioral interventions and toward conditions affecting their implementation.

**3.01: Responsibility to Clients.** Behavior analysts have a responsibility to act in the best interest of their clients and do no harm. Above and beyond this, though, they are responsible for ensuring the maximum benefits from their behavioral interventions. As Taylor et al. (2019) have identified, lack of rapport with clinicians is one main reason why behavioral interventions do not reach their maximum potential; the onus is on all behavior analysts to recognize that their responsibility to providing the maximum benefit to clients includes learning how to build rapport with families.

## Conclusion

While the importance of rapport with mediators/clients is starting to become more widely recognized within the field of Applied Behavior Analysis, how to operationally define, measure, and train individuals in skills such as active listening, empathy, and compassion has remained thus far elusive. The workshop discussed in this monograph (along with the individually defined rapport-building skills in Monograph 2.1) shows that training these behaviors, previously thought to be mentalistic and not definable from a behavioral perspective, appears to be possible. The preliminary results from the three groups of trained clinicians show that an increase in the skills of active listening, empathy and compassion occurred during the workshop.

Whether these gains generalize into real-world application with mediators, or are maintained over time, remains to be explored. The larger question also still remains—even if behavior analysts acquire and maintain these rapport-building skills through this workshop or something similar, does their use in everyday clinical practice increase adherence with treatment programs, fidelity with treatment interventions and, ultimately, result in increased behavioral changes for their clients?

The description of this workshop, its components, and the preliminary results are provided with optimism that it inspires clinicians and researchers alike to further explore these important unanswered questions. If nothing else, it is hoped that by providing the details of the workshop along with the breakdown of rapport-building skills in the previous monograph, behavior analysts will recognize that empathy and compassion are not outside the realm of behavior analytic training and measurement. This opens the door for a wealth of exploration and clinical practice that will only serve to further benefit our clients and increase the quality of service we provide.

Acknowledgements: We would like to give a special thanks to Surrey Place (Toronto) as they provided us with the opportunity to develop and facilitate our workshop as part of our role as Behavior Therapists in the Adult Program. We are grateful for the assistance of the other workshop facilitators who helped us implement the workshop. We would also like to give thanks to all of the behavioral clinicians at Surrey Place who participated in the workshops and provided us with feedback.

## References

Allen, K. D., & Warzak, W. J. (2000). The problem of parental nonadherence in clinical behavior analysis: Effective treatment is not enough. *Journal of Applied Behavior Analysis, 33*(3), 373–391.

Behavior Analyst Certification Board. (2020). *Ethics code for behavior analysts.* https://bacb.com/wp-content/ethics-code-for-behavior-analysts/

Biran, S., Santo, A. L., & Taylor, K. (2021a). "Mediator rapport building: Trainers' manual and toolkit for behavioral clinicians" [Unpublished Manuscript]. Surrey Place, Canada.

Biran, S., Santo, A. L., & Taylor, K. (2021b, November 18–19). *Mediator rapport building—A practical training curriculum and toolkit to increase anti-oppressive sensitivity and compassionate care skills for behavioral clinicians* [Presentation]. Ontario Behavior Analysis Conference, Toronto.

Burke, B., & Harrison, P. (1998). Anti-oppressive practice. In *Social work* (pp. 229–239). Palgrave, London.

Canon, L. F. & Gould, E. R. (2021). A preliminary analysis of the effects of clicker training and verbal instructions on the acquisition of relationship-building skills in two applied behavior analysis practitioners. *Behavior Analysis in Practice, 15*(2), 383–396.

Fallon, L., Collier-Meek, M., Sanetti, L., Feinberg, A. & Kratochwill, T. (2016). Implementation planning to promote parents' treatment integrity of behavioral interventions for children with autism. *Journal of Educational and Psychological Consultation, 26*, 1–23.

Fingerhut, J., & Moeyaert, M. (2022). Training individuals to implement discrete trials with fidelity: A meta-analysis. *Focus on Autism and Other Developmental Disabilities, 37*(4), 239–250.

Hastings, R., Kovshoff, H., Brown, T., Ward, N., Degli Espinosa, F., & Remington, B. (2005). Coping strategies in mothers and fathers of preschool and school-age children with autism. *Autism: The International Journal of Research and Practice, 9*(4), 377–391.

Hayes, S. C., Strosahl, K. D., & Wilson, K. G. (2003). *Acceptance and commitment therapy: An experiential approach to behavior change.* Guilford Publications.

Hock, R., Kinsman, A. & Ortaglia, A. (2015). Examining treatment adherence among parents of children with autism spectrum disorder. *Disability and Health Journal, 8*(3), 407–413.

Hojat, M., Louis, D. Z., Markham, F. W., Wender, R., Rabinowitz, C., Gonnella, J. S. (2011). Physicians' empathy and clinical outcomes for diabetic patients. *Academic Medicine, 86*(3), 359.

Kelley, J. M., Kraft-Todd, G., Schapira, L., Kossowsky, J., Riess, H. (2014). The influence of the patient-clinician relationship on healthcare outcomes: A systematic review and meta-analysis of randomized controlled trials. *PLoS One, 9*(4), e94207. doi: 10.1371/journal.pone.0094207

Lown, B. A. (2016). A social neuroscience-informed model for teaching and practicing compassion in health care. *Medical Education, 50*, 332–342.

Martin, D. J., Garske, J. P., & Davis, M. K. (2000). Relation of the therapeutic alliance with outcome and other variables: A meta-analytic review. *Journal of Consulting and Clinical Psychology, 68*(3), 438–450.

Miltenberger, R. (2004). *Behavior modification: Principles and procedures* (3rd ed.) Belmont, CA. Wadsworth Publishing.

Rohrer, J. L. & Weiss, M. J. (2022). Teaching compassion skills to students of behavior analysis: A preliminary investigation. *Behavior Analysis in Practice,* Oct 11: 1–20. doi: 10.1007/s40617-022-00748-y. Epub ahead of print. PMID: 36249892; PMCID: PMC9553076.

Taylor, B. A., LeBlanc, L. A., & Nosik, M. R. (2019). Compassionate care in behavior analytic treatment: Can outcomes be enhanced by attending to relationships with caregivers? *Behavior Analysis in Practice, 12*(3), 654–666.

# 3
# Collaboration

# 3.1 No More "Train and Train More"

## A Functional, Contextual Approach to Collaboration with Families

*Cressida Pacia, Ciara Gunning,
Aoife McTiernan and Jennifer Holloway*

### Introduction

Caregiver involvement is a key component of effective services for children with autism (e.g., Division for Early Childhood, 2014; Hyman et al., 2020). It is essential that caregivers' perspectives are incorporated within the development and implementation of behavioral interventions. Caregivers provide information on the context within which intervention will be implemented, and this facilitates the development of socially valid interventions and strategies that fit within the family's unique context (Seligman & Benjamin-Darling, 2017). For example, family expectations, values, and culture will vary across each family we work alongside in our practice.

The intensity of caregiver involvement can run on a continuum from exclusively caregiver-delivered intervention to fully therapist-delivered intervention, with varying levels of involvement in between (e.g., therapist-delivered intervention with occasional caregiver coaching). Even in interventions without an explicit caregiver coaching component (e.g., a fully center-based early intensive behavioral intervention program), best practice would still include collaboration with caregivers (e.g., Strauss et al., 2013). This could be to gain information regarding the child's skills and natural environment, developing socially valid goals that are important to the child and family, ensuring the child attends treatment consistently, participating in planning meetings, and implementing strategies at home to promote skill generalization. On the other hand, an intervention may be exclusively implemented by caregivers with no direct contact with the client, such as in a caregiver-mediated intervention delivered over Telehealth. In these scenarios, effective collaboration is even more central to the intervention, particularly collaborative practice-planning to help caregivers effectively incorporate the intervention strategies into their daily lives.

When parent coaching is included in behavioral interventions, behavior analysts tend to focus on parents' treatment fidelity as a primary concern.

Treatment fidelity refers to the extent to which the intervention strategies are implemented correctly (Carroll et al., 2007; Gresham et al., 1993). Certainly, without treatment fidelity and thus treatment effectiveness, we are unlikely to produce meaningful outcomes. However, there are other factors to consider. Treatment adherence, for example, refers to the extent to which strategies are implemented outside the treatment context, that is, whether the strategies are implemented in daily life, not just during sessions (Meichenbaum & Turk, 1987; Sackett & Haynes, 1976). Treatment fidelity is necessary for treatment adherence, but high treatment fidelity does not guarantee treatment adherence. Furthermore, other related fields, such as medicine, have taken it a step further and advanced the use of a different term, "concordance," to replace adherence (e.g., Snowden et al., 2013). Concordance is more in line with caregiver-clinician collaboration, as it necessitates agreement between both parties, rather than compliance with one party's instructions. This thus supersedes the old expert model and sets the stage for a more family-centered approach.

Establishing concordance requires fostering a collaborative, caring partnership within a client-practitioner relationship, also known as therapeutic alliance (TA) (Lejuez et al., 2005). While there are various definitions of TA, at its core, TA is a measure of the extent to which the therapist and the client are engaged and working together toward mutually acceptable goals (Bordin, 1979). Although behavior analytic research on the relationship between TA and important factors such as intervention uptake, adherence, satisfaction, and outcomes are scarce (e.g., Chadwell et al., 2019), studies outside behavior analysis have demonstrated correlation (e.g., Brewe et al., 2020; Escudero et al., 2021). Unfortunately, in a survey of parents' impressions of behavior analysts' compassionate care skills (similar to TA skills; Taylor et al., 2018), behavior analysts were rated low on important skills such as compromising, inquiring about satisfaction, role clarification, demonstrating caring about the entire family, acknowledging mistakes or treatment failures, and being patient and reassuring. Additionally, when asked about behaviors that could harm a therapeutic relationship, parents listed major concerns, including behavior analysts having their own agenda about programming, having an authoritarian demeanor when discussing programming, and underestimating the child's ability.

Given the importance of these skills, the authors advocate for teaching TA skills alongside technical skills when training behavior analysts. Although not specifically developed for behavior analysts, tools such as the Parent and Caregiver Active Participation Toolkit (PACT) can provide helpful guidance in this area. PACT provides resources to help both caregivers and clinicians promote engagement, and outlines ten evidence-informed strategies to build Alliance, Collaboration, and Empowerment (ACEs), as listed below (see Table 3.1; Haine-Schlagel et al., 2020):

*Table 3.1* Evidence-Informed Strategies to Build Alliance, Collaboration, and Empowerment (ACEs)

| Evidence-Informed Strategies to Build Alliance, Collaboration, and Empowerment (ACEs) |
| --- |
| 1. Reflectively listen.
2. Convey parent–therapist partnership.
3. Communicate positive regard.
4. Give suggestions, not directions.
5. Ask for input on intervention strategies.
6. Incorporate input into sessions.
7. Involve parent in session activities.
8. Collaboratively plan homework.
9. Focus on strengths and effort.
10. Jointly identify and problem solve barriers |

These strategies are drawn from the principles and strategies of Motivational Interviewing (MI). MI is an approach to behavior change that is defined by collaborative, goal-oriented communication that balances following (good listening) and directing (giving good information and advice) (Miller & Rollnick, 2012). While MI is derived from social psychology and client-centered counselling, a behavior analytic account of MI has been proposed (see Christopher & Dougher, 2009). A thorough description of this account is beyond the scope of the present chapter. However, this helps to illustrate that, while specific TA skills are essential and practitioners would certainly benefit from direct training in effective caregiver-clinician interactions, TA is aligned with behavioral principles and applied practices that behavior analysts already specialize in: the functional, contextual approach to behavior.

## The Functional, Contextual Approach to Behavior

While many behavior analytic programs might lack formal training in TA skills, these skills are not in juxtaposition to the science of Applied Behavior Analysis (ABA). In fact, they are a core component. In 1978, Wolf championed for the "heart" of ABA and articulated the importance of social validity: the social significance of our goals, the social appropriateness of our procedures, and the social importance of our effects. Establishing a positive working relationship and effective collaboration with the family are core to the tenets of social validity, for who, along with the client themselves, could be better suited to define which goals, procedures, and outcomes would be the most meaningful? In his summary of lessons learned in 25 years of ABA, Foxx (1996) noted that there can be important differences in the outcomes achieved by interventionists delivering ABA services in a strictly traditional way (i.e., "Behavioral Technologists") and those who demonstrate important humanistic, interpersonal behaviors (i.e., "Behavioral Artists"). While behavioral artistry primarily refers to characteristics, skills, and behaviors practitioners demonstrate toward clients (e.g., being able to establish rapport; demonstrating concern; paying careful attention to small, subtle, gradual behavioral indicators; having a sense of humor; maintaining objectivity and positivity), like other behavioral principles, they can be extended toward caregivers as well. Establishing rapport with a child with autism may involve observing their interests, unobtrusively joining in their play, and providing assistance. Similarly, establishing rapport with a caregiver may involve smiling and making eye contact, making positive comments on the child's and parent's behavior, and using humor when appropriate.

Recently, Friman (2021) cogently outlined the foundational philosophy of ABA as the Circumstances View of behavior. This view guides us away from blaming the individual and toward addressing the circumstances impacting behavior. Thus, when there is a breakdown in the collaboration process, taking the Circumstances View leads the practitioner to investigate the reasons for the challenges and functionally address those reasons, rather than blaming the caregiver or applying a one-size-fits-all approach to collaboration. Behavior analysts are well-trained to apply the Circumstances View–the functional, contextual approach to behavior–when assessing and providing intervention directly to a client. This approach is not always taught within the framework of collaboration with other stakeholders, such as caregivers. For example, when providing direct intervention to clients, behavior analysts carefully consider behavioral principles such as motivating operations, stimulus generalization, response acquisition, and competing contingencies, all of which can be applied to interactions with caregivers (Allen & Warzak, 2000).

Motivating operations can have an establishing or abolishing effect on reinforcers, such as when a period of low or diverted attention increases the value of attention as a reinforcer for a young child. For caregivers, consider how setting clear expectations and developing short-term goals can help to establish intermediate outcomes as reinforcers. When we skip this step, we run the risk of inadvertently putting caregiver adherence on extinction when their application of strategies does not produce immediate, noticeable changes in their child's behavior. Similarly, skilled clinicians consider how to program for generalization by harnessing current functional contingencies, training diversely, and incorporating functional mediators when developing skill acquisition programs for children with autism. These considerations are important when considering how a caregiver who demonstrates high treatment fidelity during sessions may struggle to implement the strategies in their daily life. It is important to ask oneself: Have current functional contingencies been harnessed by recruiting both parents to provide praise and encouragement to each other? Was training diverse by coaching caregivers outside of the clinic setting and in more challenging, relevant environments, such as the grocery store? Were functional mediators incorporated by setting up ways for the caregiver to reach out for advice or assistance; for example to their peers in a group intervention? Table 3.2 outlines some behavioral principles and applied strategies and how they could be implemented with clients and caregivers.

Needless to say, extending familiar applications of ABA to caregivers, (i.e., adults who typically have advanced verbal repertoires) is not always a straightforward transfer of commonly used strategies, especially for behavior analysts accustomed to working with young children with autism (probably wouldn't advise using a token board!). When collaborating with caregivers, harnessing the power of rule-governed behavior and exploring private events may be especially helpful. Moore and Amado (2021) argued that, in some cases, matching one's behavior to a corresponding rule is itself a reinforcer—perhaps one of the most powerful ones. Engaging in shared decision-making to establish effective rules (i.e., rules that are simple, explicit, and collaboratively developed) can further increase the probability of matching behavior to the rule. However, contextual conditions (environmental conditions and/ or private events) may make it difficult to stick to the plan, no matter how well-developed.

Fryling (2014) noted that when behavior analysts apply a blanket "train and train more" solution to breakdowns in fidelity or adherence, they risk exacerbating collaboration difficulties if the reason for non-adherence is a contextual variable such as caregiver stress. Rather than increasing stress and making their own presence aversive by placing even more demands, practitioners can work with parents to develop potential solutions that address the problem, such as by connecting the family with respite services. When considering stress as a contextual setting factor, it is important to consider the effect of the intervention itself on stress. For instance, Rivard and colleagues (2017) found that their parent coaching program effectively reduced behavior problems in children with autism, yet parents reported significantly more stress after 12 months of treatment. Mothers who were more involved in their child's intervention program reported a more personal strain (Schwichtenberg & Poehlmann, 2007).

There is a growing interest in the impact of stress and private events in the parent training literature, particularly around incorporating Acceptance and Commitment Training (ACTr) into behavioral interventions (e.g., Cameron et al., 2021; Gould et al., 2018; Yi & Dixon, 2021). ACTr is a contemporary behavior analytic intervention that aims to address the unhelpful ways that private events influence overt behavior,

Table 3.2 Implementing and Collaborating Behavioral Principles with Caregivers

| | Direct implementation | Collaboration with caregivers |
|---|---|---|
| Philosophical framework | Functional, contextual approach and the Circumstances view | Functional, contextual approach and the Circumstances view |
| Effective intervention | Use evidence-based strategies to teach socially valid skills | Use evidence-based adult learning strategies to teach socially valid skills |
| Contextual conditions | Help alleviate stress by reducing aversive stimuli and conditions (avoid punishing conditions) in their environment<br>Help improve quality of life by increasing opportunities to engage in meaningful and enjoyable activities (contact reinforcing conditions) throughout the day<br>Consider the client's culture, religion, and values | Help alleviate stress by connecting caregiver with respite and other practical resources<br>Help reduce social isolation by connecting with other caregivers (e.g., caregiver groups)<br>Consider the family's culture, religion, and values |
| Learning history | Consider how client's history with the skill/environment (e.g., social skills training with school peers) may impact their expectations of the current intervention (e.g., positive vs negative peer interactions)<br>Consider how client's previous experiences with behavioral interventions (e.g., high rates of demands, use of escape extinction) might impact their response to the current intervention (e.g., crying, running away from practitioner) | Consider how caregiver's previous experiences with other professionals (e.g., physicians) might impact their expectations of the current intervention (e.g., immediate and predictable relief of symptoms after taking medication vs slower, less predictable skill acquisition in behavioral intervention)<br>Consider how caregiver's previous experience with behavioral interventions might impact their response to the current intervention (e.g., "I already tried reinforcement–it didn't work") |

(Continued)

Table 3.2 (Continued)

|  | Direct implementation | Collaboration with caregivers |
|---|---|---|
| Assessment | Interview caregivers and other stakeholders (and client if possible) on current skills repertoire and environmental conditions (e.g., FAI) Conduct direct skills assessment (e.g., ABLLS-R, VB-MAPP) | Interview caregivers on their goals for their child and for themselves Conduct an assessment to identify family barriers and facilitators (e.g., PAIRS) Observe a parent-child interaction to identify parent's current interaction style and baseline use of intervention strategies |
| Goal-setting | Use assessment results and input from stakeholders to develop concrete, observable goals for the client (e.g., at least 5 single-word mands in a 15-minute motivating activity) | Collaborate with caregivers using assessment results and their own priorities to set explicit goals for child and themselves (e.g., setting aside 15 minutes a day to practice social communication strategies) |
| Pairing and therapeutic alliance | Build rapport with client by observing their interests, joining unobtrusively in with their play, honoring stated and implied requests, and providing assistance as needed. | Build rapport with caregivers by smiling and making eye contact, making positive comments and the child's and parent's behavior, and using humor when appropriate. |
| Motivating operations | Establish items and activities within the learning context as reinforcers through environmental arrangement and communicative temptations Disestablish (abolish) items and actions outside the learning context as reinforcers through antecedent interventions such as providing time-based access to competing reinforcers (satiation). | Collaborate with caregiver to establish intermediate outcomes as reinforcers by setting clear expectations and developing small, achievable goals Collaborate with caregiver to disestablish (abolish) competing responses (e.g., social approval, escape from aversive child behavior) as reinforcers through open and honest communication |

*Table 3.2* (Continued)

| | Direct implementation | Collaboration with caregivers |
|---|---|---|
| | Use intervention strategies that are less likely to contact competing reinforcers (e.g., provide a dense schedule of reinforcement to avoid escape behaviors) | Collaborate with caregiver to develop strategies that are less likely to contact competing reinforcers (e.g., avoid social disapproval in public and aversive child behavior by reinforcing child's precursor behavior) |
| Behavior momentum | Provide SDs for high-probability acquired behaviors to create behavior momentum for more difficult acquisition targets | Develop small, achievable goals and provide reinforcement for meeting those goals to create behavior momentum for long-term goals that require consistency and effort |
| Discriminative stimuli | Provide clear SDs for the target behavior<br>Minimize SDs for competing behavior | Collaborate with caregivers to plan effective SDs for their own established goals.<br>Collaborate with caregivers to minimize SDs for behaviors that are incompatible with their goals |
| Prompting | Provide effective, responsive prompts (e.g., use least-to-most prompting) | Collaborate with caregivers to plan effective prompts as needed (e.g., pre-planned reminder emails, strategy cheat sheets) |
| Shaping | Shape successive approximations of the target behavior (e.g., saying buh -> buh-buh -> bubbles) | Collaborate with caregivers to shape successive approximations of the target behavior/pinpoint (e.g., asking a lot of questions -> making declarative statements -> commenting) |
| Chaining | Develop a task analysis for the target behavior and teach in a logical sequence (e.g., backward chaining tying shoelaces) | Develop a task analysis for the target behavior and teach in a logical sequence (e.g., forward chaining the Project ImPACT pyramid) |

(*Continued*)

Table 3.2 (Continued)

|  | Direct implementation | Collaboration with caregivers |
|---|---|---|
| Response effort | Reduce response effort for engaging in the target behavior (e.g., set out clothes the night before when targeting fluency with the morning routine) | Collaborate with caregivers to reduce response effort for implementing strategies (e.g., coach them to write out a schedule on a post-it rather than making and maintaining a laminated visual schedule with multiple pieces) |
| Response acquisition | Ensure appropriate skill complexity (e.g., single-word mands) Use effective teaching strategies (e.g., provide teaching opportunities in an A-B-C format) and responsively adapt teaching strategies based on client performance Use data and observation to identify how they best learn (e.g., video modelling, script fading, DTT, NET) | Ensure appropriate skill complexity (e.g., tell-show-help) Use effective adult teaching strategies (e.g., providing clear explanations and rationale for goals and strategies) and responsively adapt teaching strategies based on caregiver performance and feedback Collaborate with parents to identify how they best learn (e.g., group vs individual intervention, live vs video models, etc.) |
| Data collection and analysis | Develop a robust data collection plan with appropriate parameters Graph data to monitor and analyze progress | Collaborate with parents to develop a feasible data collection plan for parent and/or child behavior Develop parent-friendly graphs to show progress |
| Program revision | Plan for and identify when skill acquisition is low or decreasing, and respond using function-based solutions (e.g., by changing prompting strategy or going back to address pre-requisite skills) | Plan for and identify when collaboration or engagement is low or decreasing, and respond using function-based solutions (e.g., by using the PAIRS) |

Table 3.2 (Continued)

| | Direct implementation | Collaboration with caregivers |
|---|---|---|
| Reinforcement | For some clients, natural reinforcement within the learning activity or context is enough to increase behavior and teach new skills. For those who require more contrived or socially mediated reinforcement, conduct a reinforcer assessment, develop a reinforcement schedule, and respond to changes in motivation in the moment. | For some caregivers, matching the behavior to the rule (rule-governed behavior) is its own reinforcer. For those who require more contrived or socially mediated reinforcement, collaborate with caregivers to develop a plan to recruit reinforcement (e.g., self-management system; praise from spouse; check-ins with BCBA) and assist with self-monitoring motivation over time. |
| Extinction | Avoid inadvertently putting appropriate behavior on extinction (e.g., by intermittently providing immediate reinforcement for appropriate mands while working on increasing delay tolerance) | Avoid inadvertently putting adherence behavior on extinction (e.g., by setting clear and realistic expectations at the outset, setting smaller goals, and reinforcing meeting those goals) |
| Punishment | Avoid inadvertently punishing appropriate behavior (e.g., by rewarding attempts rather than always implementing error correction procedures) | Avoid inadvertently punishing adherence behavior (e.g., by collaboratively planning for a potential extinction burst if the intervention plan involves some extinction) |
| Generalization | Program for generalization by capitalizing on natural contingencies, training diversely, and recruiting functional mediators. Use sequential modification if generalization is challenging | Collaborate with caregivers to program for generalization by capitalizing on natural contingencies, training diversely, and recruiting functional mediators. Use sequential modification if generalization is challenging |

Source: Adapted from Allen & Warzak, 2001; Fryling, 2014; Moore & Amado, 2021.

particularly behaviors maintained by escape or avoidance from aversive private events (Gould et al., 2018; Hayes et al., 1991). When collaborating with caregivers, this could look like helping caregivers identify and move toward their values (e.g., quality joyful moments together as a family) through committed action (e.g., setting aside ten minutes a day for child-directed play), while accepting uncomfortable private events that may arise (e.g., feelings of hopelessness that things may not improve). Certainly, private events are an essential and previously under-examined component of parent engagement. However, care must be taken to ensure that practitioners have the skills necessary to incorporate these strategies, and that behavior analysts work within their scope of practice (Tarbox et al., 2020).

## Barriers to Collaboration

One way to incorporate the functional, contextual approach into caregiver collaboration is to individualize behavioral interventions based partly on the family's barriers and facilitators to engagement (i.e., their circumstances). While caregivers may be motivated to collaborate with practitioners on their child's intervention, they can experience difficulties that impact their ability to consistently engage. A growing body of literature illustrates parents' experiences and perspectives on variables that affect their engagement with behavioral interventions. There are common barriers across a range of domains, as demonstrated by the following examples (Pacia et al., 2022):

- **Logistical barriers**—this can include difficulties with access (e.g., transportation, scheduling, cost) as well as administrative difficulties (e.g., managing multiple therapists, managing materials)
- **Child barriers**—a mismatch between the child's skills/behaviors and the intervention goals and strategies (e.g., implementing an intervention designed for more advanced learners with emerging communicators, challenging behavior impacting intervention efficacy)
- **Sibling barriers**—difficulties with siblings, such as perceived lack of attention for sibling or increased challenging behavior from sibling
- **Parent barriers**—this can include individual or cultural concerns, difficult circumstances, treatment burden too high, training not the right fit, low motivation or belief in effectiveness
- **Intervention barriers**—variations in treatment efficacy (e.g., progress slower than expected, regression in skills), difficulties with generalization

By jointly identifying and addressing barriers and how they impact each family, behavior analysts can cultivate and maintain a strong collaborative

*An Approach to Collaboration with Families* 139

therapeutic relationship and promote best outcomes of the intervention. A tool developed by the authors, the Parent-coaching Assessment, Individualization, and Response to Stressors (PAIRS), suggests specific strategies for addressing barriers within the five steps below:

*Use Good-Practice Strategies with All Families*

While assessment and individualization of specific family variables are important, there are universal strategies that can be used with all families, such as using effective adult- learning strategies. Examples include creating learning experiences that are practical and goal- focused, allow for choice and self direction, capitalize on their experience, and are internally motivating (Ingersoll & Dvortcsak, 2010).

*Identify Each Family's Unique Strengths and Barriers*

Use a structured open-ended interview to learn about contextual variables that may impact collaboration. Familiarity with common barriers can be helpful, but don't forget to ask probing questions to understand how the barrier presents for each family!

*Engage in Collaborative Problem-Solving to Address the Barriers or Minimize Their Impact*

This can look like proactively making adaptations to the intervention or connecting the family with other services (e.g., respite). The PAIRS provides suggested solutions for specific barriers as a starting point for collaborative problem-solving.

*Monitor Collaboration and Engagement Throughout the Intervention*

Consider caregiver's attendance, participation during sessions, and implementation of strategies outside of sessions as one progresses through the intervention.

*Respond to Challenges Using a Functional, Contextual Approach*

Rather than responding with a "train and train more" approach when difficulties with collaboration and engagement arise, reconsider the specific barriers or stressors that might be impacting engagement. For example, is the intervention too demanding, complex, or irrelevant to the family's current needs? Match the behavior analyst's response to the barrier, for example, see Figure 3.1.

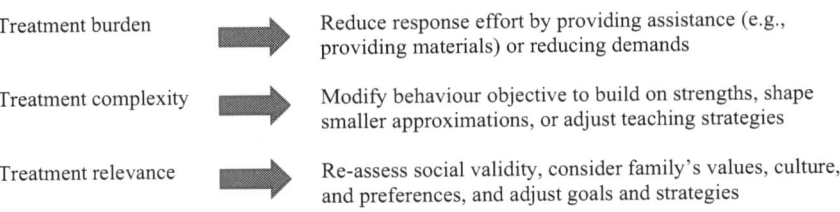

*Figure 3.1* Behavior analyst's responses to various treatment barriers.

## Take-Home Points

Effective collaboration with caregivers is essential to meaningful outcomes in ABA. While the numerous theories, behaviors, and skills to improve collaboration can feel overwhelming, it can be summarized as:

1 Prioritize the therapeutic relationship
2 Utilize one's expertise in the functional, contextual approach to behavior
3 Proactively design and responsively adapt the intervention based on client *and* family variables.

When behavior analysts build positive relationships, take the Circumstances View, and place the family at the core of the intervention: The clinician is more likely to a make meaningful impact on clients' lives.

## References

Allen, K. D., & Warzak, W. J. (2000). The problem of parental nonadherence in clinical behavior analysis: Effective treatment is not enough. *Journal of Applied Behavior Analysis*, 33(3), 373–391. https://doi.org/10.1901/jaba.2000.33-373

Bordin, E. S. (1979). The generalizability of the psychoanalytic concept of the working alliance. *Psychotherapy: Theory, Research & Practice*, 16(3), 252–260. https://doi.org/10.1037/H0085885

Brewe, A. M., Mazefsky, C. A., & White, S. W. (2020). Therapeutic alliance formation for adolescents and young adults with autism: relation to treatment outcomes and client characteristics. *Journal of Autism and Developmental Disorders*, 51(5), 1446–1457. https://doi.org/10.1007/S10803-020-04623-Z

Cameron, M. J., Moore, T., Bogran, C., & Leidt, A. (2021). Telehealth for family guidance: Acceptance and commitment therapy, parent-focused preference assessment, and activity-based instruction for the support of children with autism spectrum disorder and their families. *Behavior Analysis in Practice*, 14(4). https://doi.org/10.1007/s40617-020-00443-w

Carroll, C., Patterson, M., Wood, S., Booth, A., Rick, J., & Balain, S. (2007). A conceptual framework for implementation fidelity. *Implementation Science*, 2(1), 1–9. https://doi.org/10.1186/1748-5908-2-40

Chadwell, M. R., Sikorski, J. D., Roberts, H., & Allen, K. D. (2019). Process versus content in delivering ABA services: Does process matter when you have content that works? *Behavior Analysis: Research and Practice, 19*(1), 14–22. https://doi.org/10.1037/BAR0000143

Christopher, P. J., & Dougher, M. J. (2009). A behavior-analytic account of motivational interviewing. *Behavior Analyst, 32*(1), 149–161. https://doi.org/10.1007/BF03392180

Division for Early Childhood. (2014). *DEC recommended practices in early intervention/early childhood special education 2014*. www.dec-sped.org/recommendedpractices

Escudero, V., Friedlander, M. L., Kivlighan, D. M., Abascal, A., & Orlowski, E. (2021). Family therapy for maltreated youth: Can a strengthening therapeutic alliance empower change? *Journal of Counseling Psychology*. https://doi.org/10.1037/COU0000574

Foxx R. M. (1996). Twenty years of applied behavior analysis in treating the most severe problem behavior: Lessons learned. *The Behavior Analyst, 19*(2), 225–235. https://doi.org/10.1007/BF03393166

Friman, P. C. (2021). There is no such thing as a bad boy: The Circumstances View of problem behavior. *Journal of Applied Behavior Analysis, 54*(2), 636–653. https://doi.org/10.1002/jaba.816

Fryling, M. J. (2014). Contextual interventions for caregiver non-adherence with behavioral intervention plans. *Child and Family Behavior Therapy, 36*(3), 191–203. https://doi.org/10.1080/07317107.2014.934172

Gould, E. R., Tarbox, J., & Coyne, L. (2018). Evaluating the effects of Acceptance and Commitment Training on the overt behavior of parents of children with autism. *Journal of Contextual Behavioral Science, 7*, 81–88. https://doi.org/10.1016/j.jcbs.2017.06.003

Gresham, F. M., Gansle, K. A., & Noell, G. H. (1993). Treatment integrity in applied behavior analysis with children. *Journal of Applied Behavior Analysis, 26*, 257–263. https://doi.org/10.1901/jaba.1993.26-257

Haine-Schlagel, R., Fettes, D. L., Finn, N., Hurlburt, M., & Aarons, G. A. (2020). Parent and caregiver active participation toolkit (pact): Adaptation for a home visitation program. *Journal of Child and Family Studies, 29*(1), 29–43. https://doi.org/10.1007/s10826-019-01659-3

Hayes, S. C., Kohlenberg, B., & Hayes, L. J. (1991). The transfer of specific and general consequential functions through simple and conditional equivalence relations. *Journal of the Experimental Analysis of Behavior, 56*(1), 119–137. https://doi.org/10.1901/JEAB.1991.56-119

Hyman, S., Levy, S., Myers, S., Kuo, D., Apkon, S. D., Davidson, L. F., Ellerbeck, K. A., Noritz, G. H., O'Connor Leppert, M., Saunders, B. S., Stille, C., Yin, L., Weitzman, C. C., Childers, D. O., Levine, J. M., Peralta-Carcelen, A. M., Poon, J. K., Smith, P. J., Blum, N. J., ... Bridgemohan, C. (2020). Identification, evaluation, and management of children with autism spectrum disorder. *Pediatrics, 145*(1). https://doi.org/https://doi.org/10.1542/peds.2019-3447

Ingersoll, B., & Dvortcsak, A. (2010). *Teaching social communication to children with autism: A practitioner's guide to parent training*. Guilford Press.

Lejuez, C. W., Hopko, D. R., Levine, S., Gholkar, R., & Collins, L. M. (2005). The therapeutic alliance in behavior therapy. *Psychotherapy, 42*(4), 456–468. https://doi.org/10.1037/0033-3204.42.4.456

Meichenbaum, D., & Turk, D. C. (Eds.). (1987). *Facilitating treatment adherence: A practitioner's guidebook*. Plenum Press.

Miller, W. R., & Rollnick, S. (2012). *Motivational interviewing: Helping people change*. Guilford Press.

Moore, T. R., & Amado, R. S. (2021). A conceptual model of treatment adherence in a behavior analytic framework. *Education and Treatment of Children, 44*(1), 1–17. https://doi.org/10.1007/s43494-020-00032-0

Pacia, C., Gunning, C., McTiernan, A., & Holloway, J. (2022). Developing the Parent-Coaching Assessment, Individualization, and Response to Stressors (PAIRS) tool for behavior analysts. *Journal of Autism and Developmental Disorders*, 1–24. https://doi.org/10.1007/s10803-022-05637-5

Rivard, M., Morin, M., Mercier, C., Terroux, A., Mello, C., & Lepine, A. (2017). Social validity of a training and coaching program for parents of children with autism spectrum disorder on a waiting list for early behavioral intervention. *Journal of Child and Family Studies, 26*(3), 877–887. https://link.springer.com/content/pdf/10.1007/s10826-016-0604-5.pdf

Sackett, D. L., & Haynes, R. B. (Eds.). (1976). *Compliance with therapeutic regimens*. Johns Hopkins University Press.

Schwichtenberg, A. J., & Poehlmann, J. (2007). Applied behavior analysis: Does intervention intensity relate to family stressors and maternal well-being? *Journal of Intellectual Disability Research, 51*(8), 598–605. https://doi.org/10.1111/j.1365-2788.2006.00940.x

Seligman, M., & Darling, R. B. (2017). *Ordinary families, special children: A systems approach to childhood disability*. Guilford Publications.

Snowden, A., Martin, C., Mathers, B., & Donnell, A. (2013). Concordance: A concept analysis. *Journal of Advanced Nursing, 70*(1), 46–59. https://doi.org/10.1111/jan.12147

Strauss, K., Mancini, F., Fava, L., & SPC Group. (2013). Parent inclusion in early intensive behavior interventions for young children with ASD: A synthesis of meta-analyses from 2009 to 2011. *Research in Developmental Disabilities, 34*(9), 2967–2985. https://doi.org/10.1016/j.ridd.2013.06.007

Tarbox, J., Szabo, T. G., & Aclan, M. (2020). Acceptance and Commitment Training within the scope of practice of Applied Behavior Analysis. *Behavior Analysis in Practice, 15*(1). https://doi.org/10.1007/s40617-020-00466-3

Taylor, B. A., LeBlanc, L. A., & Nosik, M. R. (2018). Compassionate care in behavior analytic treatment: Can outcomes be enhanced by attending to relationships with caregivers? *Behavior Analysis in Practice, 12*(3), 654–666. https://doi.org/10.1007/s40617-018-00289-3

Wolf, M. M. (1978). Social Validity: The case for subjective measurement or how Applied Behavior Analysis is finding its heart. *Journal of Applied Behavior Analysis, 11*(2), 203–214. https://doi.org/10.1901/jaba.1978.11-203

Yi, Z., & Dixon, M. R. (2021). Developing and enhancing adherence to a telehealth ABA parent training curriculum for caregivers of children with autism. *Behavior Analysis in Practice, 14*(1), 58–74. https://doi.org/10.1007/s40617-020-00464-5

# 3.2 Playing Nicely in the Interdisciplinary Sandbox

## A How-to Guide on Effective Collaboration with Various Professionals

*Taylor Slobozian*

### Introduction

Within the last decade, it has become common for behavior analysts to participate in several service team delivery models through consultative practices or agencies expanding to build their team to help support the client in a "wraparound" approach. This has resulted in working with fellow colleagues with little to no behavioral experience, such as: teachers, occupational therapists, physiotherapists, speech-language pathologists, prescribing professionals, and counsellors, to name a few. Not only have behavior analysts seen a surge in participating in service delivery models with non-behavioral colleagues, they have also seen a shift in learning environments where their services are offered from a team approach (Slim & Reuter-Yuill, 2021). With these changes, it is imperative that behavior analysts start to integrate the "people skills" into their practice in order to become an effective collaborator with fellow professionals. Within the last decade, behavior analysts have seen an increase in their involvement in various collaborative service-delivery models that have included multidisciplinary, transdisciplinary, and interdisciplinary teams within many settings—that is, education, clinical, and family homes (Slim & Reuter-Yuill, 2021). Each team model brings with it its own set of advantages and disadvantages. The Behavior Analyst Certification Board (BACB) Ethics Code for Behavior Analysts states that behavior analysts are to collaborate with colleagues for the best interest of the client (Guideline 2.10; BACB, 2020). Behavior analysts are often known for their scientific rigor, which is often seen in their high standards placed on using evidence-based treatments (EBTs) as well as the high level of precision and accuracy in treatment implementation (Slim & Reuter-Yuill, 2021). This high level of standards has made it difficult for behavior analysts to be involved in an interdisciplinary team with colleagues from other governing bodies, where they are practicing from a different perspective outside one's scope of competence or practice in a field where evidence-based treatments may not be as crucial or valued (Guideline 2.10; BACB, 2020). The following will explore the concept of how behavior analysts can "play nicely in the interdisciplinary sandbox" in order to meet their client's needs while practicing the "people skills" associated with applied behavior analysis that will lead to becoming an effective collaborator. Key areas of focus will include: the breakdown of the interdisciplinary team model and its benefits, professionals who make up an interdisciplinary team that behavior analysts will commonly interact with, ethical, legal, and moral considerations and, lastly, foundational skills for becoming an effective collaborator.

### The Interdisciplinary Model

The interdisciplinary team model has often been considered one of the more inclusive models for service delivery (Slim & Reuter-Yuill, 2021). Within this model, professionals work independently of each another during the assessment period, and formal meetings are often utilized to bring the team together to present results where an exchange of ideas often occurs amongst team members. Based on each individual's assessment results, the team then collaborates on decision-making in order to move forward in terms of treatment plans and goal selection (Slim & Reuter-Yuill, 2021). It is important to note that, within this model, roles are often decided at the onset of the team being formed, and that each professional brings their own area of expertise and contributes to the "big picture" (see Figure 3.2).

While working in an interdisciplinary team, behavior analysts may often struggle with how goals are communicated and aligned to their own goals based on common, non-technical language used as well as varied techniques from within each profession. Behavior analysts, like other professionals, may also see the challenge of time being a main factor on implementing intervention strategies due to the fact that there is a significant amount of time required to collaborate with fellow team members in order to share information accurately (Slim & Reuter-Yuill, 2021). Behavior analysts may also struggle with the goal-selection process. Since each professional is acting independently of each other during the assessment phase, goals may either overlap with other professions or be contradictory to goals set by various team members (Beirne & Sadavoy, 2022). With that being said, it would be of benefit that criteria be established while working within this type of team model as a behavior analyst.

### Common Team Members, Overarching Goals, and Supporting the Team

As outlined in Figure 3.2, there is a wide variety of professionals whom behavior analysts may see themselves collaborating with, and this can

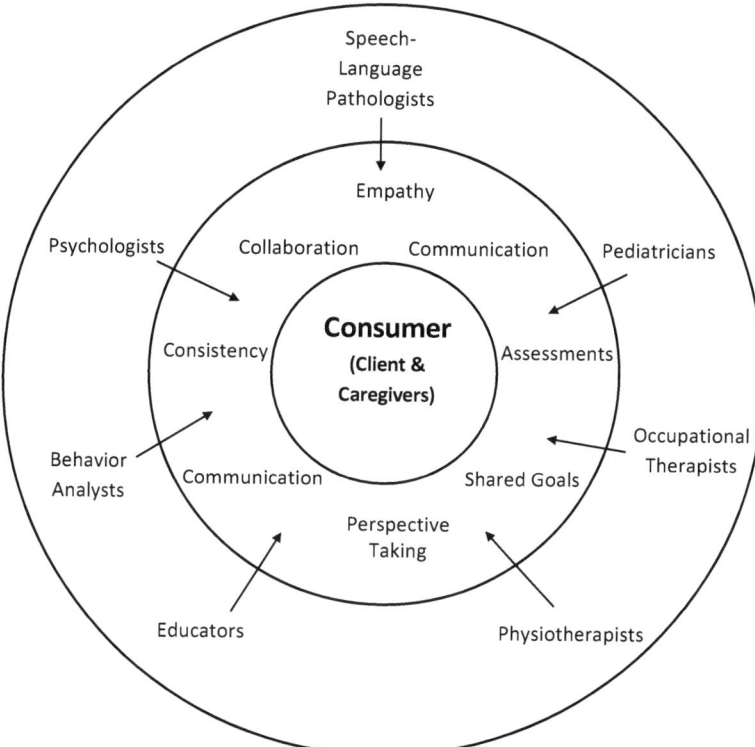

*Figure 3.2* Interdisciplinary team skills and role. A model outlining the various service providers often involved in an interdisciplinary team that behavior analysts may commonly interact with (outer ring). Within the middle ring are skills and actions that take place during the collaborative part of the team model, often referred to as the information sharing phase which is core component of this team model in order to meet the needs of the consumer/clients (inner ring).

result in overlapping of services and goal selections for clients. Speech-language pathologists, occupational therapists, counsellors, prescribing professionals, and teachers often share similar goals to those of behavior analysts who often target communicative deficits, reducing challenging behaviors, and skill acquisition, but intervention techniques differ. Speech-language pathologists often include the use of reinforcement to be built into their treatment plans and have a similar ethics code, which holds them to standards comparable to that of a behavior analyst (Donaldson & Stahmer, 2014). This further results in a common ground of using similar if not the same intervention strategies as seen in the field of applied behavior analysis. In doing so, a single intervention strategy, like the use of the Picture Exchange Communication System (PECS) can be implemented between both professionals, who share a common understanding to teach skills that both professionals have been trained in and include discrete trial training (DTT) and pivotal response training (PRT) (Donaldson & Stahmer, 2014). However, a challenge between both professions can come when trying to determine the client's current communication skills, the client's preferred communicative modality (based on response effort), when to introduce other communicative modalities and their use (i.e., total communication approach), as well as when to fade prompts as the client's verbal language starts to increase (Donaldson & Stahmer, 2014).

Occupational therapists focus on skill acquisition for their clients, which is directly related to skills to be utilized throughout their daily routine, also known as vocational skills. Many of these vocational skills are target behaviors for behavior analysts as well. However, occupational therapists are further trained in sensory integration and this is the understanding that behavior is controlled by sensory processing issues (Beirne & Sadavoy, 2022). With that being said, there may be areas where goals overlap, but the intervention strategies may differ due to what is considered to be EBTs. In this case, it would be important to focus on similarities of each profession, such as: determining how environmental factors are influencing behavior, increasing and decreasing the rate of target behaviors, and focus on generalization and maintenance of the newly acquired skill(s) to other environments. Furthermore, a behavior analyst would not simply dismiss the use of a sensory profile that may be offered by an occupational therapist but, rather, help support an operational definition that is quantifiable versus being subjective (Beirne & Sadavoy, 2022). Common ground that is often viewed between both occupational therapists and behavior analysts is the use of assessments that try to determine what is reinforcing for a client. With the above noted, a behavior analyst and occupational therapist can further support one another in an interdisciplinary team model by conducting a preference assessment and incorporating it into an intervention plan to increase the intervention's success, the client's overall success, and ultimately, their independence (Summers et al., 2022).

Counsellors and prescribing professionals provide support to clients who seek or have been referred to for mental health, pharmacology supports, and addiction services and supports. When working with these professionals, a behavior analyst can support these professionals by incorporating various interventions that are based on graduated exposure, skill building, self-monitoring, and reinforcement systems (Summers et al., 2022). Behavior analysts can support the team by operationally defining behaviors so they can be measured accurately as well as presenting individualized data in the form of visuals (graphic displays). They can support the work by applying their knowledge of being able to examine the relationship between the target behavior(s) and the environment in which they are observed—manipulating the antecedent and consequence variables to see if behavior change occurs (Summers et al., 2022). Examining the relationship of the behaviors and the environment in which they are observed, a behavior analyst can then incorporate the use of single subject research designs to evaluate treatment effectiveness and even conduct functional analyses and manipulate environmental variables to see specific changes influencing the behavior (Summers et al., 2022). Contingency management can also be supported, which is a commonly used ABA technique, based on principles of both, operant conditioning and behavioral pharmacology (i.e., providing a "token" or "voucher" to a client who sustains from drug use, a token that can later be exchanged for a reward or a prize). Through this technique, behavior analysts have been able to successfully support prescribing professionals and their clients by increasing the client's retention in treatment, increasing client sobriety and abstinence, as well as increasing client commitment to continue with treatment medication (Summers et al., 2022).

Educators share goals similar to those of a behavior analyst in the areas of school readiness, academics, and routines. However, when it comes to implementing intervention strategies, there is often a divide between practices. It would be important for a behavior analyst to have their goals clearly defined so that they can easily be measured, observed, and quantified as well as presented in common non-technical language. By doing so, this will support teachers in understanding what is being measured as well as how data will be recorded to make data informed decisions (Beirne & Sadavoy, 2022). For example, educators often use scaffolding or praise, and behavior analysts will use teaching steps and verbal reinforcement. It would be beneficial for a behavior analyst to normalize the use of intervention strategies in the natural learning environment where the skill will be taught and goals assessed, such as: creating natural learning opportunities (teachers often use this term), increasing the teaching opportunities through friendly sabotage (teachers also often use this when getting children to label or request items, known to behavior analysts as manding and tacting), and arranging the environment (i.e., staging a classroom and ensuring only the required items required to teach a skill are present and that no distractors are available) (Beirne & Sadavory, 2022). Behavior analysts would further benefit by recognizing the "business" of the school routine and work with the educator to find a treatment strategy that can easily be implemented in a classroom setting with limited to no additional resources or acquired costs.

## Legal, Ethical, and Moral Obligations While Working in an Interdisciplinary Team

While finding common ground in goal selection and intervention strategies is important, it is equally important that behavior analysts outline their ethical, legal, and moral obligations to all clients (i.e., consumers, team members, third parties, etc.). While working within an interdisciplinary team, communication is crucial, and behavior analysts should use their ABA skills to operationally define and identify the behavior goals for the team in terms of roles, effective communication amongst team members, and effective communication modalities (i.e., phone calls, emails, etc.) (Guideline 2.09; BACB, 2020). The team should also discuss ethical and legal responsibilities surrounding maintaining client confidentiality and permission for sharing client information. This relates to how documents are stored or "housed," and ensures that documentation is meeting current jurisdictional standards. This may differ amongst professions and between agency policies; however, these key components should be brought forward with the onset of service (i.e., consents to obtain or share information, videos, data, etc.). Different services may require different procedures for billing, and clear and consistent documentation, billing procedures can be accurately reflected of services provided (Guideline 2.09; BACB, 2020). Ethically speaking, behavior analysts promote an ethical culture in work environments and make others aware of this obligation while working within an interdisciplinary team (Guideline 1.02; BACB, 2020). This is important when working with diverse populations as well as with different professionals (Guideline 1.07, 1.08; BACB, 2020). From a moral standpoint, behavior analysts should put the needs of their clients above all else and advocate for additional services if required (Guideline 3.12; BACB, 2020).

## Effective Skills for Collaboration in an Interdisciplinary Team

While providing a sufficient explanation of the profession's legal, ethical, and moral obligations, behavior analysts can continue to be effective collaborators by practicing the skills as identified in Figure 3.2. As a behavior analyst working in an interdisciplinary team, it would be important that all forms of communication (i.e., verbal, written, etc.) are responsive and respectful of fellow team members' opinions and expertise. It would be equally important to encourage the use of commonly used language that is consistent across professions when explaining shared ideas and intervention strategies, limiting the use of jargon where possible. If jargon is to be used, it would be beneficial to further define using non-technical terms so team members receive a clear and consistent understanding of the information (Guideline 2.08; BACB, 2020).

A "people skill" that often flies under the radar is being empathetic. Empathy can be described as understanding another individual's experiences by placing yourself in that individual's situation (Slim & Reuter-Yuill, 2021). Becoming an interdisciplinary team by becoming an active listener and eliciting a response based on similar experiences and emotions helps one understand how others feel and allows them to respond appropriately in a given situation (Slim & Reuter-Yuill, 2021). This helps in relating to fellow team members based on their own unique experiences and further build a healthy working relationship.

Somewhat related, the skill of "perspective taking" is the act of recognizing a situation or concept and understanding it from an alternative point of view, often from another individual (Slim & Reuter-Yuill, 2021). It assists in establishing a healthy working relationship and resolving conflict. With varied opinions, expertise, and/or overlapping goals, behavior analysts can often be faced with conflict in terms of how goals and intervention strategies are decided upon. As noted above, behavior analysts often work with diverse populations and with various professionals, and perspective taking can further assist when feeling frustrated with a different discipline and meeting the consumer's needs (Beirne & Sadavoy, 2022). In addition, one can appreciate the importance of the varied skills within an interdisciplinary team, become an effective collaborator, and practice within one's own scope of competency.

## Conclusion

Within the last decade, behavior analysts have continued to see an increase in their involvement in team models and having to collaborate with various professionals across diverse disciplines. Working alongside an interdisciplinary team will allow opportunities to debunk the common misconceptions associated within the field of applied behavior analysis. Myths like, data collection is always burdensome and intensive, reinforcement procedures are the same as a "bribe," ABA does not promote individual autonomy, and newly acquired skills are always taught in controlled environments can be dispelled by working alongside other professionals to promote accurate information and understand the overlaps between groups. As outlined above, working with various professionals and practicing people skills associated with applied behavior analysis, are important to consider when it comes to playing nicely in the interdisciplinary sandbox and becoming an effective collaborator.

## References

Behavior Analyst Certification Board. (2020). *Ethics code for behavior analysts.* https://bacb.com/wp-content/ethics-code-for-behavior-analysts/

Beirne, A., & Sadavoy, J. A. (2022). *Understanding ethics in applied behavior analysis: Practical applications.* Routledge.

Donaldson, A. L., & Stahmer, A. C. (2014). Team collaboration: The use of behavior principles of serving students with ASD. *Language, Speech & Hearing Services in Schools, 45*(4), 261–276. https://doi.org/10.1044/2014_LSHSS-14-0038

Slim, L., & Reuter-Yuill, L. M. (2021). A behavior-analytic perspective on interprofessional collaboration. *Behavior Analysis in Practice, 14*(4), 1238–1248. https://doi.org/10.1007/s40617-021-00602-7

Summers, J., Busch, L., Kako, M., & Lau, C. (2022). The role of the behavior analyst on interprofessional mental health teams: Opportunities for collaboration and enhancing patient care. *Journal of Interprofessional Care, 36*(3), 434–440. https://doi.org/10.1080/13561820.2021.1969345

# 3.3 Who's the King of the Castle?

## Collaborating with Educators and Schools—as a Behavior Analyst

*Karen Manuel and Brittany Davy*

### Introduction

As young children, we are taught fundamental skills to work collaboratively within a team. We are taught that there are unwritten rules and guidelines to follow that assist one in becoming a good teammate as well as make the team function smoothly. As adults, we continue to apply those same rules and guidelines, yet we encounter challenges that require us to use all of our knowledge and expertise to manage. As behavior analysts, when working within and/or consulting alongside educational teams, including but not limited to teachers, administrators, psychologists, occupational therapists, and speech language pathologists, viewpoints can sometimes vary to the point of disagreement (Brown, et al, 2021; Zume, 2019). The requirement of a behavior analyst is to determine how to best manage/support the client under potentially challenging circumstances (Brodhead, 2015). One must focus on how to contribute to the team, ensure viewpoints are accepted and deemed valid, while also ensuring ethical standards are upheld. This task is not necessarily easy, but success is achievable when behavior analysts effectively use the many tools that have been provided through the Behavior Analyst Certification Board (BACB) and government resources. Work within school teams can be conducted collaboratively, respectfully, and ethically. Yet, why does it feel like sometimes there has to be a King of the Castle?

### Joining the Team

At the beginning, gaining access to the treatment conversation may be the first hurdle to overcome. School board structures can vary and can include behavior analysts as part of the internal structure or depend solely on external consultation. As a school board member, getting to the table may require internal referral, service request, school team meetings, and/or any additional method of making a formal service request. Depending on the

DOI: 10.4324/9781003300465-15

knowledge and acceptance of the direct school staff, a behavior analyst may experience a delay in acceptance into the team discussion, although typically the invitation is inevitable if the student's needs require additional treatment. Whereas, when contributing as an external consultant to the school board, the invitation typically is initiated through parent/caregiver discussion and brought to the school team. Parents/Caregivers hold the right to invite anyone to case conferences as advocates, although the openness at the table to allow the behavior analyst to engage in the conversation may still be limited by the school team.

It is important to continue to remember that every member of an interdisciplinary team, including behavior analysts, educators, and other professionals, are crucial to facilitating effective support within the school environment. Collaboration can be challenging as each professional is certified by a regulatory body with a unique code of ethical guidelines (Newhouse-Oisten, Peck et al., 2017). Therefore, regular reviewing of the BACB Ethics Code in relation to collaboration is fundamental to the success of behavior analysts navigating these spaces. For example, it describe the obligation of behavior analysts in interdisciplinary teams to contribute information driven by the philosophical assumptions and principles of ABA to serve clients well (Guideline 1.02, 2.03; BACB, 2020). Behavior analysts should speak only to the areas that they are competent in and only provide recommendations of intervention that are based on ABA principles.

One must also ensure they are prepared with the documentation that is required. The Ethics Code, as well as school regulations, require consent forms to be signed for information sharing prior to engagement of professionals supporting the client/student (Guideline 3.06; BACB, 2020). Making oneself familiar with other special education documents can further one's confidence. Many individual school boards have created resource guides outlining the best ways to navigate special education within their board (e.g., Toronto District School Board, 2020). Figure 3.3 outlines the common reasons conflicts occur in special education.

### At the Table

This is the moment to shine. Assessment of data to determine baseline requires advocating for the most effective collection method while also taking into consideration response effort for team members. Collecting data within a school setting rather than in a clinical setting has an important difference because non-behavioral team members collect data. These same team members may also be responsible simultaneously for large groups of children. To address this, select methods that are clear, concise, require little staff response effort, and use few resources that are easily received

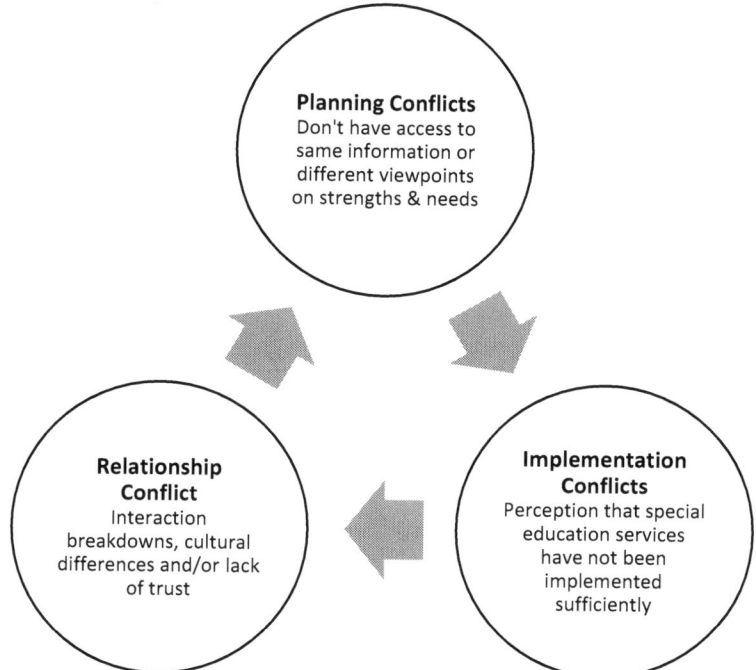

*Figure 3.3* Common reasons for conflict in special education (Ontario Ministry of Education, 2007).

(Zube, 2019). If possible, immerse oneself in the process to be present and engaged in whichever role can be most supportive to the collection process within the school. Be considerate of the client, classroom, and staff needs while developing the plan.

Providing reinforcement to staff for completion of tasks is crucial (Brodhead, 2015). This requested task is additional work for staff, who are already overworked. When presenting data findings, be sure to use both ABA terminology and insight while also ensuring full comprehension through use of laymen's terms. Their ability to thoroughly understand the findings will assist in the team's ability to implement the treatment plan. When presenting a treatment plan, behavior analysts ensure the plan is based on the assessment findings and behavioral principles, and they use evidence-based treatment strategies and positive reinforcement procedures (Guideline 2.14; BACB, 2020). While advocating for the least intrusive effective method, speed of success will be a factor in whether the recommendations are likely to be heard and implemented. Continue to be

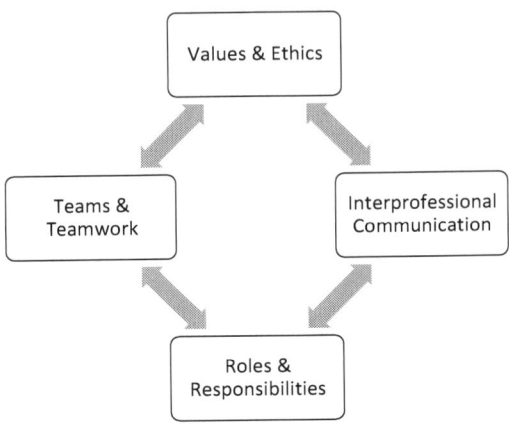

*Figure 3.4* Interprofessional Education Collaborative (IPEC) Framework.

considerate of the non-behavioral members, as their buy-in will be key to your client's success.

Collaboration models created to guide efficient and effective interdisciplinary teamwork encourage the use of respect, ethical integrity, and communication such as the Interprofessional Education Collaborative (IPEC) framework. Slim and Reuter-Yuill (2021) interpreted the IPEC framework through a behavior analytic perspective that emphasizes four domains: values and ethics, roles and responsibilities, interprofessional communication, and teams and teamwork (see Figure 3.4). The teams and teamwork domain specifically explains the importance of considering controlling variables such audience, motivational and contextual variables, as they "…allow for the prediction and control of the behavioral responses of each member and the 'team' as a unit" (Slim & Reuter-Yuill, 2021, p. 1245). While the team goals are to promote the most effective and therapeutic environments for clients, it is important to engage in perspective taking to understand the motivations and ethical guidelines of the others at the round table.

### Decision-Making Challenges: "Because I Said so" Just Doesn't Cut It

Not every collaboration runs smoothly. There may be some moments where play reviews are needed. For behavior analysts, a common challenge is when non-evidence-based behavioral treatments are suggested. The need to support evidence while also maintaining positive team relations can be difficult. Again, resources exist to assist, such as the use of a decision-making model to evaluate when to address such concerns developed by Brodhead (2015). He suggests that one steps up and takes a stand against the use of non-evidence-based treatments primarily when the client's safety is at risk. If there are no safety concerns, then further evaluate the elements of the plan to determine whether the risks of the treatment outweigh the benefits of maintaining the professional relationship. This can also include further evaluation of whether the non-behavioral treatment can be implemented using behavioral principles, such as using the suggestions as antecedent controls, discriminative stimuli, scheduled reinforcement, and more (Brodhead, 2015). Although it is important to engage all of the viewpoints and not suppress any member of the interdisciplinary team, keeping an open mind and a listening ear when participating is important (Slim & Reuter-Yuill, 2021).

Being offered a spot in an interdisciplinary team with educators, although challenging, may be far easier than maintaining that seat at the table. An interdisciplinary team is the coming together of two or more disciplines with a singular goal, acknowledging that those disciplines are all contributing and producing plans towards the goals. As various professionals attempt to juggle different assessment techniques, ethical guidelines, intervention strategies, and areas of competency, maintaining a seat at the table is inherently challenging. To be successful in a interdisciplinary team, the group must facilitate a system that prioritizes the uniqueness of each professional's role and competency as opposed to suppressing these areas (Slim & Reuter-Yuill, 2021).

Work by King et al. (2009) discusses three components of a successful interdisciplinary team that are crucial tips of effective collaboration: arena assessment, role release, and intensive ongoing interaction between group members.

1 Arena assessment consists of the team working together to collect data about a behavior being observed at the same time (Boyer & Thompson, 2013). In this approach, one person takes on the role of assessor and the others observe synchronously. This technique minimizes breakdowns in communication that occur when professionals observe different dimensions and measurements of a singular behavior of interest, which is common in independent observations.
2 Role release refers to the procedure of each professional on the team releasing an aspect of their services, informed by their profession area, to one person on the team who amalgamates them to provide a comprehensive support to the client. The remainder of the team supports that professional though consistent interactions and training (King et al., 2009).
3 Intensive ongoing interaction is essential to the success of any interdisciplinary team (Boyer & Thompson, 2013). Research has shown that the

use of clear and easy to understand descriptions as well as commitment to quality and timely interactions are important to increasing the willingness of professionals to collaborate respectfully.

## Final Game Play: Use Your Power Skills to Keep Your Seat at the Table and Keep Out of the Principal's Office

Power skills are those hidden strengths or tools that are inherent to behavior analysts who aid in supporting successful communication in an interdisciplinary school team. Using these skills can not only get behavior analysts to the table, but also maintain that seat.

Looking to the example of Jane, one can see how to use the skills described. Jane is a behavior analysts and clinical supervisor at a privately funded ABA clinic. Jane's client is Amil. Amil attends a public school, in a regular education grade 3 classroom with limited educational assistant support during larger transitions throughout the day. Amil has been reported by the education and clinical team as a very physically active child, with higher than typical levels of fidgeting than other children his age as well as out of seat behavior accompanied by a low verbal toned sound production that some have described as "humming." Amil's parents have asked for a case conference and invited Jane to attend; and all consents to share information have been signed prior to the case conference.

At the meeting, the school administrator leads the conversation asking for updates and input from all around the table. Amil's classroom teacher shares a thorough update on his mathematics and language, and then shares that, although active, Amil is a joy to work with but causes disruptions to other classmates throughout the day. The Occupational Therapist shares that the team will purchase specialized seating options, sensory toys, and a system for leveled breaks. Jane provides her update regarding the skill acquisition programs based on Tarzan's last assessment results, currently in his treatment plan. She also quickly shares data on the behaviors that have been observed within therapy as well as the positive reinforcement strategies that have been successful. She then discusses how it is understandable that these observable behaviors could add to a challenge within a school classroom, ensuring that she makes suggestions that fit within the teacher's current structure, including a structured schedule, self-regulation strategies learned, and a discrimination training that was started in therapy and increases Amil's awareness of his "humming." Although in the clinical setting the goal was teaching Amil to use differential reinforcement with extinction techniques, Jane explains to the school team when and where this humming is reinforced and when and where it is creating disruption, that the team would not give attention to it. This training would include teaching him the perspective of taking elements supporting how impactful his humming is on the staff and students in his class. The plan is presented collaboratively, respectfully, and ethically, and the result is the full team establishing a successful plan that is manageable and, more importantly, successful for Amil.

So, learning from Jane and Amil, if we as behavior analysts come prepared, use power skills, and refer to the many resources that are available, teams can recognize that there really is no competition at all. The winner of this game, the King of the Castle, is always the client. Working collaboratively, respectfully, and ethically, one can assist clients to achieve success in their primary learning environments. Remembering that the requirement is to always support the client's needs to the best of one's ability, no matter which court they are playing on.

## References

Behavior Analyst Certification Board. (2020). *Ethics code for behavior analysts*. https://bacb.com/wp-content/ethics-code-for-behavior-analysts/

Boyer, V. E., & Thompson, S. D. (2013). Interdisciplinary model and early intervention: Building collaborative relationships. *Young Exceptional Child*, 17(3), 19–32.

Brodhead, M. T. (2015). Maintaining professional relationships in an interdisciplinary setting: Strategies for navigating nonbehavioral treatment recommendations for individuals with autism. *Behavior Analysis in Practice*, 8(1), 70–78. https://doi.org/10.1007/s40617-015-0042-7

Brown, T. J., Schaible, L., Samantha, D., Moscatelli, A., & Linke, N. (2021). *Wearing both hats: Finding fulfillment as a school psychologist and BCBA*. Bethesda, MD: National Association of School Psychologists. Retrieved from www.proquest.com/other-sources/wearing-both-hats-finding-fulfillment-as-school/docview/2515586835/se-2

King, G., Strachan, D., Tucker, M., Duwyn, B., Desserud, S., & Shillington, M. (2009). The application of a transdisciplinary model for early intervention services. *Infants and Young Children*, 22(3), 211–223.

Newhouse-Oisten, M. K., Peck, K. M., Conway, A. A., & Frieder, J. E. (2017). Ethical considerations for interdisciplinary collaboration with prescribing professionals. *Behavior Analysis in Practice*, 10(2), 145–153. https://doi.org/10.1007/s40617-017-0184-x

Ontario Ministry of Education. (2007). *Shared Solutions. A guide to preventing and resolving conflicts regarding programs and services for students with special education needs*. https://files.ontario.ca/edu-shared-solutions-2007-en-2022-01-28.pdf

Slim, L., & Reuter-Yuill, L. M. (2021). A behavior-analytic perspective on interprofessional collaboration. *Behavior Analysis in Practice*, 14(4), 1238–1248. https://doi.org/10.1007/s40617-021-00602-7

Toronto District School Board. (2020). *Guide for special education for parents and guardians: Special education plan 2020.* www.tdsb.on.ca/Portals/0/docs/SpecialEducationPlan_ParentGuide.pdf

Zube, M. (2019). The behavior analyst and "the team": Interdisciplinary teams and working in schools. In *Understanding ethics in applied behavior analysis* (pp. 312–341). Routledge.

# 3.4 Overcoming the Aversive
## Handling Difficult Conversations with Professionalism and Compassion

*Brianna M. Anderson and Dana Kalil*

### Introduction

From managing personal boundaries to mediating differences between others, professionals in supervisory roles regularly engage in critical conversations that have the potential to be challenging. These conversations can include delivering bad news, evaluating supervisees' performance, addressing unethical behavior, and even asking for help (Greeny et al., 2022; Sellers et al., 2019). Though the subject matter of difficult conversations can vary, the commonality across all topics that makes them so difficult is the strong negative emotions they evoke—dread, frustration, embarrassment (Bradley & Campbell, 2016).

### What Are Difficult Conversations?

No one conservation topic is universally difficult for every person. What one behavior analyst finds difficult, another might not. However, topics can be generally categorized by three overarching themes: *self*, *relationship*, and *task* (Kofman, 2014). Conversations involving the *self* occur when the issue is internal and involves only oneself. For example, a behavior analyst may feel inadequate in their role; therefore, they may decide to schedule a meeting with their manager to discuss their performance. *Relationship* conversations occur when an issue arises between two or more individuals. For example, a behavior analyst and a speech-language pathologist may have conflicting opinions about a client's communication program; therefore, they may engage in a conversation to determine how to proceed. Finally, *task* conversations occur when the attention of all parties is directed toward a specific issue. For example, a behavior reduction plan may no longer be effective; therefore, a meeting may need to be scheduled to address that protocol.

DOI: 10.4324/9781003300465-16

## Why Are These Conversations So Difficult?

While the abovementioned conversation topics are an important and often necessary part of behavior analysts' roles, why are they so difficult? Given the principles of operant conditioning, addressing certain topics may be difficult as a result of learning history, namely positive punishment. If an individual has previously engaged a difficult conversation that led to an aversive outcome, the behavior analyst may avoid engaging in similar conversations in the future (Cooper et al., 2019). For example, a behavior analyst might delay sharing these assessment results with a parent because the last time they had to share similar results, the behavior analyst experienced guilt related to that parent's emotional response.

The anticipation of such aversive outcomes alone can lead a behavior analyst to avoid having difficult conversations, even if they have never contacted the punisher directly. For instance, a behavior analyst might have witnessed their mentor sharing the assessment results and simply observing the contingency could be a sufficient punisher for the behavior analyst. Individuals may also avoid difficult conversations due to rule-governed behavior (Cooper et al., 2019). Rule-governed behaviors are controlled by "contingencies that may not be explicitly stated or that have never been contacted directly" (Bradley & Noell, 2022, p. 434). For example, a behavior analyst who has been taught to circumvent conflict at all costs may postpone difficult conversations to avoid the *possibility* of punishment for breaking this rule.

Whether due to learning history, rules, or both, behavior analysts may avoid having these conversations with the hope that the issue will resolve itself without interference. They may also choose to address the issue directly, but fail to provide explicit expectations and solutions. Unfortunately, this approach can lead to detrimental outcomes for everyone involved.

## Why Do Behavior Analysts Need to Have Difficult Conversations?

Engaging in potentially challenging conversations is a crucial part of behavior analysts' roles. As per the Ethics Code for Behavior Analysts, behavior analysts must "develop, communicate, implement, and evaluate an improvement plan with clearly identified procedures" when supervisee performance issues arise (4.08 Performance Monitoring and Feedback; Behavior Analyst Certification Board [BACB], 2020, p. 15). Behavior analysts must also "actively identify and address environmental conditions […] that may interfere with or prevent service delivery" (2.19 Addressing Conditions Interfering with Service Delivery; BACB, 2020, p. 12). Further, it is a behavior analyst's responsibility to "consider discontinuing services when [...] the client is not benefiting" from it (3.15 Appropriately Discontinuing Services; BACB, 2020, p. 14).

These are only some of the circumstances in which behavior analysts are ethically *required* to have difficult conversations. There are also reasons why behavior analysts might *want* to engage in them. Behavior analysts can use these conversations as opportunities to establish clear expectations and avoid potentially detrimental misunderstandings. For instance, delivering performance feedback is an essential component of implementer training and is associated with improved implementer treatment adherence, which can in turn lead to greater improvements for the learner as well (Anderson et al., 2023).

When done sensitively and compassionately, having these conversations can also foster further important and often necessary discussions (Farrell, 2015). They can create an open line of communication, leading to more opportunities for collaboration, particularly when behavior analysts encourage others to provide their input and develop joint solutions (Farrell, 2015). They can also lead to a more positive work environment by providing opportunities for individuals to explicitly discuss their values and expectations (Farrell, 2015). Lastly, engaging in difficult conversations can build stronger relationships through rapport-building with implementers, families, and other professionals.

## What Strategies Are Beneficial for Having Difficult Conversations?

### Before the Conversation

Training and Mentorship
Obtaining training or mentorship is a proactive way to develop the skills needed to engage in difficult conversations (Polito, 2013). Training can involve case discussions, enhanced role-playing, and performance feedback, all of which can occur in vivo before the actual conversation takes place (Epner & Baile, 2014). When seeking out a mentor, it can be helpful to look for someone who is comfortable engaging in difficult conversations and has the people skills needed to do this work in a compassionate manner.

Develop an Outline
Using an outline to guide the conversation can help behavior analysts maintain focus when emotions are heightened. Outlines can include important talking points as well as relevant information, related ethical standards, agency policies, and/or resources (BACB, 2020; Farrell, 2015). They can help behavior analysts preemptively consider challenging points in the conversation and prepare possible solutions. They can also help behavior analysts prepare the language needed to engage in the conversation in a

non-accusatory, non-threatening manner. If there is a possibility that boundaries may be tested during the conversation, it can be helpful to write down where this testing might occur, which boundaries are firm, and where some flexibility can be provided. Lastly, it can be helpful to consider how the other person might react and how to respond in a professional manner.

Schedule the Conversation
Both timing and the location of the conversation are important to consider. A conversation that takes place in an inappropriate location (e.g., delivering negative individual feedback in front of others) is inappropriately timed (e.g., delivering difficult feedback in a moment of frustration) or is timed too late (e.g., waiting until the next team meeting three weeks from now to deliver feedback about someone's performance when they'll be continuing to work with clients between now and then) can be more detrimental than not having the conversation at all (Farrell, 2015). Before scheduling the conversation, consider the questions provided in Table 3.3.

*During the Conversation*

Stay Focused
Having a pre-established list of topics that need to be addressed can be beneficial for guiding the conversation (Farrell, 2015). That said, the list should not be exhaustive, as this can be overwhelming, can devalue the most important topic(s) of the discussion, and reduce the other party's capacity to process the information delivered. Narrowing the focus to one or two items can improve the effectiveness of the conversation and leave time for discussion to occur (Farrell, 2015). If considerable amounts of information need to be shared, consider providing the information in written format, creating a cohesive summary of the information, or dividing the conversation up into multiple meetings. Lastly, some conversations are best had at regular intervals, such as routine performance reviews or monthly parent meetings, and doing so can prevent an over-accumulation of information requiring conversation.

Be Professional *and* Compassionate
Maintaining professionalism is critical when having difficult conversations (Andzik & Kranak, 2021). In some cases, it can be helpful to model composure and control of emotions during the conversation, particularly when either individual may become escalated (Polito, 2013; Price & Baker, 2012). It is also important to acknowledge the sensitivity of the conversation and treat the other individual with compassion. Compassion involves having "a sense of common humanity, mindfulness, and lessened indifference" toward others (Pommier et al., 2020, p. 22). Compassionate behaviors can be verbal (providing reassurance, using affirmative statements, offering help), or nonverbal (smiling, nodding; Beck et al., 2002; Rohrer et al., 2021).

Table 3.3  Question to Consider When Scheduling a Difficult Conversation

| Question | Answer | |
|---|---|---|
| | Yes | No |
| Does the conversation need to happen immediately? | If emotions are heightened, deep breathing (Perciavalle et al., 2017), guided imagery (e.g., visualizing a positive outcome of the conversation; Bigham et al., 2014), and progressive muscle relaxation (e.g., progressively tensing, holding, and relaxing different muscle groups; Ozgundondu & Metin, 2019) are just some of the empirically supported techniques that can promote relaxation in a stressful moment. | Schedule the conversation at a more appropriate time. |
| Do you have the emotional capacity to have the conversation in a professional manner at the scheduled time? | Proceed with the conversation as scheduled. | Consider rescheduling the conversation to a more appropriate time. If this is not possible, develop an outline to help you prepare and/or bring a colleague with you to the conversation. |
| Do you anticipate the other person will feel embarrassed, ashamed, sad, or other internalizing (inwardly directed) emotions? | Find a private location where other people cannot see or hear the conversation. Consider ventilation systems, adjacent rooms, and the space beneath doors that can cause sound to carry even when in a private location. | If possible, schedule the conversation in any appropriate location. |
| Do you anticipate the other person will experience anger, aggression, or other externalizing (outwardly directed) emotions? | Schedule the meeting in a safe location close to others who can be called upon if assistance is required, such as next to a high foot traffic area or in a public location. If this is not possible, bring a colleague to the meeting. When in the meeting, position yourself near the exit and have a means of contacting the appropriate authorities if needed. | Depending on the content being shared and the individuals involved, the meeting may be able to take place in a more private location. |

Active listening is another type of compassionate behavior that can be beneficial during difficult conversations. It involves paying attention to the other person, suspending judgment, engaging in reflection, and clarifying, summarizing, and sharing as needed throughout the conversation (Hoppe, 2006). Not only can active listening allow the behavior analyst to obtain important information, it can also help the other person feel understood and validated. Further, it signals to the other person that they are a full participant in the conversation and allows both parties to build rapport with each other (Jahromi et al., 2016).

Summarize
At the end of the meeting, it can be helpful to verbally summarize what has been discussed, clarify expectations, set an agreed-upon goal, and review any required action items (e.g., a follow up conversation or additional training). At this time, both parties should clarify any misunderstandings to prevent confusion and ask questions if anything is not fully understood.

*After the Conversation*

Documentation
After the meeting, the conversation should be documented in detail, including the agreed-upon expectations, solutions, and any follow up tasks that need to occur. A copy of this document should then be shared with the other person. This step can be done formally (e.g., meeting minutes, formal report) or informally (e.g., email, hand-written clinician notes). Not only does this uphold the Ethics Code for Behavior Analysts (3.11 Documenting Professional Activity; BACB, 2020), it provides a relatively objective record of the conversation that both parties can refer to at a later date. Documentation can also hold both parties accountable for completing any agreed-upon action items.

Follow Up
It can be beneficial to follow up with the other person after having the difficult conversation. During the follow up, the behavior analyst can inquire about the other person's mental well-being, answer any questions that may have arisen after the conversation ended, or clarify any areas of confusion. Of note, it is important to recognize when supporting another individual's mental well-being is outside of one's scope of practice and refer the individual to a qualified professional (BACB, 2020). Finally, it is important to monitor the individual's progress and schedule additional meetings as needed, for there may not be a significant change in behavior after one conversation (Polito, 2013).

### Self-Reflection

It can be helpful to engage in self-reflection after the conversation (Epner & Baile, 2014). This time can be used to reflect on what went well, what didn't go well, and what could be done differently. It can also be helpful to plan for any skill-development opportunities before the next difficult conversation. For instance, it may be beneficial to seek assistance from a mentor to aid in interpreting why some parts of the difficult conversation did not go well and practice using an alternate approach.

### Reinforcement

Difficult conversations are not always reinforcing—in fact, they are often punishing—and therefore, it may be necessary to find other sources of reinforcement. The more a behavior analyst can obtain their own means of reinforcement, the easier these conversations will become. Reinforcement looks different for everyone, but here are some examples of possible reinforcers:

- Writing down the benefits of having difficult conversations and reviewing this list before and/or after these conversations.
- Schedule a desired activity immediately following the difficult conversation.
- Take a break, go for a walk, or talk to a friend about a desired (and unrelated) topic.

See Table 3.4 for the Difficult Conversations Preparation and Self-Reflection Checklist.

### Summary

Difficult conversations are—inherently—not easy. They are often linked to a history of punishment and, therefore, can be aversive. Although these conversations can evoke unpleasant emotions in the moment, it is important to recognize the potential short- and long-term benefits of having them. Not only are difficult conversations necessary for behavior analysts to maintain an ethical and professional practice, they can also allow them to build rapport with others, set clear expectations, and develop a professional and compassionate treatment environment.

Table 3.4 Difficult Conversations Preparation and Self-Reflection Checklist

| Questions | Answer | Notes |
|---|---|---|
| **Before the Conversation** | | |
| Do I require training or mentorship before having this conversation? | Y / N | |
| If yes, have I sought this out? | Y / N | |
| Do I have all of the information needed to have the conversation? | Y / N | |
| If applicable, do I have the evidence and/or supporting documentation necessary to have the conversation? | Y / N | |
| Have I determined the purpose of the conversation and the ideal outcome? | Y / N | |
| Have I considered possible outcomes of the conversation and what my response would be? | Y / N | |
| Have I developed an outline for the conversation? | Y / N | |
| Have I chosen an appropriate location in which the conversation will take place? | Y / N | |
| Have I chosen an appropriate time for the conversation? | Y / N | |
| Should a mentor or colleague join me in the conversation? | Y / N | |
| **During the Conversation** | | |
| Did I remain focused for the majority of the conversation? | Y / N | |
| Did I follow my original outline with some flexibility when appropriate? | Y / N | |
| Did I maintain objectivity, separating my own emotions from the conversation? | Y / N | |
| Did I maintain professionalism during the conversation? | Y / N | |
| Was I compassionate to the party(ies) involved? | | |
| Did I arrive at my ideal outcome? | Y / N | |
| If no, was the outcome agreed upon by both/all parties? | Y / N | |
| Did I have the opportunity to summarize the key takeaways at the end of the conversation? | Y / N | |
| **After the Conversation** | | |
| Have I documented the conversation appropriately? | Y / N | |
| Have I shared a copy of the conversation with the necessary parties (e.g., the other party, supervisor, manager)? | Y / N | |
| Have I reflected on what went well during the conversation and in what areas I could improve? | Y / N | |
| If improvements are necessary, have I sought out training or mentorship in this area? | Y / N | |
| Have I chosen a method of self-reinforcement? | Y / N | |

## References

Anderson, B. M., Kozluk, A., Morgan, M.-C., MacDonald, M., Friedel, J., & Cox, A. D. (2023). Exploring characteristics of implementer training associated with improved learner outcomes. *Journal on Behavioral Education*. [Advanced online publication] https://doi.org/10.1007/s10864-022-09504-2

Andzik, N. R., & Kranak, M. P. (2021). The softer side of supervision: Recommendations when teaching and evaluating behavior-analytic professionalism. *Behavior Analysis: Research and Practice, 21*(1), 65–74. https://doi.org/10.1037/bar0000194

Beck, R. S., Daughtridge, R., & Sloane, P. D. (2002). Physician-patient communication in the primary care office: A systematic review. *Journal of the American Board of Family Practice, 15*(1), 25–38.

Behavior Analyst Certification Board. (2020). *Ethics code for behavior analysts*. https://bacb.com/wp-content/ethics-code-for-behavior-analysts/

Bigham, E., McDannel, L., Luciano, I., & Salgado-Lopez, G. (2014). Effect of a brief guided imagery on stress. *Biofeedback, 42*(1), https://doi.org/10.5298/1081-5937-42.1.07

Bradley, G. L., & Campbell, A. C. (2016). Managing difficult workplace conversations: Goals, strategies, and outcomes. *International Journal of Business Communication, 53*(4), 443–464. https://doi.org/10.1177/2329488414525468

Bradley, R. L., & Noell, G. H. (2022). Rule-governed behavior: Teaching social skills via rule-following to children with autism. *Developmental Neurorehabilitation, 25*(7), 433–443. https://doi.org/10.1080/17518423.2021.2018735

Cooper, J. O., Heron, T., & Heward, W. (2019). *Applied behavior analysis* (3rd ed.). Pearson.

Epner, D. E., & Baile, W. F. (2014). Difficult conversations: Teaching medical oncology trainees communication skills one hour at a time. *Academic Medicine, 89*(4), 578–584. https://doi.org/10.1097/ACM.0000000000000177

Farrell, M. (2015). Difficult conversations. *Journal of Library Administration, 55*(4). https://doi.org/10.1080/01930826.2015.1038931

Greeny, K., Rosenberg, N., Fettig, A., & Schwartz, I. (2022). Common struggles of behavior analysts and their relation to the professional and ethical compliance code. *Behavior Analysis: Research and Practice*. Advanced online publication. https://doi.org/10.1037/bar0000254

Hoppe, M. H. (2006). *Active listening: Improve your ability to listen and lead*. Center for Creative Leadership.

Jahromi, V. K., Tabatabaee, S. S., Abdar, Z. E., & Rajabi, M. (2016). Active listening: The key of successful communication in hospital managers. *Electronic Physician, 8*(3), 2123–2128. https://doi.org/10.19082/2123

Kofman, F. (2014). *Authentic communication: Transforming difficult conversations in the workplace*. Sounds True: Boulder.

Ozgundondu, B., & Gok Metin, Z. (2019). Effects of progressive muscle relaxation combined with music on stress, fatigue, and coping styles among intensive care nurses. *Intensive & Critical Care Nursing, 54*, 54–63. https://doi.org/10.1016/j.iccn.2019.07.007

Perciavalle, V., Blandini, M., Fecarotta, P., Buscemi, A., Di Corrado, D., Bertolo, L., Fichera, F., & Coco, M. (2017). The role of deep breathing on stress. *Neurological Sciences, 38*(3), 451–458. https://doi.org/10.1007/s10072-016-2790-8

Polito, J. M. (2013). Effective communication during difficult conversations. *Neurodiagnostic Journal*, *53*(2), 142–152. https://doi.org/10.1080/21646821.2013.11079899

Pommier, E., Neff, K. D., & Tóth-Király, I. (2020). The development and validation of the compassion scale. *Assessment*, *27*(1), 21–39. https://doi.org/10.1177/1073191119874108

Price, O., & Baker, J. (2012). Key components of de-escalation techniques: A thematic synthesis. *International Journal of Mental Health Nursing*, *21*(4), 310–319. https://doi.org/10.1111/j.1447-0349.2011.00793.x

Rohrer, J. L., Marshall, K. B., Suzio, C., & Weiss, M. J. (2021). Soft skills: The case for compassionate approaches or how behavior analysis keeps finding its heart. *Behavior Analysis in Practice*, *14*(4), 1135–1143. https://doi.org/10.1007/s40617-021-00563-x

Sellers, T. P., Valentino, A. L., Landon, T. J., & Aiello, S. (2019). Board Certified Behavior Analysts' supervisory practices of trainees: Survey results and recommendations. *Behavior Analysis in Practice*, *12*(3), 536–546. https://doi.org/10.1007/s40617-019-00367-0

Svarovsky, T. (2013). Having difficult conversations: The advanced practitioner's role. *Journal of the Advanced Practitioner in Oncology*, *4*(1), 47–52. https://doi.org/10.6004/jadpro.2013.4.1.5

# 3.5 Running Efficient and Effective Meetings with Collaboration and Compassion

*Olivia Ng and Kerry-Anne Robinson*

## Introduction

There is a long-running joke that meetings are places where minutes are kept but hours are lost. How often do behavior analysts spend their time sitting in meetings wishing they could be doing other more productive work with clients instead? And yet, meaningful behavior-change programs are impossible to implement without a team-based approach and, yes, meetings. Although technology has allowed people to communicate efficiently via emails, text messages, and shared documents, nothing can replace face-to-face human interaction (whether virtual or in-person) and the in-the-moment opportunities to read and respond to others' body language and micro-expressions. Face-to-face interaction enables people to make more meaningful personal connections that are difficult to establish via written communication.

Because behavior analytic work takes place in a broad variety of settings and with diverse groups of people, behavior analysts' skill sets need to take on chameleon-like properties depending on the contexts in which we work. Social-communication skills needed for consulting to schools and hospitals may look vastly different from those required to effectively work with families in their homes and communities. Across all environments, efficient and effective meetings are regularly required for support teams to successfully plan and implement behavioral interventions. Here, we present a task analysis for meetings, with an emphasis on using effective interpersonal skills that can be used across a variety of contexts.

## Why Use This Approach?

To the author's knowledge, there has been little to no literature exploring the topic of meetings specific to behavior analysts, with the exception of LeBlanc and Nosik (2019), who provide several important considerations for planning and leading effective meetings. Among several key points, they emphasize the importance of a clear agenda, limiting distractions and interruptions to the attendees (i.e., asking everyone to turn off their phones), tips for addressing off-task meeting behavior, and advice for managing interpersonal conflicts. As well, Rohrer and colleagues (2001) developed the Compassionate Collaboration Tool that contains 25 skills that have been shown to improve treatment outcomes, treatment adherence, and client perceptions of service providers. As an extension to LeBlanc and Nosik (2019), drawing from Rohrer and colleagues (2021), and referring to the health administration and human services literature, a task analysis for behavior analysts is presented below to run efficient and effective meetings with an emphasis on social-communication strategies that are essential for strong, collaborative support teams.

## A Task Analysis for Meetings

Table 3.5 depicts a suggested task analysis for meetings led by behavior analysts. Note that this is not a task analysis for effective, efficient, and pleasant meetings but merely a basic framework for organizing the duration of a respectful meeting from beginning to end. Rather, the discussion that follows in this chapter will provide specific recommendations and examples of running a meeting with powerful social-communication skills where each attendee can walk away from the meeting feeling heard, valued, and that they have made a positive contribution to the life of the consumer of behavior analysis in question.

## Planning a Successful Meeting

Table 3.5 outlines several pre-meeting behaviors that should take place. These include ensuring that all attendees are aware of the day, time, and location of the meeting, ensuring virtual meeting invitations have been sent out beforehand, and sending out a proposed agenda at least a day before the meeting. Prior to the meeting is our first opportunity to set a compassionate and collaborative tone with our meeting attendees.

When scheduling the meeting, we should consider the attendees' schedules, lifestyle, and any other barriers that may impact how they are able to participate. For example, scheduling a 7 p.m. meeting with a therapist who has a 7 a.m. client session shows a lack of consideration for the therapist's schedule and the need for a work-life balance. It may also be difficult for parents and caregivers to attend meetings during work hours or while their child is at home and needs their attention. Instead, behavior analysts should collaborate with meeting attendees to determine

*Table 3.5* Task Analysis for Meetings

*Meeting Task Analysis*

1. Ensure that all meeting attendees are aware of the day/time/location of the meeting.
2. Ensure that virtual meeting invitations have been sent out (if the meeting is virtual).
3. Prepare a proposed agenda prior to the meeting and send it out to meeting attendants the day before.
4. Arrive to the meeting approximately 5-10 minutes early. Ensure you are dressed appropriately.
5. Begin the meeting on time.
6. Welcome attendants to the meeting.
7. Engage in small talk for 1–2 minutes.
8. Review the agenda and ask if there are any additions to the agenda.
9. Confirm duration of meeting (end time).
10. Begin moving through the agenda items.
11. Effectively (and politely) redirect off-task behaviors/discussion.
12. Make notes during your discussions and write down any outstanding questions that people may have.
13. Speak clearly and with an appropriate voice volume.
14. Continue moving through agenda items until all items have been discussed. Ensure that enough is allocated for discussion of all agenda items.
15. Ask the attendants of the meeting if they have any questions and address them if possible.
16. Conclude the meeting–review and confirm action items and designated person.
17. Thank everyone for their time and participation.
18. Set a future meeting date/time/location if necessary.
19. Send out a copy of the meeting notes/minutes to all meeting attendees.

Data Review

20. Orient audience to the graph.
21. Provide a concise summary of data and trends.
22. Provide an analysis of data summary.
23. Discuss questions that may arise or need to be addressed based on the data.
24. Discuss next steps based on the data.
25. Ask the attendees if they have any questions.

the best possible dates and times for meetings to ensure everyone is able to fully attend and participate. When meeting attendees are included in the planning of the meeting, including the timing of the meeting, they are more likely to feel like the behavior analyst is taking their needs into account and that their time and involvement are needed and valued.

When sending out invitations and pre-meeting agendas, behavior analysts should be sure to solicit input from the attendees about the goals of the meeting. The meeting invitation and agenda should be written in language that is accessible to each of the attendees. Behavior analytic technical jargon should be avoided if the meeting will include caregivers or non-behavioral professionals. If there are potential language barriers for attendees, translation apps (e.g., Google Translate) or translation services may be beneficial. When the attendees are provided with the agenda and encouraged to collaborate on the goals of the meeting, they are more likely to feel like an integral part of the team and ready to participate and collaborate during the actual meeting.

**Ensuring Everyone Feels Welcome**

It is important for the meeting leader to arrive early to ensure they can set up as needed, test the technology, and welcome the attendees as they arrive. When setting up for the meeting, the leader should consider any accessibility needs of the attendees as well as any possible cultural considerations. It is important for meeting leaders to prepare for potential individual differences and try to anticipate how they may impact the meeting (Knight, 2015). Communication styles can vary between individuals, as do notions of authority and hierarchy. If the meeting leader fails to consider the potential differences impacting meeting behavior of attendees, it may lead to fewer opportunities for participants to contribute, misunderstandings, and hurt feelings. Individual differences may impact how comfortable people are participating in the meetings, providing feedback, advocating for their needs, voicing disagreements, and the style, demeanor, and tone with which they communicate during the meeting. Receiving ongoing education, training, or reading about diversity, equity, and inclusion can also be helpful and a priority in the Ethics Code for Behavior Analysts (Guideline 1.07; BACB, 2020; see Levy et al., 2021, for several recommendations for promoting inclusive, anti-racist, and culturally humble environments in the workplace).

At the beginning of each meeting, it is important to create expectations related to how the meeting will run and the type of participation the meeting leader would appreciate from the group. This may set the stage to allow for some to move out of their comfort zone, while also providing the leader with guidelines they can follow to keep everyone focused (Knight, 2015).

Dressing for the meeting is not as simple as just looking professional. The behavior analyst's attire should be professional, while also ensuring the attendees will feel comfortable and confident participating and collaborating with them. If the meeting is with a caregiver in their home, for example, who typically wears sweatpants and t-shirts, it may be uncomfortable for them if the behavior analyst shows up in a suit. The differences in dress may unintentionally create an unwanted power differential and may make it more difficult for the caregiver to collaborate, ask questions, and advocate for their child's needs. If the behavior analyst was dressed in a professional looking pair of jeans and a sweater instead, for example, the caregiver may not experience the same level of discomfort and be able to participate fully.

While waiting for meeting attendees to arrive, and prior to starting to go through the agenda, it is important to engage in small talk for 1–2 minutes (i.e., Step 7 in the task analysis). This is an excellent opportunity to build rapport with the meeting participants. The tone of the small talk is very important. The tone of the meeting leader's initial interactions lets the meeting attendees know what type of interactions are expected throughout the meeting and how everyone will be treated. Small talk may also serve as the foundation for building trust.

Even though it is beneficial, many people find small talk awkward and uncomfortable. At a meeting it is best to keep things professional, so it is probably best to avoid bringing up politics, religion, or potentially contentious topics. To engage in meaningful small talk that sets the tone for a collaborative effective meeting, pay close attention to your tone. Talk as if speaking with a friend. Avoid being so literal or serious. Having a relaxed playful attitude, just as when speaking with friends, will set attendees at ease. It is okay to be lighthearted, funny, and even sarcastic during the first few minutes of a meeting before going through the agenda items. Superficial small talk topics such as the latest weather or how the local sports teams are doing may also be less likely to establish feelings of connection. If a behavior analyst is meeting with caregivers, small talk could involve a brief anecdote about an interaction they or their colleague had with the client earlier that week. As uncomfortable as it is, do not avoid elephants in the room during this time! If a team member or caregiver arrives to the meeting looking completely frazzled, tired, and stressed, provide a brief check-in, and validate how they may be feeling before moving on.

When welcoming meeting participants and doing introductions, the meeting leader should model what each attendee's introduction should look like. It should be brief and should attempt to minimize any emphasis on authority or power. For example, a leader in the role of Senior Behavior Analyst in a meeting with caregivers and group home staff may simply state her name and role within the team as opposed to her title. For example, compare these introductions:

A "Hi, I'm Robin. I'm the Senior Behavior Analyst with Best ABA Group, Inc. I will be clinically overseeing all the work that the team is doing."
B "Hi, I'm Robin. I'm really excited to talk about the next steps for Mohammed. I'll be here to provide direction, support, and answer questions."

Introduction A is factual and less friendly. It also does not provide the caregivers or the group home staff information about what her involvement will entail. Introduction B, on the other hand, is much warmer, client and team focused, and lets all the meeting attendees know how she can support the team and client. It also avoids giving the impression that the behavior analyst believes their role is more valuable than the other team members.

## Tackling the Agenda

Next, the meeting leader will guide the group through the key items on the agenda (Table 3.5, Items 8–15). The leader should provide an opportunity for attendees to add any last-minute agenda items before proceeding—this is an additional way for the meeting leader to communicate that they are open to the group's ideas, and it continues to foster an environment of collaboration and open discussion. Confirm the duration of the meeting so that everyone has a clear expectation of when it will end. Meetings should be no longer than one hour in length to hold the attendees' attention and participation to a meaningful degree (Gallo, 2016). If meetings need to exceed an hour, scheduled breaks are recommended. Meetings that are excessive in duration may lead to a variety of off-task behaviors and ultimately become conditioned aversive events that attendees may potentially dread the next time a meeting agenda gets circulated.

On the other hand, a fine compromise must be made to ensure that sufficient time is allotted for discussing each of the proposed agenda items. This is where the meeting leader's role becomes the most important, as they are responsible for skillfully chairing and moving agenda items along for the duration of the meeting (McConnell, 1997). When attendees engage in any off-task behaviors or discussion, the leader should politely but effectively redirect the conversation back to the relevant topics. Readers are encouraged to refer to LeBlanc and Nosik (2019), who provide several useful redirection phrases in response to off-task meeting behaviors. The meeting leader should pay particular attention to their own personal biases and, when comments are not in line with their own perspective, be careful not to dismiss, as off-task behavior, potentially valuable input from attendees. One way of acknowledging attendee input that is not relevant to the current agenda is to make note of the discussion and document any outstanding questions or topics that can be addressed later (i.e., a follow up email or discussion with the relevant individuals). If a question can be addressed quickly within the allocated time frame of the meeting, it may be best to address it immediately and reinforce attendee participation rather than chastising the individual for swaying from the agenda.

Rohrer and colleagues' (2021) Compassionate Collaboration Tool offers several recommendations for language and communication strategies that should be used while moving through agenda items in meetings. As with planning for the meeting and ensuring everyone feels welcome, the meeting leader and attendees should be sure to use language accessible to all while moving through the agenda items and avoid technical jargon as much as possible. Meeting leaders should engage in active listening behaviors (e.g., nodding, mirroring facial expressions, body language, and vocal language), and framing what attendees are trying to communicate ("It sounds like you're saying …"), to ensure that everyone has the same understanding of the discussion. Meeting leaders should also reflect on the content being presented by others ("It sounds like you're worried about the number of things you have to do"). This is a great way to summarize and redirect potential conversational tangents or off-task behavior, as well as to bring the meeting back to a more focused, action-orientated state (Henggeler et al., 2009). As well, it provides the attendees with validation that their ideas are being heard and understood. It is recommended that emotions being heard are identified and calibrated, whether it be in tone of voice or what is being communicated. This may be aversive for many behavior analysts, but it is an essential part of compassionate collaboration. If a supervisee, for example, begins speaking about their never-ending to-do list and all the barriers in the way, the meeting leader could say something like, "It sounds like you're worried you won't be able to put that plan in place with everything else you have on your plate right now." This validates how the supervisee is feeling and shifts the meeting back to the topic at hand. When going through agenda items, it is likely helpful to promote feelings of hope and optimism to the attendees through discussion of potential positive outcomes (Rohrer et al., 2021). A meeting leader should not only talk about the tasks that need to be completed, but also the potential positive outcomes of doing so. This may increase motivation to complete action items and create an atmosphere of collaboration and participation since all attendees are aware of the long-term goals and benefits of their actions.

The meeting leader should continue moving through each of the agenda items until they all have been discussed. They should also check in intermittently with attendees to ensure that everyone is following along, and whether they have any questions. Rohrer and colleagues (2021) suggest requesting and accepting corrections ("Did I get that right? Did I miss anything?"). An open, safe, friendly, and conversational tone should be maintained throughout the meeting as much as possible. This type of environment may be better *experienced* than merely *understood* through reading this text.

### Presenting Data at Meetings

Items 20 to 25 on the task analysis in Table 3.5 outline a concise process for reviewing data at meetings. As mentioned earlier, it is wise to eliminate the use of technical jargon when meeting with caregivers or non-behavioral professionals, and this applies to conversations on data analysis as well.

What the data looks like should vary based on the attendees at the meeting. When meeting with other behavior analysts, a supervisor or supervisee, it would be appropriate, maybe even expected, that the graph be presented using APA format. When meeting with caregivers or other professionals, the visual presentation of the graph should be made more accessible to all attendees. After a brief orientation to the visual display, all attendees should be able to understand what the graph is depicting. Use colors, a clear legend, and language that is accessible to all attendees. Use a type font and graph size that is appropriate. Scaling should be accurate and representative of the social significance of actual behavior change.

After drawing the attendees' attention to the graph, provide a concise summary of your analysis of the data and any trends via visual inspection or any calculations that might have been completed prior to the meeting. It is best to describe data in contextual and applicable terms to the client's real-world experience. For example, if data is trending in a desirable direction, it may be clinically appropriate to discuss whether these trajectories have been observed to a significant, noticeable, or meaningful degree by the client or others. On the other hand, if the data suggest that behavior is deteriorating, variable, or that no significant improvement has been observed, at this point an opportunity (and, perhaps, the most impactful, opportunity) for collaborative discussion can take place among all attendees at the table. Each participant should be given an opportunity to provide their input to help determine the next steps for the client. Finally, once next steps are determined, provide an opportunity for attendees to provide any additional feedback or ask any questions they might have.

As a rule of thumb, always provide ample opportunity for everybody's input on topics that relate to critical clinical decision making for the client. Behavior analysts would benefit from genuinely and actively listening to others from a place of humility (Kirby et al., 2022). Not all perspectives and opinions on intervention may make it to the final plan of action, given the risks, benefits, and evidence base, but all should be respectfully heard and carefully considered as part of the client's context.

### Wrapping Up the Meeting

Steps 16 through 19 in Table 3.5 involves wrapping up and concluding the meeting. It is important to review and confirm action items with each designated person to ensure all attendees clearly understand the next steps, their individual responsibilities, and all due dates for tasks that need to be completed. If it has not already been done by this point in the meeting, check in with the attendees regarding potential barriers to getting action items completed by the proposed due date and any support they may need to get the item completed. Depending on the meeting attendees present, this can be done during the meeting or following the meeting as a one-to-one conversation. Doing this will increase the likelihood of the action item being completed, while also demonstrating compassion, thereby providing opportunities to validate or address challenges an attendee may experience—utilizing a collaborative approach to support them in getting tasks done and overcoming obstacles.

Be sure to thank everyone for their participation and contributions. This will have attendees leaving the meeting feeling appreciated and valued. It is also recommended that you plan any follow-up meetings prior to ending the current meeting. This increases accountability to get any action items done. It is also often easier to schedule a meeting with several people while they are all in the same room as opposed to sending emails back and forth to try to find a time that works best for everyone.

Finally, following the meeting, your notes (i.e., minutes) should be sent out to all participants. Again, consideration should be taken to ensure that the meeting notes are presented in plain language that is accessible to all attendees. If there were suspected language barriers during the meeting, it is highly recommended that the notes be provided in the attendee's first language, perhaps by using a translation service. This is especially important if treatment planning decisions or consent is needed for next steps to occur.

## Conclusion

Several recommendations for the types of social-communication skills needed for running the most successful meetings have been provided, namely, being flexible, approachable, collaborative, considerate, and thoughtful regarding individual differences. It should be acknowledged here that positive behavior change is more likely to occur with a combination of support that includes instruction, modeling, rehearsal, and feedback (Parsons et al., 2012). Reading about social-communication skills from a book is highly unlikely to lead one to suddenly engage in exceptional meeting leader behavior. As such, it is recommended that behavior analysts continue to operationally define what it means to engage in ideal meeting behaviors and espouse the above-mentioned values. A few authors have developed quizzes, checklists, and other tools that behavior analysts could use to evaluate their own and others' social-communication skills and meeting behaviors (see LeBlanc & Nosik, 2019; Rohrer et al., 2021;

Sachs, 2000). In turn, meetings might just become less of a chore or a burden to endure and, rather, a more enjoyable and outcome-focused process for all.

## References

Behavior Analyst Certification Board (BACB). (2020). Ethics code for behavior analysts. Author.

Gallo, A. (2016, July 6). The condensed guide to running meetings. *Harvard Business Review*. https://hbr.org/2015/07/the-condensed-guide-to-running-meetings

Henggeler, S. W., Schoenwald, S. K., Borduin, C. M., Rowland, M. D., & Cunningham, P. B. (2009). *Multisystemic therapy for antisocial behavior in children and adolescents*. Guilford Press.

Kirby, M. S., Spencer, T. D., & Spiker, S. T. (2022). Humble behaviorism redux. *Behavior and Social Issues*. Advance online publication. https://doi.org/10.1007/s42822-022-00092-4

Knight, R. (2015, December 4). How to run a meeting of people from different cultures. *Harvard Business Review*. https://hbr.org/2015/12/how-to-run-a-meeting-of-people-from-different-cultures

LeBlanc, L. A., & Nosik, M. R. (2019). Planning and leading effective meetings. *Behavior Analysis in Practice, 12*, 696–708. https://doi.org/10.1007/s40617-019-00330-z

Levy, S., Siebold, A., Vaidya, J., Truchon, M. M., Dettmering, J., & Mittelman, C. (2021). A look in the mirror: How the field of behavior analysis can become anti-racist. *Behavior Analysis in Practice*, 1–14. Advance online publication. https://doi.org/10.1007/s40617-021-00630-3

McConnell, C. R. (1997). The chairperson's guide to effective meetings. *Health Care Supervision, 15*(3), 1–9.

Parsons, M. B., Rollyson, J. H, & Reid, D. H. (2012). Evidence-based staff training: A guide for practitioners. *Behavior Analysis in Practice, 5*(2), 2–11. https://doi.org/10.1007/BF03391819

Rohrer, J. L., Marshall, K. B., Suzio, C., & Weiss, M. J. (2021). Soft skills: The case for compassionate approaches or how behavior analysis keeps finding its heart. *Behavior Analysis in Practice, 14*, 1135–1143. https://doi.org/10.1007/s40617-021-00563-x

# Part II
# Specialized Skills

# 4
# Consultation and Training

# 4.1 Don't Blame the Mediator

Keeping Applied Behavior Analysis Doable for Best Possible Outcomes

*Kendra Thomson, Reghann Munno, Louis P. A. Busch, and Maurice Feldman*

**Introduction**

Practice in applied behavior analysis (ABA) has grown substantially in the last decade. As of July 1, 2022, there were 56,961 Board Certified Behavior Analyst (BCBA) certificants worldwide (Behavior Analyst Certification Board; BACB, 2020) compared to 22,274 certificants in 2016 (Carr & Nosik, 2017). This exponential growth has been accompanied by positive developments, including professionalization of the discipline as well as some associated growing pains such as needing to further develop infrastructure and policy (Carr & Nosik, 2016). Although ABA is applicable across diverse contexts (e.g., Heward et al., 2022), the majority (70%) of BCBA certificants provide services to people with autism and those around them such as family members and paid caregivers. This proportion also mirrors the quantity of related ABA research, demand for services, and associated funding. Recently, members of the neurodiverse and/or autistic community have raised several serious concerns with ABA-based intervention (e.g., it is abuse, it is overly rigid), some of which have been discussed in peer-reviewed papers with recommendations for behavior analysts (e.g., Leaf et al., 2022). The impact of these discussions on practice and the experience of service users is unclear. Although challenging, this moment is an opportune time for behavior analysts to reflect—from a place of compassion, humility, and curiosity—on the potential causes of the criticism. Some behavior analysts have begun to engage in critical self-reflection, prompted largely by anti-ABA sentiments and other social-justice movements about systemic issues such as ableism, racism, and prejudice. Overdue discussions are taking place in the field about important issues that impact the success of the discipline, such as compassion (e.g., Taylor et al., 2019) and cultural humility (e.g., Wright, 2019). Behavior analysts also need to consider how the field can be more approachable by those seeking services and those who support them, as well as by other professionals.

DOI: 10.4324/9781003300465-20

Three decades ago, Neuringer (1991) wrote that "humility will prove to be functional" for behavior analyst researchers (p. 1). Kirby et al. (2022) recently revisited this topic in the context of behavior analyst practitioners working inter-professionally. They provided suggestions for how behavior analysts can engage in cultural humility and reciprocity (versus disciplinary centrism and hubris) to promote inter-professionalism and scalability of behavioral interventions. Conceivably, behavior analysts can and should generalize the same humility when working with clients and mediators as per Item H-9, "Collaborate with others who support and/or provide services to clients" of the BCBA *5th Edition* Task List (herein referred to as the Task List; BACB, 2017). They should also strive to develop strong inter- and intrapersonal skills and therapeutic alliance since "behavioral artists" (Foxx, 1998, p. 14) are preferred over those who may be solid technicians but lack people skills and do not form a therapeutic alliance (e.g., Callahan et al., 2019). Behavior analysts should also be flexible in adjusting procedures to suit dynamic environments while remaining technological and analytic versus reductionist or rigid (e.g., Leaf et al., 2016; Shook et al., 2002).

Most ABA practitioners are aware of the seven dimensions of ABA described in the inaugural edition of the *Journal of Applied Behavior Analysis* (Baer et al., 1968), which were reiterated as "still current" 20 years later (Baer et al., 1987). Cooper, Heron, and Heward (2019) cite additional characteristics of ABA (accountable, public, doable, empowering, optimistic), which have received less attention and can be traced to an earlier article. In *Reasons Applied Behavior Analysis is Good for Education and Why those Reasons Have Been Insufficient*, Heward et al. (2005) cited twelve benefits of ABA in education (i.e., it is meaningful, effective, focused, broadly relevant, self-correcting, accountable, public, doable, replicable, empowering, optimistic, and knows motivation). Although Heward's paper focused on why these reasons had only limited impact in education, we argue that if behavior analysts carefully consider whether their practice is aligned with these characteristics, the answer may serve as an antidote to some of the reported concerns with ABA. We will focus this brief discussion on ways behavior analysts may be able to make ABA "doable" for mediators and illustrate why this is both ethically and practically important based on the literature and our combined clinical and research experience.

Keeping ABA services doable aligns with the BACB *Ethics Code for Behavior Analysts* (2022, herein referred to as "the Code"), which describes four guiding principles for behavior analysts (i.e., benefit others; treat others with compassion, dignity, and respect; behave with integrity; and ensure their own competence). Item 2.09, *Involving Clients and Stakeholders*, states that, "behavior analysts make appropriate efforts to involve clients and relevant stakeholders throughout the service relationship, including selecting goals, selecting and designing assessments and behavior-change interventions, and conducting continual progress monitoring" (BACB, 2020, p. 11). As a scientific discipline, ABA applies established principles to make socially valid improvements in behavior. As such, a hallmark of ABA is data-based decision-making. Depending on the nature of the services and the context where services are provided, implementation of the procedures and data collection may involve others in the client's environment. Mediators can include caregivers, school personnel, direct support professionals, and even athletic coaches. Item 2.14 of the Code; *Selecting, Designing, and Implementing Behavior-Change Interventions,* states that behavior analysts

> select, design, and implement behavior-change interventions that "best meet the diverse needs, context, and resources of the client and stakeholders. Behavior analysts also consider relevant factors (e.g., risks, benefits, and side effects; client and stakeholder preference; implementation efficiency; cost effectiveness) and design and implement behavior-change interventions to produce outcomes likely to maintain under naturalistic conditions."
>
> (BACB, 2020, p. 14)

Caregivers of children with autism and other Neurodevelopmental Disabilities (NDDs) frequently act as mediators to promote generalization of skills, to empower caregivers, and to decrease reliance on resource-intensive, expensive services (e.g., Cheng et al., 2022; Feldman & Werner, 2002). Importantly, caregivers of children with NDDs experience higher levels of stress than families without children with autism (e.g., Hayes & Watson, 2013). Therefore, behavior analysts should consider the contextual factors, such as caregiver values and preferences that may influence the success of mediator-implemented interventions to ensure that whatever they are expecting and training mediators to do, is, well, doable!

Skinner (1977) stated an organism "does what it is induced to do by its genetic endowment or by the prevailing conditions" (p. 1007). The organism "not being wrong" has become ubiquitous in ABA. In other words, don't blame the learner if they are not behaving as expected. Rather, look for reasons in their environment. Unfortunately, however, we are likely not the only ones who have heard trainees or colleagues say, "(mediator) is not implementing the procedure" or "(mediator) will not collect the data." Issues such as these, which are sometimes egregiously referred to as "non-compliance" are not likely due to some inherent problem with the mediator, but rather due to a convergence of environmental variables including the behavior analyst's behavior. Mediator behavior should be assessed with the same functional approach and careful consideration of environmental variables as is done with clients through analysis of the four-term contingency, which includes motivation (e.g., Michael, 1993) and interlocking contingencies (e.g., Busch et al, 2020; Vichi et al., 2009). For example, a caregiver's behavior may contact negative reinforcement if they attend to their child's immediate needs (e.g., provide a snack to terminate crying). The snack may serve as positive reinforcer of the child's crying versus reinforcing a functional communicative response. Demands in the caregiver's environment, such as other children or competing advice, may also impact their motivation (e.g., serve as an abolishing operation) for implementing ABA programming. The caregiver's previous attempts to implement an ABA extinction program may have led to aversive consequences (e.g., child crying, their own feelings of frustration), or alternative responses (e.g., playing with their child) may be contacting a denser schedule of putative reinforcement than implementing procedures and collecting data. As per Item H-3 of the Task List (2017), behavior analysts "recommend intervention goals and strategies based on such factors as client preferences, supporting environments, risks, constraints, and social validity." Given that the consistency and effectiveness of mediator-implemented ABA programming may be impacted by how much those involved support the goals, procedures, and outcomes (Kazdin, 1977; Smith, 2013, Wolf, 1978), a focus on keeping ABA procedures doable and acceptable to those involved seems highly warranted. Surprisingly, social validity of processes and outcomes are under-reported in the ABA literature (e.g., Carr et al., 1999; Dixon, 2018; Ferguson et al., 2019; Kennedy 1992). Therefore, it may not be surprising if also under-assessed in practice. To our knowledge, no empirical studies have attempted to assess the relation between how doable mediators find procedures to be and how often and effectively they implement them or whether it impacts withdrawal from service. It seems plausible that a positive relation exists between how doable and how frequently and effectively procedures are implemented. Relatedly, behavior analysts are also obligated to "evaluate the validity and reliability of measurement procedures" (Task List Item C-8; BACB, 2017, p. 3), which can only be accomplished if data are collected. Engaging mediators from the start, when choosing a measurement system may lead to increased adherence and more successful outcomes for all involved.

To help behavior analysts assess whether the procedures they are expecting the mediator to implement are doable, we suggest they humbly ask themselves the following questions about their own behavior:

- Have I broken down the procedures into manageable steps, or are they overly complex?

- Is the measurement system suitable for the environment? Are the expectations for data collection realistic?
- Did I provide sufficient training on procedures and associated data collection systems?
- Did I consider the mediator's preferences and exercise cultural humility?
- Have I reinforced successive and sufficient approximations of desired mediator behavior, or has extinction of the trained responses occurred?
- Has the mediator's behavior contacted any punishment contingencies?
- Am I aware of the competing contingencies and the impact of motivating operations in the mediator's environment?
- Have I discussed the program and/or data collection system in sufficient detail with the mediator to adequately assess their preferences and problem solve any barriers to successful implementation?
- Am I engaging in avoidance responses such as telling the caregiver what to do vs. modeling it for them?

Behavior analysts should also keep in mind that they may need to be flexible in their approach to adapt to contexts, while still providing the most effective, evidence-based intervention to produce meaningful behavior change. Meaningful changes are those that are long-lasting, generalizable, and endorsed by the person receiving services and those around them. Most ABA research is conducted in analogue (laboratory) or highly controlled settings (e.g., treatment centers), conditions that may not be realistically generalizable to real-world settings (homes, schools, playground, office, etc.).

The following scenario is hypothetical, yet the target behaviors, living environment, and assessment and intervention discussed are common in adult behavioral services for persons with NDDs and comorbid mental health conditions based on our combined experience. We have included points in the vignette ("Spot the Issue") for the reader to consider whether the behavior analyst's behaviors are aligned with the Code and the Task List and reflect on what the behavior analyst could have done differently.

Harry is a 25-year-old autistic man who lives with his father, Michael (his substitute decision maker), in their family home. Harry was diagnosed with obsessive compulsive disorder (OCD) by his psychiatrist. Harry has many skills and independently completes most of his self-care routines (e.g., getting dressed). Harry's primary method of communication is via a portable augmentative communication device. Harry loves to spend time with his family, especially going for walks. Harry also enjoys gathering all the soap bottles in the house, bringing them into his bathroom, and emptying all the soap into his sink while turning on the water to form bubbles to play with. Although Harry smiles and laughs while engaging in this activity, Michael has requested support from a behavior analyst to help attempt to replace this activity with something else that Harry enjoys. The behavior analyst met with Michael to discuss goals for services. Michael mentioned that Harry was having some other challenges at his day program. The behavior analyst suggested that they target those behaviors first given that she didn't feel the soap activity was that harmful.

> Spot the Issue: To be consistent with the Task List and the Code, the behavior analyst should have meaningfully involved Harry and Michael in selecting and prioritizing the goals for intervention.

The behavior analyst continued an indirect functional assessment through interviewing Michael, which revealed that Michael found the large amount of soap Harry used to be costly as a single income family, and cleaning up the mess was difficult for him because of limited mobility in his wrists due to an injury. He said that Harry's behavior also interfered with their religious practices out of the home in which water is often used. Michael reported that he usually ignored these behaviors because when he previously tried to prevent Harry from accessing some of the soap bottles by locking them in cupboards, Harry yelled loudly and kicked the door. He said he also had an "outburst" at their place of worship. The behavior analyst obtained informed consent for baseline data collection and told Michael to also collect descriptive (antecedent–behavior–consequence; ABC) data. She gave a brief description of what it was, handed Michael a data sheet that she used with another client, and asked him to collect as much data as possible.

> Spot the Issue: To be consistent with the Task List and the Code, the behavior analyst should have involved Harry and Michael in selecting the data collecting system. They should have created specific data sheets and provided behavioral skills training (e.g., Schaefer & Andzik, 2021) to ensure that Michael was using the data sheets appropriately and comfortable doing so. She should have also discussed collecting data on a reasonable schedule to ensure it was doable for him. The behavior analyst should have shown cultural humility regarding the family's religion and asked if and how they wanted to integrate it into the intervention plan.

The behavior analyst followed-up with Michael about his ABC data in an email. Michael sent it back indicating that he had only collected data twice, but Harry's behavior was occurring more frequently with more serious outcomes. On a phone meeting the behavior analyst asked Michael if he could commit to collecting more data. He indicated that he would try. ' Michael shared that he had also received some recommendations from an occupational therapist (OT), and he was having trouble balancing everything. The behavior analyst replied that he should just try his best to collect the data and he could ignore the OT's strategies due to lack of evidence. She recommended that he attempt to interrupt the behavior at home and ignore it in the place of worship.

> Spot the Issue: To be consistent with the Task List and the Code, the behavior analyst should have explored the barriers to data collection with Michael to come up with a doable system. She should not have recommended that Michael interrupt the behavior without reliable and valid data from a comprehensive functional behavior assessment and an appropriate intervention in place. This puts Harry and Michael at risk. The conflicting recommendations are also confusing. The behavior analyst should have made efforts to collaborate with other professionals with specialization in OCD (i.e., psychologist or psychiatrist) and automatic reinforcement/sensory needs (i.e., OT). The behavior analyst should not have commented about the OT's recommendations without more information. The behavior analyst could have also reflected on her own behavior asking the questions listed earlier in the chapter.

As the number of behavior analysts continues to exponentially climb, the field needs to critically reflect on increasing concerns with service delivery. In addition to teaching theoretical foundations of the science and the development of technical skills, academic and clinical training should focus on intra- and interpersonal skills and the development of therapeutic alliance. Wolf argued decades ago that for behavior analysis "to find its heart" (1978, p. 1) the quality of the goals, procedures, and outcomes of ABA would need to be assessed from the service users' perspectives, which may be best captured using subjective measures. The field has yet to embrace this advice consistently, nor has it done justice to measuring experience (e.g., quality of life). Acknowledging the concerns presented by the autistic community and other ABA service recipients, we echo the calls for behavior analysts to focus on how to be more approachable, culturally informed, and flexible in their practice. Behavior analysts are ethically obliged to deliver services that are highly effective, acceptable, and feasible for those receiving them and others who participate in implementing ABA procedures. Behavior analysts should consider available resources, client and mediator preferences, culture, and personal values, and engage in ongoing and humble self-reflection and communication with all involved to ensure that the procedures are doable.

# References

Baer, D. M., Wolf, M. M., & Risley, T. R. (1968). Some current dimensions of applied behavior analysis. *Journal of Applied Behavior Analysis, 1*(1), 91–97. https://doi.org/10.1901/jaba.1968.1-91

Baer, D. M., Wolf, M. M., & Risley, T. R. (1987). Some still-current dimensions of applied behavior analysis. *Journal of Applied Behavior Analysis, 20*(4), 313–327. https://doi.org/10.1901/jaba.1987.20-313

Behavior Analyst Certification Board. (2017). *BCBA task list* (5th ed.). Author.

Behavior Analyst Certification Board (2020). *Ethics code for behavior analysts.* https://bacb.com/wp-content/ethics-code-for-behavior-analysts/

Busch, L. P., Porter, J., & Barreira, L. (2020). The untapped potential of behavior analysis and interprofessional care. *Journal of Interprofessional Care, 34*(2), 233–240. https://doi.org/10.1080/13561820.2019.1633292

Callahan, K., Foxx, R. M., Swierczynski, A., Aerts, X., Mehta, S., McComb, M.-E., Nichols, S. M., Segal, G., Donald, A., & Sharma, R. (2019). Behavioral artistry: Examining the relationship between the interpersonal skills and effective practice repertoires of applied behavior analysis practitioners. *Journal of Autism and Developmental Disorders, 49*, 3557–3570. https://doi.org/10.1007/s10803-019-04082-1

Carr, J. E., Austin, J. L., Britton, L. N., Kellum, K. K., & Bailey, J. S. (1999). An assessment of social validity trends in applied behavior analysis. *Behavioral Interventions: Theory & Practice in Residential & Community-Based Clinical Programs, 14*(4), 223–231. https://doi.org/10.1002

Carr, J. E., & Nosik, M. R. (2017). Professional credentialing of practicing behavior analysts. *Policy Insights from the Behavioral and Brain Sciences, 4*(1), 3–8. https://doi.org/10.1177/2372732216685861

Cheng, W.M., Smith, T.B., Butler, M., Taylor, T. M., & Clayton, D. (2023). Effects of parent-implemented interventions on outcomes of children with Autism: A meta-analysis. *Journal of Autism and Developmental Disorders, 53*, 4147–4163. https://doi.org/10.1007/s10803-022-05688-8

Cooper, J., Heron, T., & Heward, W. (2019). *Applied behavior analysis* (3rd ed.). Pearson.

Dixon, M. R., Belisle, J., Rehfeldt, R. A., & Root, W. B. (2018). Why we are still not acting to save the world: The upward challenge of a post-Skinnerian behavior science. *Perspectives on Behavior Science, 41*(1), 241–267. https://doi.org/10.1007/s40614-018-0162-9

Feldman, M. A., & Werner, S. E. (2002). Collateral effects of behavioral parent training on families of children with developmental disabilities and behavior disorders. *Behavioral Interventions: Theory & Practice in Residential & Community-Based Clinical Programs, 17*(2), 75–83. https://doi.org/10.1002/bin.111

Ferguson, J. L., Cihon, J. H., Leaf, J. B., Van Meter, S. M., McEachin, J., & Leaf, R. (2019). Assessment of social validity trends. In the Journal of Applied Behavior Analysis. *European Journal of Behavior Analysis, 20*(1), 146–157. https://doi.org/10.1080/15021149.2018.1534771

Foxx, R. M. (1998). Twenty-five years of applied behavior analysis: Lessons learned. *Discriminanten, 4*, 13–31.

Hayes, S. A., & Watson, S. L. (2013). The impact of parenting stress: A meta-analysis of studies comparing the experience of parenting stress in parents of children with and without autism spectrum disorder. *Journal of Autism and Developmental Disorders, 43*(3), 629–642. https://doi.org/10.1007/s10803-012-1604-y

Heward, W. L. (2005). Reasons applied behavior analysis is good for education and why those reasons have been insufficient. In W. L Heard, T. E. Heron, N. A. Need, S. M. Peterson, D. M. Sainato, G. Cartledge, R. Gardner III (Eds.), *Focus on behavior analysis in education: Achievements, challenges, and opportunities* (pp. 316–348). Merrill/Prentice Hall.

Heward, W. L., Critchfield, T. S., Reed, D. D. et al. (2022). ABA from A to Z: Behavior science applied to 350 domains of socially significant behavior. *Perspectives on Behavioral Science, 45*, 327–359. https://doi.org/10.1007/s40614-022-00336-z

Kazdin, A. E. (1977). Assessing the clinical or applied importance of behavior change through social validation. *Behavior Modification, 1*, 427–452. https://doi.org/10.1177/014544557714001

Kennedy C. H. (1992). Trends in the measurement of social validity. *The Behavior Analyst, 15*(2), 147–56. https://doi.org/10.1007/BF03392597

Kirby, M. S., Spencer, T. D. & Spiker, S. T.(2022). Humble behaviorism redux. *Behavior and Social Issues, 31*, 133–158. https://doi.org/10.1007/s42822-022-00096-0

Leaf, J. B., Cihon, J. H., Leaf, R., McEachin, J., Liu, N., Russell, N., Unumb, L., Shapiro, S., Khosrowshahi, D. (2022). Concerns about ABA-based intervention: An evaluation and recommendations. *Journal of Autism and Developmental Disorders, 52*, 2838–2853. https://doi.org/10.1007/s10803-021-05137-y

Leaf, J. B., Leaf, R., McEachin, J., Taubman, M., Ala'i-Rosales, S., Ross, R. K., Smith, T., Weiss, M. J. (2016). Applied behavior analysis is a science and, therefore, progressive. *Journal of Autism and Developmental Disorders, 46*, 720–731. https://doi.org/10.1007/s10803-015-2591-6

Michael, J. (1993). Establishing operations. *The Behavior Analyst, 16*(2), 191–206. https://doi.org/10.1007/BF03392623

Neuringer, A. (1991). Humble behaviorism. *The Behavior Analyst, 14*(1), 1–13. https://doi.org/10.1007/BF03392543

Schaefer, J. M., & Andzik, N. R. (2021). Evaluating behavioral skills training as an evidence-based practice when training parents to intervene with their children. *Behavior Modification, 45*(6), 887–910. https://doi.org/10.1177/0145445520923996

Shook, G. L., Ala'i-Rosales, S., & Glenn, S. (2002). Certification and training of behavior analyst professionals. *Behavior Modification, 26*(1), 27–48. https://doi.org/10.1007/s10803-015-2591-6

Skinner, B. F. (1977). Herrnstein and the evolution of behaviorism. *American Psychologist, 32*(12), 1006–1012. https://doi.org/10.1037/0003-066X.32.12.1006

Smith, T. (2013). What is evidence-based behavior analysis?. *The Behavior Analyst, 36*, 7–33. https://doi.org/10.1007/BF03392290

Taylor, B. A., LeBlanc, L. A., & Nosik, M. R. (2019). Compassionate care in behavior analytic treatment: Can outcomes be enhanced by attending to relationships with caregivers? *Behavior Analysis in Practice*, *12*(3), 654–666. https://doi.org/10.1007/s40617-018-00289-3

Vichi, C., Andery, M. A. P. A., & Glenn, S. S. (2009). A metacontingency experiment: The effects of contingent consequences on patterns of interlocking contingencies of reinforcement. *Behavior and Social Issues, 18*(1), 41–57. https://doi.org/10.5210/bsi.v18i1.2292

Wolf, M. (1978). Social validity: The case for subjective measurement or how applied behavior analysis is finding its heart. *Journal of Applied Behavior Analysis*, 11(2), 203–214. doi: 10.1901/jaba.

Wright P. I. (2019). Cultural humility in the practice of applied behavior analysis. *Behavior Analysis in Practice, 12*(4), 805–809. https://doi.org/10.1007/s40617-019-00343-8

# 4.2 Beyond the PowerPoint

*Carmen Hall*

**Beyond the PowerPoint: Becoming an Engaging Presenter**

Behavior analysts coach and train others regularly as clinicians providing behavior analytic services. Recently, Applied Behavior Analysis (ABA) critics indicated language used by behavior analysts can be a barrier for consumers (Critchfield et al., 2017). Graduate programs in behavior analysis focus on the task list itself—a list of competencies that behavior analysts are required to demonstrate to become Board Certified Behavior Analysts (BCBA) (BACB, 2017). More recently, Personnel Selection and Management was added to the fifth-edition task list to focus on supervision and training skills for the behavior analyst (BACB, 2017), and although an item specifies "train[ing] personnel to competently perform assessment and intervention procedures" (p. 5), actual presentation skills are often not addressed in graduate training programs. Many students present as a learning activity in post-secondary and then join supervisors and the academic community in presenting at academic conferences. Although conference presentations are common across disciplines, behavior analysts are often asked to present to audiences beyond the conference. Consumers of this information can include educators, parents, and community members who are looking to apply the behavior analytic principles with their children or students, campers, and clients they support.

Since Microsoft's creation of the PowerPoint software in 1990, the presentation landscape in conferences, classrooms, and online training has been dominated by the use of slides. Many individuals who prepare a presentation begin with a slide outline and the creation of the individual slides before formulating their presentation. Craig and Yewman (2013) state that this leads to "dense slides, low energy, 'um-ah' verbal tics, and bored audiences" (p. 5). Unfortunately, more time and energy may be spent on the actual PowerPoint itself than the discussion of who the audience is, what they need to learn, and how the audience members will most effectively learn and demonstrate the skills taught.

DOI: 10.4324/9781003300465-21

### What Is the Function of that Presentation?

Critchfield (2014) highlights that every behavior has a function, including the behavior of one's audience. An effective presenter will understand the function of the audience's behavior before the presentation, just as one would before implementing a treatment plan. For people to engage in behavior analytic presentations, it is important to understand the audience's learning history and environmental variables that are affecting their knowledge and motivation coming into the presentation, including potential preconceived notions of behavior analytic topics (Critchfield, 2014). Although it is not possible to know each audience member, knowing what types of professionals, parents, or students are in the group (and thus their common educational and experiential backgrounds) will help tailor the presentation to meet their needs. This may include using simplified language and avoiding the use of jargon so the audience can fully grasp the intended message. When a topic meets a person's needs, it is a more effective reinforcer—more applicable to one's situation and thus seen as more valuable. Learners will be more focused and attend when they perceive a presentation as reinforcing and useful.

Many business-presenting guidebooks focus on the concept, What's in it for me? (Craig & Yewman, 2013; see Figure 4.1). The audience wants to know the relevance to their lives, and the presentation needs to be simple to understand: focusing on the concept of "What's in it for me?" Since many behavior analysts were taught the theory of what they are teaching others via high-level theoretical textbooks and presentations, it could be the case that they, in turn, are teaching others in the same fashion. Although these are appropriate to learn the concepts to apply as a practicing behavior analyst, most clients do not need that level of knowledge or technicality. Critchfield (2014) states that as a behavior analyst, it is their job to change their behavior for "what the other person needs to better understand" (p. 141). If that level of detail is required, the behavior analyst has to ask themselves if they should be relaying this information in a presentation, or if it is more appropriate for the behavior analyst to implement and work with the family to implement the strategies hands-on in the environment. Since behavioral analysis relies on evidence-based strategies, future behavioral analysts can enhance the value of their presentations by balancing between evidence-based interventions and attention-grabbing presentation skills.

### What Effective Speakers Do

Effective speakers understand the concept that *they* are the presenter—not their slides (Craig & Yewman, 2013). Great communicators make it look easy, and great presenters transform audiences, all because they have

*Figure 4.1* Getting to know your audience: What's in it for me?

rehearsed the concepts to resonate with the audience (Duarte, 2010). The slides act as visual back-up and helps the learner visually understand the concept, turning back to the presenter to tell the story of their presentation (Craig & Yemen, 2013). Too often, presenters start their presentation by creating a slide deck. By doing this, the presentation is focused on content and not on presenting. If one is doing this, they should question whether this is really a presentation, or if it would be better suited as a printed handout (for making effective handouts see Chapter 6.4).

The presenter's behavior needs to change to meet the needs of the audience and to enable them to better understand the concepts (Critchfield, 2014). Put in another way, resonance, a physics concept, occurs when an object responds to an external stimulus that has the same frequency (Duarte, 2010). As presenters change their frequency to that of their audience, Duarte (2010) talks about resonating deeply with them. Thus, using overly technical terms, talking about theory, and lecturing at them rather than engaging with the techniques does not meet the needs of the audience. "Great communicators make it look easy and great presenters transform audiences, all because they have rehearsed the concepts to resonate with the audience" (Duarte, 2010, p. 14). Presentations have the ability to create change by using human contact and communication—the emotion is what makes the audience feel the same passion as the behavior analyst—something that no other medium can do (Duarte, 2010).

## Organizing the Presentation

In daily life, connections with people around us occurs by communicating in a real manner by sharing common beliefs and emotions. Presentations need to be organized like a story, not a report—a report is sharing information and a presentation is sharing an experience (Duarte, 2010). A presentation is one way of communicating. This involves applying the storytelling skills used in one's daily life to assist behavior analysts in successfully engaging audiences and conveying messages—much more effectively than *just* a slide deck (Yewman & Craig, 2013). For behavior analysts, a blend of evidence-based information combined with storytelling creates a presentation over facts being talked about. Just like screenplays and stories, presentations have a beginning, middle, and an end, where the middle is longer. There is an inherent structure, and it includes the clinical part, where the information captures the audience's attention as the turning point (Duarte, 2010). In essence, the turning point is the strategy or information that answers the question, What is in it for me?

During a presentation, storytelling can aid the listener in the visualization of the information and is an effective way of relaying data (Kosara & Mackinlay, 2013). The martini glass approach is one approach by stating a broad introduction and further narrowing it down to make a point, ending the presentation by allowing the listener to explore the topic (Segel & Heer, 2010). Storytelling tools can also assist in describing complex materials as it makes them easier to understand while engaging the audience and sparking their interest by building a deeper connection to the topic. Similar to the plot of a story, with a build up of characters and desires, action with conformation, and a resolution—the presentation needs to follow a similar structure (Duarte, 2010). As behavior analysts, when presenting, provide an idea for the audience to adopt and take forth and generalize to their environment. However, in order to do that there needs to be a conflict or issue (seen as a problem) that the behavior analyst is offering a solution to—just like a plot (Duarte, 2010). See Figure 4.2 for Duarte's (2010) Audience Journey.

## Openings, Endings, and Transitions

Presenters have a short time to gain their audience's attention. The first two minutes are crucial and is when audiences are listening the most (Medina,

196  *Carmen Hall*

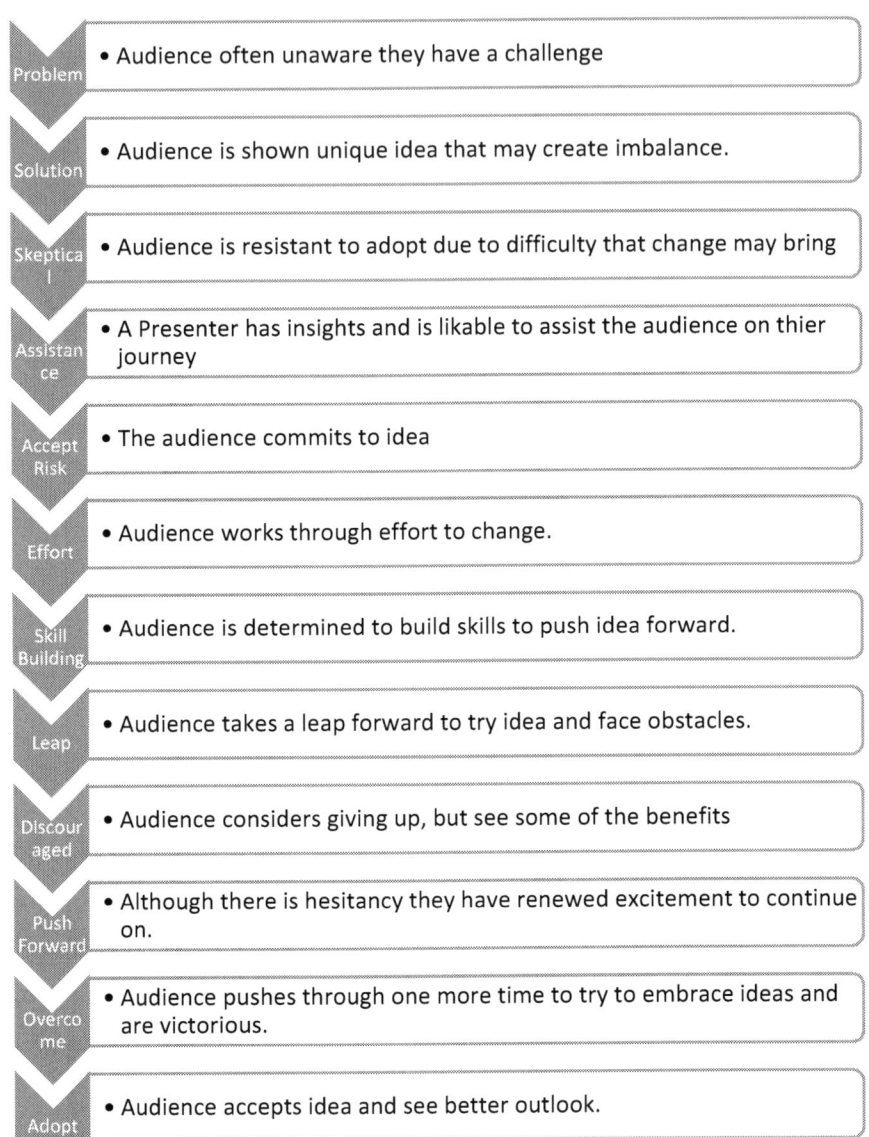

*Figure 4.2* The audience's journey.

2014). Otherwise, failure to grab their attention may mean the remainder of the presentation may be lost (Craig & Yewman, 2013). Presenters are often competing with many other reinforcers in the environment—someone's Facebook feed, an interesting video, a text message from a friend, and so forth. However, if the opening is multisensory and reinforcing, the likelihood of catching the audience's attention and keeping them engaged is higher and, in turn, attending to the rest of the presentation. Simply stated, the opening is not about the presenter—it's about the audience and what the behavior analyst can do for the audience—with no translation needed! (Craig & Yemna, 2013). No jargon is needed, but the opening constitutes the critical moment to make a true testament to the audience on what this presentation will do for them. For example, the research behind the functions of behavior is not what a parent wants to hear: they want to hear how looking at their child's behavior through a functional lens will make their job easier. Thus, more time should be spent on planning the opening than any other part of the presentation! See Table 4.1 for a list of possible opening ideas.

Don't stop at the opening. But think about how the opening will flow into the first slide, the next activity, and the next presenter. Introduce the next presenter and move around the room. As one activity ends, think about how to transition into drawing the audiences' attention back to the front. Nothing looks more unprofessional than unrehearsed transitions.

Likewise, presenters often get to the end, and they exhale from all of the knowledge they successfully relayed to the audience. A presentation should not end with, "That's all … and now questions." Think about how to will leave the audience—what is their call to action? Table 4.1 also illustrates a number of ways a presentation can end. When imagining good movies, they have good endings. Thus, if the presentation is a story, think about a good ending. It is suggested that the beginning and the ending are rehearsed. You know the middle (the content), often what the behavior analyst was trained in.

## Pausing

Intentionally incorporating an active pause while talking is a vital way to attract the attention of the audience. Pausing can assist the audience in processing the points raised or increasing the expectation of what is going to be said next. Pausing can be difficult, however, as it can be used to get the attention of some audience members if used effectively; but, it can also distract and frustrate others if used too much. Therefore, presenters need

*Table 4.1* Effective Presentations: Opening and Closing Ideas

| Creative Strategy | Example |
| --- | --- |
| 1. Important Points | News article |
| 2. Ice Breaker | Related or unrelated activity |
| 3. Quiz, Activity, Game | General topic or to highlight the take home message |
| 4. Story | Similar story of success, overcoming challenges, etc. |
| 5. Questions | Polling the audience (how many … have you ever …) |
| 6. Audio/ Visuals | Short video, demonstration, commercial, etc. |
| 7. Quote | Sentimental quote to ask audience to reflect |
| 8. Cartoons or Pictures | Explain situation or highlight main message |
| 9. Jokes | Start or end with tasteful humor to engage audience |
| 10. Statistics | Prevalence or agreement on a topic |
| 11. Audience Knowledge | Quiz or pre- and post-test to test the audience's knowledge |
| 12. Anecdotes | Short story to highlight experiences and relate to audience |

to be intentional and very selective about when and how long to pause. Most public speakers often pause for two to three seconds to attract the audience they are speaking to (Rosemarie, 2020).

Deliberately using pauses can aid in the presentation's success in two ways. If a pause occurs after presenting an important point, the audience will know that the points mentioned are worthy of their attention and it can give them time to process those points. For instance, pausing just after the key point will allow the audience to digest the concept and create anticipation of what will come next. The audience will be more likely to listen when pausing just before highlighting crucial points or conclusions (Craig & Yewman, 2013).

### Voice

To effectively deliver a message, a lively voice, pitch, and pace are essential to maintain audience engagement. The power of the voice is that it communicates energy by using an appropriate volume and proceding at a reasonable pace. The right amount of volume lets others know that the presenter believes in what they are presenting (Chambers, 2016). It is very common to lose the audience's attention in a few seconds if the presenter's voice is too low. A helpful technique to practice projecting is delivering the presentation with headphones on and ensure one is concentrating on body language and chest to confirm they are assisting their vocal cords and lungs to project their voice (Yewman & Craig, 2013).

When presenters are nervous, they tend to speak faster than normal, and this can cause the audience to get lost. Fast talking can help presenters share the excitement with the audience; however, it doesn't always work. It is a fine line to share the excitement while not exhausting the audience. On the contrary, if the pace is too slow, audiences may become bored, and presenters then lose attention. Thus, speaking at a normal speed and changing the pace according to the points made ensures the audience understands the message delivered (Rosemarie, 2020).

Using filter words like "um," "uh," "well," "so," "you know," "er," and "like" are very common in everyday speech. However, when presenting, these words can be very distracting and minimize the ability for the audience to listen and understand the presentation. More filter words often occur when the presenter is unprepared and disorganized, and thus knowing the slides thoroughly and focusing on transitions, activities, and pace will help presenters speak fluently (Yewman & Craig, 2013).

**Body Language**

The presenter's physical stature, posture, gestures, movement, or nonverbal cues are impactful and influence the spoken words in a presentation (Craig & Yewman, 2013). Studies have shown that messages passed in coordination with gestures are more persuasive than those with limited gestures (Clough & Duff, 2020). Furthermore, a mismatch between a person's speech and gestures creates communication problems for the listener (Knapp et al., 2021). Therefore, during a presentation, presenters should be mindful not only to reduce unnecessary body movement but also to use appropriate body language such as eye contact and smiling (Chambers, 2016).

Similarly, cultural values can also guide when, where, and how much eye contact one may give when presenting (Knapp et al., 2021). If a speaker is nervous, this eye contact can decrease further. However, despite the above factors, eye contact is vital in face-to-face communication. Throughout the presentation, eye contact maintained with the audience shows interest and assists in keeping their attention. If eye contact is too long, it can create discomfort, and if it is too short, it will not feel like genuine contact. It is recommended to hold one or two seconds of eye contact with each member of the audience (Craig & Yewman, 2013). Smiling between phrases will also send a message of friendliness. When smiling, a message is expressed that the presenter is happy to be presenting the topic and they come with enough energy, as long as the smile is sincere and true, not artificial (Craig & Yewman, 2013).

### Activities and Active Learning

When creating active learning opportunities for audiences, it is commonplace to believe in the visual, kinesthetic, or auditory learner. The theory is based on the premise that educators should match teaching styles to the person in the audience's learning style (Fury, 2020). However, there is little scientific research to support this, and although people may have strengths and weaknesses in certain areas, teaching to one's learning style has not been proven to accelerate learning (Fury, 2020). Instead, presenters need to think about multiple factors, "where accessible presentations tend to be effective presentations" that are based on the theory of universal design (Vitullo, 2008, para 4). Bugstahler (2015) highlights that

> making facilities, information resources, and instruction welcoming to, accessible to, and usable by everyone is called universal design (UD). Universal design means that rather than designing something for the average user, you design it for people with a broad range of characteristics such as native language, gender, race, ethnic background, age, sexual orientation, learning style, and ability.
>
> (para. 2)

Combining the need for universal design as well as research in neurocognition, activities, and multimodal presentations are required to be effective in order to disseminate behavior analysis principles. Medina (2014) found that audiences are more likely to remember emotionally charged events rather than neutral events. Thus, presenting on an ABA topic where the audience is practicing with one another, while laughing and correcting each other as they practice, will increase memory of the event (Medina, 2014). Universal design "incorporate[s] a variety of instructional methods that use a variety of senses" and "illustrate[s] key points with a variety of examples, real-life experiences, or stories that appeal to multiple demographic groups" (Bugstahler, 2015, para 9). Thus, to keep the energy consistent, practice, practice, and practice the introduction of the exercise, how it will run, and then the activity's debriefing. Within this domain, the presenter needs to consider how to transition to the next slide. As Medina (2014) mentions, the incredible inefficiencies of text-based information along with the power of images highlights the need for activities and other creative media to assist the presenter in making the presentation active and thus memorable (see Medina's website, www.brainrules.net, for videos and research to support this rule).

### PowerPoints Tips and Tricks

Taken together, if PowerPoint will still be used in a presentation, it is important to have accessible fonts and colors that can be read, be

*Table 4.2* PowerPoint Tips and Tricks Based on Universal Design

| Category | Description |
| --- | --- |
| 1. Font | 24-pt font minimum |
|  | San Serif Fonts (e.g., Helvetica) |
| 2. Background Colors | High contrast background to text colors |
| 3. White Space | Leave large amounts of white space with little clutter |
| 4. Charts | Large, simple charts and tables |
| 5. Organize. | Organize material to bring audience on journey |
| 6. Sentences | Simple language and keyword phrases rather than full sentences |
| 7. Abbreviations | Spell out abbreviations and acronyms on first use |
| 8. Videos | Caption videos shown |

multimodal, and accommodate those with disabilities, thereby allowing everyone to access the content, materials, and activities (Bugstahler, 2015). The number of words per slide should be minimal and just outline what the presenter is talking about. When preparing slides, remember that the speaker is the presenter, not the PowerPoint (Craig & Yewman, 2013). Break up text slides with slides that just include images. As Segel and Heer state, "data stories appear to be most effective when they have constrained interaction at various checkpoints within a narrative, allowing the user to explore the data without veering too far from the intended narrative" (p. 1147).

Using visuals to support one's story can help the audience follow along. See Table 4.2 for a list of universal design PowerPoint tips to ensure the PowerPoint is accessible, easy to understand, and reinforcing for the audience.

Gratitude and appreciation are expressed to Sarah Talebi and Rediet Debe for their research assistance and contributions to this article.

## References

Behavior Analyst Certification Board. (2017). *BCBA task list* (5th ed.). Author.

Bugstahler, S. (2015). *Equal access: Universal design of your presentation.* Disabilities, Opportunities, Internetworking, and Technology. www.washington.edu/doit/equal-access-universal-design-your-presentation

Chambers, S. (2016). *Body language: How to develop effective nonverbal communication skills to empower your personal and professional life.* E. C. Publishing.

Clough, S., & Duff, M. C. (2020). The role of gesture in communication and cognition: Implications for understanding and treating neurogenic communication disorders. *Frontiers in Human Neuroscience, 14,* 323. https://doi.org/10.3389/fnhum.2020.00323

Craig, A., & Yewman, D. (2013). *Weekend language: Presenting with more stories and less powerpoint*. DASH Consulting Incorporated.

Critchfield, T. S. (2014). Ten rules for discussing behavior analysis. *Behavior Analysis in Practice, 7*(2), 141–142. https://doi.org/10.1007/s40617-014-0026-z

Critchfield, T. S., Doepke, K. J., Kimberly Epting, L., Becirevic, A., Reed, D. D., Fienup, D. M., Kremsreiter, J. L., & Ecott, C. L. (2017). Normative emotional responses to behavior analysis jargon or how not to use words to win friends and influence people. *Behavior Analysis in Practice, 10*(2), 97–106. https://doi.org/10.1007/s40617-016-0161-9

Duarte, N. (2010). *Resonate: Present visual stories that transform audiences*. John Wiley & Sons Publishing.

Furey, W. (2020). The stubborn myth of "learning styles": State teacher-license prep materials peddle a debunked theory. *Education Next, 20*(3), 8–12.

Knapp, M., Hall, J., & Horgan, T. (2021). *Nonverbal communication in human interaction* (9th ed.). Kendall Hunt Publishing.

Kosara, R., & Mackinlay, J. (2013). Storytelling: The next step for visualization. *Computer, 46*(5), 44–50. https://doi.org/10.1109/MC.2013.36

LeBlanc, L. A., Heinicke, M. R., & Baker, J. C. (2012). Expanding the consumer base for behavior-analytic services: Meeting the needs of consumers in the 21st century. *Behavior Analysis in Practice, 5*(1), 4–14. https://doi.org/10.1007/BF03391813

Medina, J. (2014). *Brain rules for baby: How to raise a smart and happy child from zero to five*. Pear Press.

Rosemarie, B. (2020). *Essential presentation skills speaking to inform, inspire, and invite* (1st ed.). Spotlight Publishing.

Segel, E., & Heer. J. (2010). Narrative visualization: Telling stories with data. *Transactions on Visualization and Computer Graphics, 16*(6), 1139–1148. https://doi.org/10.1109/TVCG.2010.179

Vitullo, M. W. (2008). Universal design: Creating presentations that speak to all. *Footnotes, 36*(6). www.asanet.org/wp-content/uploads/savvy/footnotes/julyaugust08/presentation.html

# 4.3 Behavior Analyst as Changemaker
## Science, Advocacy and Activism in Practice

*Louis P. A. Busch, Jaime Santana, and Mark A. Mattaini*

Introduction

As scientist-practitioners, behavior analysts work to understand the contextual variables that contribute to social challenges with a goal of making the world a better place. However, as most behavior analysts are clinicians who support individuals living with neurodevelopmental disabilities, their focus is often on changing the microsystems their clients interact within by applying functional assessment and contingency management in an effort to promote safety, autonomy, and meaningful community engagement. This important microsystem work may include exchanges within family systems, schools, peer groups, workplaces, sport and social clubs, neighborhoods, local health units, and social service organizations. But, if "saving the world" is to be the true goal of a science of human behavior (Dixon et al., 2018; Skinner, 1987; Skinner, 1953), then behavior analysts must commit to an agenda of social justice and extend its reach to collective action at the level of cultural systems (Mattaini & Roose, 2021). Unfortunately, cultural systems analysis, social justice, policy influence, and community organizing, are not topics regularly prioritized in the training of behavior analysts (BACB, 2017a; BACB 2017b). We contend that, in addition to striving to become effective scientist-practitioners, behavior analysts should strive to become effective scientist-advocates (Cihon & Mattaini, 2020). As this volume is a practical resource for behavior analysts, this chapter focuses on three primary questions: (1) Why advocate? (2) What do we advocate for? And, (3) How do we advocate effectively?

Why Advocate?

Advocacy is defined as "public support for, or recommendation of, a particular cause or policy" and activism as "the action of using vigorous

DOI: 10.4324/9781003300465-22

campaigning to bring about political or social change" (Oxford, 2015). While there are certainly causes near and dear to the hearts of many behavior analysts (e.g., disability rights movement, right to education, reduction of intrusive procedures, poverty reduction, anti-racism, environmental issues), advocacy and activism may be interpreted as being outside of the scope, responsibility, and professional decorum of the practicing behavior analyst. The same might be said at the organizational level: while the professional associations of other disciplines (medicine, psychology, social work) have long made a practice of addressing social issues in the form of public statements, calls to action, or direct political lobbying, this sort of political activity is relatively new to many behavior analytic organizations. Mattaini et al. (2020) noted, however, that many behavior analysts were deeply involved in the social justice movements of the 1960s and 1970s, with their participation typically occurring outside the purview of behavior analytic organizations and instead in conjunction with other community organizations and collectives. It should also be recognized that the ABAI Special Interest Group "Behaviorists for Social Responsibility" has recognized this responsibility since 1978 (Board of Planners, 2015.) Perhaps in response to large-scale political movements (e.g., MeToo, Black Lives Matter), calls for behavior analysts to both reform their own practices and to advocate for changes to cultural systems appear to have increased in recent years (Kirby et al., 2022; Sadavoy & Zube, 2021). Indeed, the growth of the sub-field of culturo-behavior analysis may also be an indication of an increasing political momentum (Cihon et al., 2021).

One of the first signs of public engagement from a major behavior analytic organization came with the formation of the Association for Behavior Analysis' Task force on Public Policy in 1987. Task force members noted that, while "science and advocacy are most often viewed as different arenas" (Fawcett et al, 1988, p. 28), with advocacy efforts threatening the credibility of the scientist, and a scientific orientation threatening the credibility of the advocate, the two roles could co-exist. Fawcett and colleagues (1988) suggested that behavior analysts could participate in the public policy arena by (1) conducting research relevant to policy debates; (2) involving policy makers in the design of behavioral interventions; (3) learning to analyze and report on cost effectiveness and cost benefit; and (4) promoting the idea that science and advocacy could co-exist. They also provided recommendations on disseminating policy research and on training future behavior analysts to be effective in this area. While significant political efforts have been aimed at expanding funding for behavioral services, particularly targeting insurance mandates for the treatment of autism spectrum disorder (Jacobson et al, 1998; Green & Johnston, 2009; Harvey et al; 2010; Trump & Ayres, 2019), advocacy in other areas has been relatively limited. This isn't to say however, that there haven't been significant accomplishments in other areas. In recent years, behavior analysts have produced innovative research targeting climate change (Gelino et al., 2020; Gelino et al., 2021; Bonner & Biglan, 2021), racism (Belisle et al., 2022; Gingles, 2021; Beaulieu et al., 2022), domestic violence (Amorim et al., 2022; Myers, 1995); poverty reduction (Silverman et al., 2018), youth violence (Friman, 2021; Roose & Mattaini, 2020; Aspholm & Mattaini, 2017; Mattaini, 2001), and gender-based equity (DeFelice et al., 2019; Leland & Stockwell, 2019).

Three and a half decades ago, B. F. Skinner lamented the field's lack of progress on applying the technology of human behavior in the service of prominent social issues (Skinner 1987). Since then, a number of researchers have continued to encourage behavior analysts to advocate for radical social change (Dixon et al., 2018; Mattaini et al., 2020; Mattaini & Roose, 2021). A rationale for behavior analysts to become engaged in advocacy and activism centered on social justice is evident. Although the bulk of the field's focus exists within microsystem interactions (e.g., improving communication of a child with developmental disability by training parents and teachers to apply systematic teaching procedures), those systems are clearly connected to and impacted by the broader macro-systems in which they exist. The most well-researched behavioral interventions will be of little use to society if they cannot be implemented within the contexts they are needed and for the length of time they are needed (Dixon et al, 2018). Integrity and longevity of behavioral change relies, at least in part, upon the resources and conditions within the settings in which they are applied and maintained. Environments consistently impacted by negative social contingencies (e.g., poverty, crime, war, pollution, structural violence, institutional racism) are less likely to be comprised of the conditions necessary for individuals and groups to make lasting changes in their behavior. In short, to have a meaningful and lasting impact at the level of individual and interpersonal contingencies, behavior analysts must be sensitive to contingencies at the cultural level, and work to advocate for improvements that are likely to support maintenance and generalization of behavioral changes that improve the lives of their clients. Furthermore, behavior analysts frequently work with and profit from work with marginalized populations; understanding and responding to issues of power and oppression, and adopting a value set that prioritizes social justice within these populations is a prerequisite to ethical practice.

Working to acknowledge and offset the injustices caused by current and historical applications of the science of behavior, also makes advocacy a moral responsibility, and one that the longevity of the field may be contingent upon. Recent controversies about early and current applications of behavior analysis, including the field's connection to the practice of conversion therapy and the ongoing use of contingent skin shock, illuminate the harms and, at times, conflicting values of society and the practice of behavior analysis (Cappriotti et al., 2021; Johnson, 2022; Zarcone et al., 2020). Although discussions have begun on how to examine and respond to these harms (and harms that have yet to be defined), they have not been without considerable resistance (Leaf et al., 2021).

Dr. Cindy Blackstock, a social worker and fierce advocate for Indigenous children in Canada, provides a model that may be of use for the difficult task of critically looking inward. In *The Occasional Evil of Angels'* (2009), Blackstock examined the historical and current roles of social workers in separating Indigenous children from their families. Blackstock suggested that the first step in a process of reconciliation is to hear about the harms "in a way that cannot be rationalized or abided" (p. 28). Importantly, Blackstock noted that the definition of harm and the identification of its potential solutions are not the role or right of those who have harmed, but rather those who have experienced it. Blackstock presented arguments that are commonly leveraged in defense of the field (with a particular focus on the harm caused to Indigenous children by social workers) and proceeded to meticulously and systematically dismantle those arguments. The arguments presented could easily be generalized to behavior analysts: (1) there were different standards back then; (2) we didn't know; (3) we are/were needed; (4) the issue is/was outside of the mandate/jurisdiction of the profession. Blackstock encourages undefended contact with the reality and inevitability of harm caused by well-intentioned professionals, and notes that codes of ethics, professional training and standards, and even the development of an anti-oppressive disciplinary paradigm will fail to prevent harms, and may inadvertently create a false sense of confidence that may limit sensitivity to harms while stifling corrective responses to them. Blackstock concludes with the following advice for her profession:

> [S]ocial work must look in the professional mirror to see its history from multiple perspectives including that of those who experienced the harm. We must look beyond our need to not feel blamed so we can learn and change our behavior. It sounds trivial to write about the power of blame and shame among social workers but I have seen its power. I have seen many bright and compassionate non–Aboriginal social workers raise walls of rationalization and distance to insulate themselves from it. As the doers of good, we have not been trained to stand in the shadow of our harmful actions so we ignore or minimize them. It is a privilege to put up those walls—to be able to insulate yourself from what happened.
>
> (p. 35)

Dr. Blackstock's exemplary work as an advocate, and her non-defensive and solution-focused approach to critically examining her own discipline, serve as a model of how behavior analysts may examine historical and current practices that may be in conflict with an agenda of social justice. Her work also provides a road map of how behavior analysts may move forward towards reconciliation and a more effective model of understanding, preventing, and responding to practice-based harms.

**What to Advocate For?**

In 1978, Montrose Wolf, one of the field's founders, suggested that behavior analysis could "find its heart" through an inquiry of the subjective experience of consumers of the science; a fairly controversial statement at the time. Wolf (1978) described social validity as the extent to which the goals, procedures, and outcomes of behavioral intervention have value to the individual and to society at large; and although social validity is hailed as one of cornerstones of applied behavior analysis (Baer et al, 1987), it continues to be assessed and reported on at an alarmingly low rate in behavior analytic research (Ferguson et al, 2018; Dixon et al, 2018). Further, the majority of work in this area has been in evaluating the experiences of individuals, families, and communities retrospectively, after the interventions have already been applied. Relatively little scholarly work has been done on determining the values, hopes, and expectations of individuals, families, and communities before procedures are applied. One way to ensure our goals align with the best interest of stakeholders and allies, as defined by our shared values (rather than ours exclusively), is to leverage a community-based participatory action approach to both research and clinical intervention. The American Psychological Association (VandenBos, 2013) defines participatory action research as "a form of action research that emphasizes collaboration between researchers and members of the disadvantaged community of interest" (p. 766). A participatory action approach is fundamentally about the democratization of research and its outcomes, and positions the individuals and collectives within their contexts as the experts of their own experience (Glassman & Erdem, 2014). Practically, this approach requires the development of relationships with, and partnerships with, non-academic (or non-clinical) community stakeholders to identify, examine, and respond to an issue identified by those stakeholders (rather than by the researchers). The goal of this type of research or action, is first and foremost, to promote positive social change for those individuals and communities (Leavy, 2017). A community-based participatory action approach to research and practice in behavior analysis could contribute to increased social validity, as the values and perspectives of people behavior analysts interact with would be at the forefront from the onset of our contact with consumers. One example of this sort of approach can be found in the work of Atkins and colleagues (2019) on prosocial processes within and amongst groups; the cornerstone of which is the collaborative identification of shared values at the outset of any change initiative.

Behavior analysts engaged in advocacy and activism must ask themselves repeatedly if their values and hopes for the future align with those of the community that will be impacted by that advocacy. While a collaborative process of exploring the values of those who will be most impacted by the advocacy may be the most appropriate way to determine what specifically to advocate for, there may be targets that could be considered universal. For example, the equitable distribution of resources and physical and psychological safety may be considered a general guiding goal of social justice efforts (Mattaini & Roose, 2021). Access to accessible, effective, equitable, and culturally appropriate social, health, and educational services may also provide appropriate advocacy targets for behavior analysts. Finally, behavior analysts can identify advocacy focus by targeting injustice. Mattaini and Roose (2021) suggest that scientist-advocates "identify cultural practices that create and sustain injustice, like poverty and structural racism" and that "those data offer opportunities to challenge such conditions constructionally by assisting in identifying, arranging and sustaining alternative contingencies supporting social justice, and thereby healthy communities" (p. 5). There may be instances of injustice that are specific to the practice of behavior analysis or to the individuals who receive behavior analytic services, such as the use of aversive procedures in educational, social, and healthcare settings, and practices that perpetuate unnecessary restraint, sedation, or segregation. In addition to the use of aversive procedures, behavior analysts should take care to recognize systems of coercion, inequity, and injustice that leverage positive reinforcement in order to maintain oppressive power dynamics or to control vulnerable populations. Finally, systems that value profit over human dignity, such as the growing influence of private equity in behavioral services for youth at risk, the elderly, and children with neurodevelopmental disabilities should be of particular concern for the scientist-advocate (O'Grady, 2022).

Whatever the target for change, advocates must work to identify specific actions that can be executed and are within the capability of those in positions of influence to make the changes. Advocating for world peace, gender equity, or an end to racism and poverty, although noble pursuits, are unlikely to be effective advocacy targets in such nebulous forms (Mattaini et al., 2020). Scientist-advocates may choose to identify specific policies, programs, and issues within their jurisdictions to target for action. For example, lobbying for increased funding for restorative youth justice initiatives, housing affordability programs, or community programs may be within reach for many practicing behavior analysts. Likewise, lobbying against policy or programs that perpetuate injustice (e.g., healthcare privatization, race-based carding practices, extractive and polluting industries, proposed cuts to social service programs) may be fruitful areas for advocacy efforts. Additionally, it is vital that scientist-advocates identify individuals who have the power and capacity to direct changes at each level of organizational functioning. In the Canadian legislative context this group might include ministers, deputy ministers, associate ministers, associate deputy ministers, and program and policy directors. In academic institutions, this group may include chancellors, chairs and vice-chairs, presidents, vice-presidents, provosts, deans, program administrators, and department heads. Within private and not-for-profit sectors, this group may include chief executive officers, executive directors, directors, and managers. In addition to targeting multiple key decision makers simultaneously, it is often necessary to identify the networks of influence to which those decision makers are connected; directing advocacy efforts towards the stakeholders to whom the decision-makers are beholden (funders, media, lobbyist organizations) can be an effective and practical option (Mattaini et al., 2020).

### How to Advocate?

An advocacy framework that provides strategic direction for the scientist-advocate can be a useful starting point. Sharp (2011) likens the strategic planning process in advocacy work to planning a trip, which requires a method of travel (i.e., method for change), identification of costs, required paperwork/documentation, and identifying potential unexpected situations and plan solutions for them. In Cihon and Mattaini's volume on *Behavior Science Perspectives on Culture and Community* (2020), Mattaini and colleagues provide several cultural systems frameworks for identifying important contextual variables and contingencies that are relevant to the action of decision makers. Applying an advocacy framework that maps out motivative operations, resources, and conditions, driving and inhibiting forces, positive and aversive consequences for action of decision makers and their network of influence, may lead to innovative and solution-focused approaches for scientist-advocates. Ardila Sánchez et al. (2020) suggest that completing a situation-specific cultural systems analysis may "identify interlocking variables and systemic practices within the behavioral, cultural, and ecological milieu that maintain the status quo, and from there, to navigate complex interactions and identify strategic directions for change" (p.20). Recent advances in relational frame theory also provide a number of tools that scientists-advocates can leverage in their work. Ardila Sánchez et al. (2020) described how overgeneralized relations of coordination or opposition can lead to misunderstanding, conflict, racism, stereotypes, and extreme views. They also suggested that understanding the relations at play in political messages can help to identify manipulative content and to combat disinformation. As an example, Belisle et al. (2022) applied relational frame theory to provide a theoretical model of "derived implicit bias against members of the black community," as well as a nested model of anti-black racism at individual, community, and system levels that could prove valuable in the development of "anti-racist interventions for individuals and groups" (p. 29).

Advocacy repertoires can take many forms, and some forms may be more appropriate in specific contexts and for specific issues. Gene Sharp's work on *Waging Non-Violent Struggle* (2005) identifies classes of action containing 198 methods. The first level, which Sharp describes as mild, comprises "nonviolent protest and persuasion" and may include "written declarations, petitions, leafleting, picketing, wearing of symbols, symbolic sounds, vigils, singing, marches, mock funerals, protest meetings, silence, and turning one's back" (p.4). The second class comprises non-cooperation, which may include "social boycotts, student strikes, stay-at-home initiatives, or collective disappearance" (p. 4). Mattaini et al. (2000) differentiate between constructive noncooperation (e.g., refusing to participate and creating alternatives for unjust systems) and disruptive noncooperation (e.g., withdrawing from and protesting unjust systems). There may be occasions where these milder forms of activism are ineffective and "in such cases, there remains a clear ethical obligation to advocate for justice, even where less collaborative approaches may be required" (Mattaini et al., 2000, p. 389). Sharp (2005) describes *non-violent intervention* as those that "actively disrupt the normal operation of policies or the system by deliberate interference, either psychologically, physically, socially, economically, or politically" (p. 5). These may include

> sit-ins, nonviolent raids, nonviolent obstruction, nonviolent occupation, overloading of facilities, alternative social institutions, alternative communication systems, reverse strikes, stay-in strikes, nonviolent land seizures, defiance of blockades, seizures of assets, selective patronage, alternative economic institutions, overloading of administrative systems, seeking of imprisonment, and dual sovereignty and parallel government.
>
> (p. 5)

Mattaini et al. (2000) differentiate between resource disruption (e.g., temporarily shutting down roads) and retaliation (e.g., the application of aversives).

Fawcett (1988) suggested that in operating as advocates, behavior analysts should "not work alone" and instead should "attend to the orientation, roles, and skills of others involved in the process of social change" (p. 20). In an important paper on professional humility, Kirby et al (2022) suggest that interprofessional collaboration, the ability to create respectful and reciprocal relationships with professionals from other disciplines, is a critical component of effective advocacy for behavior analysts. As mentioned earlier, disciplines such as social work and medicine have long histories of advocacy and can make effective partners in the change-making goals of behavior analysts.

In advocating for increased access to behavior analytic services, or for changes in how supports are currently implemented, behavior analysts tend to rely on persuasion through the demonstration of the scientific merits of a behavior analytic approach. Specifically, behavior analysts often produce research syntheses, meta-analyses, and systematic reviews to convince funders, policy-makers, educators and consumers of behavior analytic services that behavioral procedures can have positive social and economic impacts. Unfortunately, this approach in isolation may not always be effective. Although a graph demonstrating a contingent replication of behavioral change across settings may be convincing for other behavior analysts, without a more fulsome context it is likely to have little impact on those outside the field. Mattaini et al. (2020) suggest that "policy-makers may be unwilling to act based on well-established data for political reasons, or because of established, contradictory relational networks." (p. 389). Further, many program and policy decisions prioritize cost and utilization estimates, the opinions of political allies, and actions that are most likely to defuse public criticism and contribute to positive public impressions of the agency, department, or leader. Alternatively, policy makers may be especially cautious when making decisions that pose a risk of negative public perception. Although not a common practice for behavior analysts, leveraging the "story behind the data" can be effective. Many behavior analysts may be hesitant to explore and share the subjective and qualitative outcomes of their work and, up until recently, the major certification body discouraged the use of testimonials within its code of ethics (BACB, 2020). However, the personal and emotional impact of an intervention can be a potent form of persuasion. For example, Grant and Forrest (2020) suggested that narratives can be an important part of scientific work and a powerful "source of influence, persuasion, and behavior change" (p. 292). Ardila Sánchez et al (2020) provided ten "culturo-behavioral guidelines for strategic activism" with which scientist-advocates should familiarize themselves. The authors emphasized ethics in activism and caution that some actions meant to combat injustice may instead be damaging, and that aversive strategies may have retaliatory consequences as well as emotional and behavioral side-effects, including "anger, counter-coercion, suppression, rigidity, and immobilization"

(p. 17). Sánchez and colleagues also suggest engagement "in intentional efforts to minimize colonial or exploitative relationships," support of collective and decentralized leadership, prioritization of movement sustainability, and engagement in acceptance and relationship-based strategies to manage the fear and uncertainty often associated with advocacy and activism work.

## Conclusion

Change-making is complex and often emotional work that requires sustained effort over extended periods, and it can have profound impacts for marginalized peoples, for better or worse (Mattaini et al., 2000). As such, it is critical that the social justice efforts of scientist-advocates are built upon a solid foundation of values and an axiology that prioritizes the best interest of the communities we serve, as identified by those communities. Behavior analysts need to be flexible in their work as scientist-advocates. To this point, epistemologies that center holism, reciprocity, and reflexivity such as those found within the practice of feminist and Indigenous advocates may be of value. Collaboration with existing advocacy groups with an agenda of learning and contributing (in that order), is also paramount to a successful advocacy agenda. Research and evaluation can be powerful tools in advocacy work, and are skill sets that many behavior analysts have to offer. Likewise, advocating using popular media, in collaboration with those who have that expertise, can be invaluable to the scientist-advocate. The science of human behavior offers many tools that can contribute to social justice, and with an openness to learning, collaboration, and a healthy dose of humility, behavior analysts are well-positioned to become scientist-advocates who can make the world a better place.

## References

Advocacy. (2015). *Oxford English Dictionary*. Oxford University Press.

Amorim, V. C., Tourinho, E. Z., & Cihon, T. M. (2022). Brazilian public policies for assistance to women in situations of violence: Contributions from Culturo-Behavioral Science. *Behavior and Social Issues*. Advanced online publication. https://doi.org/10.1007/s42822-022-00095-1

Ardila Sánchez, J. G., Cihon, T. M., Malott, M. E., Mattaini, M. A., Rakos, R. F., Rehfeldt, R. A., Richling, S. M., Roose, K. M., Seniuk, H. A., & Watson-Thompson, J. (2020). Collective editorial: Ten guidelines for strategic social action. *Behavior and Social Issues*, 29(1), 15–30. https://doi.org/10.1007/s42822-020-00038-8

Aspholm, R. R., & Mattaini, M. A. (2017). Youth activism as violence prevention. *The Wiley Handbook of Violence and Aggression*, 1–12. https://doi.org/10.1002/9781119057574.whbva104

Atkins, P. W., Wilson, D. S., & Hayes, S. C. (2019). *Prosocial: Using evolutionary science to build productive, equitable, and collaborative groups.* New Herbinger Publications.

Baer, D. M., Wolf, M. M., & Risley, T. R. (1987). Some still-current dimensions of applied behavior analysis. *Journal of Applied Behavior Analysis, 20*(4), 313–327. https://doi.org/10.1901/jaba.1987.20-313

Beaulieu, L., & Jimenez-Gomez, C. (2022). Cultural responsiveness in applied behavior analysis: Self-assessment. *Journal of Applied Behavior Analysis, 55*(2), 337–356. https://doi.org/10.1002/jaba.907

Behavior Analyst Certification Board. (2017a). *BCBA/BCaBA coursework requirements.* www.bacb.com/wp-content/uploads/2017/09/170113-BCBA-BCaBA-coursework-requirements-5th-ed.pdf

Behavior Analyst Certification Board. (2017b). *BCBA/BCaBA task list* (5th ed.). www.bacb.com/wp-content/uploads/2017/09/170113-BCBA-BCaBA-task-list-5th-ed-.pdf

Behavior Analyst Certification Board. (2020). *Ethics code for behavior analysts.* https://bacb.com/wp-content/ethics-code-for-behavior-analysts/

Belisle, J., Payne, A., & Paliliunas, D. (2022). A sociobehavioral model of racism against the Black community and avenues for anti-racism research. *Behavior Analysis in Practice.* Advanced online publication. https://doi.org/10.1007/s40617-022-00702-y

Blackstock, C. (2009). The occasional evil of angels: Learning from the experiences of Aboriginal peoples and social work. *First Peoples Child and Family Review, 4*(1), 28–37. https://doi.org/10.7202/1069347ar

Board of Planners. (2015). Editorial: Behaviorists for social responsibility 2016. *Behavior and Social Issues, 24*(1), 1–3. https://doi.org/10.5210/bsi.v24i0.6425

Bonner, A. C., & Biglan, A. (2021). Rebooting behavioral science to reduce greenhouse gas emissions. *Behavior and Social Issues, 30*(1), 106–120. https://doi.org/10.1007/s42822-021-00058-y

Capriotti, M. R., & Donaldson, J. M. (2021). "Why don't behavior analysts do something?" The behavior analysts' historical, present, and potential future actions on sexual and gender minority issues. *Journal of Applied Behavior Analysis, 55*(1), 19–39. https://doi.org/10.1002/jaba.884

Cihon, T. M., Borba, A., Benvenuti, M., & Sandaker, I. (2021). Research and training in Culturo-Behavior Science. *Behavior and Social Issues, 30*(1), 237–275. https://doi.org/10.1007/s42822-021-00076-w

Cihon, T. M., & Mattaini, M. A. (Eds.). (2020). *Behavior Science Perspectives on Culture and Community.* Springer.

DeFelice, K. A., & Diller, J. W. (2019). Intersectional feminism and behavior analysis. *Behavior Analysis in Practice, 12*(4), 831–838. https://doi.org/10.1007/s40617-019-00341-w

Dixon, M. R., Belisle, J., Rehfeldt, R. A., & Root, W. B. (2018). Why we are still not acting to save the world: The upward challenge of a post-Skinnerian behavior science. *Perspectives on Behavior Science, 41*(1), 241–267. https://doi.org/10.1007/s40614-018-0162-9

Fawcett, S. B., Bernstein, G., Czyzewski, M. J., Greene, B. F., Hannah, G. T., Iwata, B. A., Jason, L. A., Mathews, R. M., Morris, E. K., Otis-Wilborn, A., Seekins, T., & Winett, R. A. (1988). Recommendations of the Task Force on Public Policy. *The Behavior Analyst*, 11(1), 27–32. https://doi.org/10.1007/bf03392451

Ferguson, J. L., Cihon, J. H., Leaf, J. B., Van Meter, S. M., McEachin, J., & Leaf, R. (2018). Assessment of social validity trends in the *Journal of Applied Behavior Analysis*. *European Journal of Behavior Analysis*, 20(1), 146–157. https://doi.org/10.1080/15021149.2018.1534771

Friman, P. C. (2021). There is no such thing as a bad boy: The circumstances view of problem behavior. *Journal of Applied Behavior Analysis*, 54(2), 636–653. https://doi.org/10.1002/jaba.816

Gelino, B. W., Erath, T. G., & Reed, D. D. (2021). Going green: A systematic review of proenvironmental empirical research in behavior analysis. *Behavior and Social Issues*, 30(1), 587–611. https://doi.org/10.1007/s42822-020-00043-x

Gelino, B. W., Erath, T. G., Seniuk, H. A., Luke, M. M., Berry, M. S., Fuqua, R. W., & Reed, D. D. (2020). Global sustainability: A behavior analytic approach. *Behavior Science Perspectives on Culture and Community*, 257–281. https://doi.org/10.1007/978-3-030-45421-0_11

Gingles, D. (2021). Igniting collective freedom: An integrative behavioral model of acceptance and commitment towards black liberation. *Behavior Analysis in Practice*. Advanced online publication. https://doi.org/10.31234/osf.io/zk7jh

Glassman, M., & Erdem, G. (2014). Participatory action research and its meanings. *Adult Education Quarterly*, 64(3), 206–221. https://doi.org/10.1177/0741713614523667

Grant, L., & Forrest, M. (2020). Can stories influence sustainable behavior? *Behavior Science Perspectives on Culture and Community*, 283–306. https://doi.org/10.1007/978-3-030-45421-0_12

Green, G., & Johnston, J. M. (2009). Licensing behavior analysts: Risks and alternatives. *Behavior Analysis in Practice*, 2, 59–64. https://doi.org/10.1007/BF03391739

Harvey, A. C., Harvey, M. T., Kenkel, M. B., & Russo, D. C. (2010). Funding of applied behavior analysis services: Current status and growing opportunities. *Psychological Services*, 7(3), 202–212. https://doi.org/10.1037/a0020445

Jacobson, J. W., Mulick, J. A., & Green, G. (1998). Cost-benefit estimates for early intensive behavioral intervention for young children with autism—general model and single state case. *Behavioral Interventions*, 13(4), 201–226. https://doi.org/10.1002/(SICI)1099-078X(199811)13:4<201::AID-BIN17>3.0.CO;2-R

Johnson, A. H. (2022). The weight of harm: A response to "Editor's note: Societal changes and expression of concern about rekers and Lovaas' (1974) behavioral treatment of deviant sex-role behaviors in a male child." *Behavior Analysis in Practice*, 1–9. https://doi.org/10.1007/s40617-022-00683-y

Kirby, M. S., Spencer, T. D., & Spiker, S. T. (2022). Humble behaviorism redux. *Behavior and Social Issues*. Advance online publication. https://doi.org/10.1007/s42822-022-00092-4

Leaf, J. B., Cihon, J. H., Leaf, R., McEachin, J., Liu, N., Russell, N., Unumb, L., Shapiro, S., & Khosrowshahi, D. (2021). Concerns about ABA-based intervention: An evaluation and recommendations. *Journal of Autism and Developmental Disorders, 52*(6), 2838–2853. https://doi.org/10.1007/s10 803-021-05137-y

Leavy, P. L. (2017). *Research design: Quantitative, qualitative, mixed methods, arts-based, and community-based participatory research approaches.* The Guilford Press.

Leland, W., & Stockwell, A. (2019). A self-assessment tool for cultivating affirming practices with transgender and gender-nonconforming (TGNC) clients, Supervisees, students, and colleagues. *Behavior Analysis in Practice, 12*(4), 816–825. https://doi.org/10.1007/s40617-019-00375-0

Mattaini, M. A. (2001). Constructing cultures of non-violence: The Peace Power! strategy. *Education and Treatment of Children, 24*(4), 430–447.

Mattaini, M. A., Esquierdo-Leal, J. L., Ardila Sánchez, J. G., Richling, S. M., & Ethridge, A. N. (2020). Public policy advocacy in culturo-behavior science. *Behavior Science Perspectives on Culture and Community*, 385–412. https://doi.org/10.1007/978-3-030-45421-0_16

Mattaini, M. A., & Roose, K. M. (2021). Emerging culturo-behavior science contributions to global justice. *Behavior and Social Issues, 30*(1), 215–236. https://doi.org/10.1007/s42822-021-00073-z

Myers, D. L. (1995). Eliminating the battering of women by men: Some considerations for behavior analysis. *Journal of Applied Behavior Analysis, 28*(4), 493–507. https://doi.org/10.1901/jaba.1995.28-493

O'Grady, E. (2022). *The Kids Are Not Alright: How private equity profits off of behavioral health services for vulnerable and at-risk youth.* Private equity stakeholder project. https://pestakeholder.org/reports/the-kids-are-not-alright-how-private-equity-profits-off-of-behavioral-health-services-for-vulnerable-and-at-risk-youth/

Roose, K. M., & Mattaini, M. A. (2020). Challenging violence: Toward a twenty-first century, science-based "constructive programme." *Behavior Science Perspectives on Culture and Community*, 307–331. https://doi.org/10.1007/978-3-030-45421-0_13

Sadavoy, J. A., & Zube, M. L. (2021). *A scientific framework for compassion and social justice: Lessons in applied behavior analysis.* Routledge.

Sharp, G. (2005). *Waging nonviolent struggle: 20th century practice and 21st century potential.* Extending Horizons Books.

Sharp, G. (2011). *There are realistic alternatives.* Albert Einstein Institution.

Silverman, K., Holtyn, A. F., & Subramaniam, S. (2018). Behavior analysts in the War on Poverty: Developing an operant antipoverty program. *Experimental and Clinical Psychopharmacology, 26*(6), 515–524. https://doi.org/10.1037/pha 0000230

Skinner, B. F. (1953). *Science and human behavior.* Macmillan.

Skinner, B. F. (1987). Why we are not acting to save the world. In B. F Skinner (Ed.) *Upon Further Reflection* (pp. 1–14). Prentice Hall PTR.

Trump, C. E., & Ayres, K. M. (2019). Autism, insurance, and discrimination: The effect of an autism diagnosis on behavior-analytic services. *Behavior Analysis in Practice, 13*(1), 282–289. https://doi.org/10.1007/s40617-018-00327-0

VandenBos, G. R. (2013). *APA dictionary of clinical psychology*. American Psychological Association.

Wolf, M. M. (1978). Social validity: The case for subjective measurement or how applied behavior analysis is finding its heart. *Journal of Applied Behavior Analysis, 11*(2), 203–214. https://doi.org/10.1901/jaba.1978.11-203

Zarcone, J. R., Mullane, M. P., Langdon, P. E., & Brown, I. (2020). Contingent electric shock as a treatment for challenging behavior for people with intellectual and developmental disabilities: Support for the IASSIDD policy statement opposing its use. *Journal of Policy and Practice in Intellectual Disabilities, 17*(4), 291–296. https://doi.org/10.1111/jppi.12342

# 4.4 Using Movement Education, Motor Learning, and Phenomenology to Inform Teaching Strategies in Other Professional Contexts

*Maureen Connolly and Elyse Lappano*

## Introduction

Two of the basic premises of phenomenology are that (1) humans studying phenomena in "the lifeworld" (that is, the world humans inhabit but in a largely taken for granted fashion) can engage in deep study of phenomena by engaging with the mundane world (the world of simple, everyday things and actions, which are also largely taken for granted or taken at face value) and, (2) that humans engaging with the phenomenon under study should attempt to suspend their already known or learned information about that phenomenon. In this way, meaning can be constructed from sensorial experience by and in the embodied consciousness (that is, perceptually) from a position of attentive wonder (almost as if experiencing the phenomenon for the first time). This process has been our experience over these many years of working with children, youth, teens, transition-aged participants, and adults who bring their complex bodies to our well-intentioned, yet sometimes woefully inadequate, efforts at activity program planning. We committed to being taught by them, by their sometimes wild and usually unconventional approaches to embodied engagement with the world. We committed to observing-noticing-looking with the disposition of attentive wonder, with the commitment to imaginative free variations, with the comportment of willingness, of "in-dwelling," to see/hear/touch/taste/smell/ feel/know from the perspective of our participants' bodies, to assume wisdom from their bodily expressive preferences and engagements, to be taught by them. It has made all the difference. These critically reimagined strategies allow us to plan, implement, analyze, explicate, and understand intense pedagogic and relational episodes between participants in our specialized individualized movement programs and the facilitators who work with them on a weekly basis. In our monograph, we describe our approaches and strategies so that practitioners can adapt our movement-based and phenomenologically oriented strategies for other learning contexts. We contend that these approaches are practical and

DOI: 10.4324/9781003300465-23

oriented to real-world situations, and that they can be learned. It will be our privilege to offer them to our readers.

## Practicing Attentive Wonder

About twenty or so years ago, Author M attended a conference where one of the keynote speakers offered comments about how "good" qualitative inquiry works. The speaker referred to the analogy of the Old Tracker teaching his young apprentice. "How do you track a fox?" the young apprentice asks. "Start tracking," the Old Tracker says, "the Fox will teach you." This has been our experience over these many years of working with children, youth, teens, and transition-aged participants who bring their complex bodies to our well intentioned, yet often woefully inadequate, efforts at activity program planning. We have made the decision to let them be our foxes. We committed to being taught by them, by their sometimes wild and usually unconventional approaches to embodied engagement with the world. We committed to observing-noticing-looking with the disposition of attentive wonder, with the comportment of willingness, of "in-dwelling," to see/hear/touch/taste/smell/ feel/know from the perspective of our participants' bodies, to assume wisdom from their bodily expressive preferences and engagements, to be taught by them. It has made all the difference. These foxes have taught us and continue to do so.

As the reader will discern from our title, we engage with a phenomenological orientation to our research and pedagogy. This commitment is not trivial in any sense. We embrace this orientation because we see the role of the adaptive physical educator as one of engaging, respecting, and facilitating embodied knowledge. This becomes especially more relevant when working with participants whose expressive repertoires are primarily embodied, as opposed to more typical expressive modalities of speech. We are not alone in this commitment. Standal (2015) argues that the phenomenological perspective, represented by the work of Maurice Merleau–Ponty (1963; 2002; 1968) and Samuel Todes (2001), shows that the knowledge objects of physical education are embodied. This means that the knowledge object and the knowing subject are umbilically connected (p. 72). Stolz (2014) in his *Philosophy of Physical Education, A New Perspective*, also takes Merleau-Ponty's phenomenological perspective in advancing a new tradition of physical education, namely "embodied learning." We are taking on the embodied assumption that unconventional bodies also deserve to participate in physical education and have developed activities and approaches to support that participation.

Two of our prominent strategies, which we have borrowed from phenomenologically oriented approaches to inquiry, are Bracketing and Imaginative Free Variations (IFV). When we attempt Bracketing, we attempt, as much as possible, to suspend the already known "certainties" of a phenomenon such that we experience it in as open and "unknowing" a sensibility as possible, what Merleau-Ponty (1962) would call "attentive wonder" to the mundane world. For Imaginative Free Variations (IFV), we work toward proposing numerous and (sometimes preposterous) elaborations, explanations, theories, hypotheses, scenarios, anticipatory sets and/or "what ifs" for how a thing is understood to reveal itself AS itself; that is, we commit to a form of engaging with eidetics unconstrained by pre-ordained expectations. For example, one of our most provocative and transforming insights from IFV is that much of the so-called *purposeless problem behavior* (such as hand flapping or rocking) exhibited by autistic people is NOT a problem, but is, rather a *solution* for their perceived reading of what the problem might be for them. Once we see their embodied expressions as solutions, then we can collaborate with them in problem-solving strategies that are relevant and resonant with their felt sense of the world.

For a therapeutic practitioner, this strategy might translate into longer observation times to develop a sense of the participant's expressive gestural repertoire and to accept that repertoire as being meaningful. Behavior is communication; it can also be problem solving. The practitioner might also accept an expressive gestural response as being as legitimate as a verbal response. Once practitioners become familiar with participants' gestural and expressive repertoires, these expressive repertoires can gradually become reliable anticipatory sets that allow practitioners to plan and respond according to what the participants communicate in and through their gestural and expressive repertoires.

## Intersected Elements: Movement Education, Station-Based Pedagogy, Embedded Curriculum

**Movement Education** is an approach to the body that influences teaching, learning, and attitude in physical/bodily-based programs. It is based on the premise that there are overarching "themes" of the moving body—body, space, quality, (effort), and relationship—which are always present, regardless of the movers, the context, or the activity. These "existential" movement themes allow the teacher to plan activities that are *inclusive*, (i.e., operate at a *conceptual* planning level), and to analyze and describe program participants' movement and embodied expression over time and across contexts. Our approach is based in activity contexts and uses a spectrum of dimensions of movement, all of which can relate to each other in scaffolded, progressive, and/or sequential ways. These dimensions are games skills, fitness and conditioning, gross motor skills, body management skills, fine motor skills, sensorimotor development. Shafir et al.

(2016) established a solid rationale for applying LMA (Laban Movement Analysis) in the analysis of motor characteristics, emotion, and bodily expression. Dott (1995) applied LMA to motor stereotypes and distortions in autism and psychosis, and Foroud and Wishaw (2006, 2012) used LMA to analyze reaching movement patterns in stroke survivors and infants. We use both LMA and less formalized applications of Laban's system of movement existentials to describe and analyze patterns of fundamental movement skills, bodily expression and idiosyncratic gait, and postural and gestural repertoires.

**Station Based Pedagogy (SBP)** is an approach to learning and teaching based in task breakdown and distributed practice. A basic skill that can be more easily learned in its entirety can be practiced at a station designed specifically for that skill. For example, a sending task like rolling a ball toward a target as a progression for bowling could be a single station task with a focus on gripping and releasing a ball close to the floor after a preliminary arm swing. A more complex skill that presents challenges can be "broken down" into smaller elements or components and each element or component can be practiced at its own station, allowing for greater overall practice of a skill that would be more elusive if practiced only in entirety. For example, many of our participants have challenges with a squat action, where they are flexing at the hips and the knees at the same time. We frequently do knee flexion at one station and hip flexion at a different station, and various stances and weight transfers at yet another station before we "put it all together" as a squat action at a compound skill station. A basic or more complex skill that requires more elaboration of difficulty can be practiced at stations devoted to increased challenge and/or variation on the skill. Skill practice can be blocked in early stages and then randomized so that participants can develop problem solving and strategy selection. We want some elements of the skill to be more routinized, so that less processing is required in ongoing ways; and we want some elements to have more unpredictability, with limits given the high predictability needs of our participants, so that there are opportunities for novel engagement with the variables of the tasks. When possible, at a given station, we chain our sub-components, especially in early learning stages of a new skill or station. We commit to constraining the environment to reduce the number of outcome errors during practice, especially in the initial stage of learning (Poolton, et al, 2005; Capio, et al, 2013), and we typically engage in teaching strategies that use an external focus of attention (Chiviacowsky, Wulf and Avila, 2013) as well as minimal verbal cueing, simple, singular instruction, modeling, coactive engagement (engaging in moving with a participant in a hand over hand, arm over arm and so forth pattern of movement support), de-escalation and re-direction strategies, and purposeful, usually self-regulated rest-pause (resting and pausing before continuing). Since there is sparse research literature on motor learning and low functioning autism (Azar, McKeen, Carr, Sutherland & Horton, 2016), we work with activity-oriented literature on higher functioning autistic participants and participants with intellectual delay and we extrapolate from conventional motor learning research with neurotypicals. We then shift to using phenomenologically oriented strategies based in the embodied "felt sense" of movement to "translate" this strategy into activity and pedagogy that makes sense for the more unconventional bodies and processing strategies of our neuro-diverse participants. For example, Herzfeld and Shadmehr (2014) propose that motor variability is not "noise" (or distraction), but that it is helpful for learning. Our participants regularly process what might be considered immense quantities of "noise" in that they have significant issues filtering and processing their sensations and perceptions, including those elements of a task that they want to engage. Does this mean that their learning is enhanced by their "high noise" variabilities? Or are they over-saturated? Or does saturation lead to other kinds of processing? We have their movement repertoires and their chosen forms of communicative expression to assist us in addressing these questions. But we cannot engage them in more traditional research designs that require neurotypical standards of attention, specificity, and control. Our insights take longer and are, by necessity, descriptive.

**An Embedded Curriculum Approach** involves embedding the movement patterns that our participants have indicated are relevant, desired, helpful, enjoyable and/or meaningful. These patterns would be those that we have discerned over longitudinal observations of the movement patterns of the participants and in conversations and other forms of communication with them (Connolly, 2008). An example of how embedding works can be seen with midline crossing. Typically, children begin midline crossing as infants (e.g., two hands grabbing one foot) and continue with thousands of repetitions over the years. Many of our participants have challenges with everyday applications of mid-line crossing (e.g., getting dressed, toileting, eating, carrying objects, using a seatbelt, engaging with peers in desired play or activity experiences) and yet have expressed preferences for doing these activities. Here, embedding can be helpful. Clearly it is not enjoyable or meaningful to do repetition of movement patterns that feel disconnected from the desired end goal; however, with an embedded curricular strategy, we can set an environment where midline crossing happens regardless of the activity that is scheduled. Likewise, we cannot expect participants to engage in high repetition of midline crossing on demand or on command. Instead, we must make it inevitable and unavoidable within the activity they are doing. Examples include pulling with two hands on a rope, pushing with two hands on a stick, deliberate cross-body reaching for objects. In this way midline crossing is embedded and does not have to be requested. Along with midline crossing, our embedded curriculum includes spinal extension and flexion, large range of motion movement in the legs and arms (e.g., bending and straightening), travelling backwards, firm and sustained qualities of movement, the 3D Cross (that is, the directions that are available for movement, forward, backward, right, left, up, down), coredistal relationships (the relationship between the center of the body and the limbs), balance loss and regain, dorsi-flexion at the ankle, weight bearing and weight transfers (moving the body from one set of weight bearing parts to another set of weight-bearing parts). Our challenge in station construction is to make the embedded ingredients invisible: to infuse as many of our embedded curriculum ingredients and motor learning strategies into the design of the station so that what unfolds is an entirely new entity, one in which the elements have been transformed into an activity that is a thing unto itself and that is also responsive to the embodied preferences of our participants. For example, one of our participants expressed an interest in improving his stair-climbing skills. We began his stair-climbing adventure by engaging in several sessions in which we emphasized spinal extension, especially on the BOSU (a half-ball flat-bottomed piece of equipment), on scooters, on wedge mats and in the aquatic environment. As his spinal extension improved, his hip girdle mobility improved, his gait lengthened and his coordination of hip flexion with knee flexion and hip extension with knee extension also improved. Back extension is not usually the "go to" progression for stair climbing, but we surmised from our embodied problem solving that once our participant was more functional and mobile in his head-to-tail and limb girdle relationships, he would be more likely to engage in step like activities with more ease at stations with which he was already familiar.

### Knowledge Transfer of the Intersected Elements

*Movement Education* approaches emphasize conceptual engagement with skill development so that there are many correct responses to a task, and there are always ways to make a task simpler and more elaborated, depending on how the learner responds to the task. Let's say we are working with the concept of balance. The specificity of a demand for a handstand or a stand on one foot or walking on a straight line means that only with those specific responses can the learner be successful. But if we ask for balances (instead of specific skills) we can get a variety of responses that fulfill what it means to be balanced: aligning the body's center of gravity (usually in the hips) and other body parts over a restricted base of support. The skills listed above are indeed balances; however, so is weight-bearing on one hand and one foot or establishing stillness on two knees without holding on to a support or lying on one's side and holding that position without falling over. These other options also fulfill the conceptual requirements of balance and are safer and simpler progressions for novice or anxious learners. Practitioners might consider the conceptual features of specific skills they are hoping to teach their participants and plan to accept an array of responses, all of which might be correct, a continuum of correctness allows more learners to succeed and enables learners to work progressively from simple to more complex.

*Station Based Pedagogy* allows not only distributed practice of a basic or complex skill, but also allows for practice of all the components of a skill in a progressive manner without the need to "do the whole skill" before the learner feels ready. This strategy allows us to engage several tried-and-true motor learning strategies. *Break it down and build it up* gives practitioners the option to take a complex skill and subdivide it into smaller parts, practice the smaller parts, and gradually build it back up again. This also exemplifies *part-whole learning*, where the whole is greater than the sum of the parts, but practicing the parts contributes to a better sense of the whole. We also get *repetition*, a key element in motor learning, and what feels like repetition of novel components (since the various subdivided parts all feel like different things) rather than the impatience that sometimes accompanies repeating the same thing over and over. Further, the build-it-up process typically follows a *chaining strategy*. Imagine a skill is ABCD and we break it into A B C D for subdivided component practice. Then we chain it back together: practice A, practice B, practice A+B; practice C, practice B+C, practice A+B+C; practice D, practice C+D, practice B+C+D, practice A+B+C+D. Voila: chaining. Skill practice can be done in *blocks*, that is, repeating a same set of movements over time with little variation, and gradually with *random* variations so that learners can feel the learned skill in different conditions. All these strategies transfer to behavior-based learning if the practitioner has developed the ability to analyze what is to be learned and break it into its components. This practice sounds easy and, eventually it is, but it, too, is a skill and practitioners can practice it in professional development settings, in informal groups, and solo.

*Embedding Curriculum* allows practitioners to engage participants in practicing a required skill within a desired activity. This strategy works best when practitioners have spent time getting to know their participants so they can work with the context of a desired activity and embed the skill or behavior the participant needs to learn. Again, embedding requires some familiarity with the desired activity so that the embedding is effective, that is, invisible. Once participants recognize the required activity, the strategy is no longer effective since they no longer participate in the (previously) desired activity because they know it is being used to "force" them to practice something else. This is another skill that practitioners can develop in professional development contexts, informal groups, and solo.

### Overall Learning

Over the several decades that we have been working with our programs and participants, we have learned what our participants enjoy. We include these insights below so that you, as practitioners, might consider if these learning experiences and outcomes might be something that you want to work toward. **The participants enjoy the experience of "challenge by choice".** Once they are familiar with an activity station, they can manipulate the elements of the station to increase or vary the challenge. This opportunity for autonomy can be quite exhilarating. **The participants enjoy the experience of problem solving and physical exertion.** Some motor tasks might seem like a straightforward problem... but can still be tricky. Balance and weight transfers are challenging for high level performers when apparatus and equipment can be modified and interchanged. Many of our participants have balance, postural control, and kinesthetic awareness challenges, so working through this movement problem is an achievement, not a tokenistic "well done" of an easy task. **The participants enjoy the variety that is possible with well-designed stations and activities.** The routines can be changed on a weekly basis; for example, variations in body position and weight-bearing parts. Or use heavier objects like medicine balls, or different shapes and textures of objects. **The participants enjoy watching each other fail and succeed.** They learn from their own and others' successes and failures. **The participants enjoy being watched by their peers and their student facilitators.** Having an audience is not necessarily off-putting; and if it is, they let us know. **The participants enjoy deciding on the duration and pace of their activity station.** They can tell when they are getting tired. They are paying attention to their bodies' clues and signals. They know when they need to break for hydration. They know when they are getting frustrated. And they also like repetition—they like doing something over and over with randomized variety; it is not boring, it is fulfilling. **The participants enjoy being silly.** They like to goof around with some of the tasks at the activity station. They like to make weird noises. They like to exaggerate their off-balance positions and their facial expressions. They have the experience of being teens or young adults or kids while they are learning fundamental movement skills without compromising their dignity. They are working on movement goals that are meaningful to them and they are doing it in a community of learners with whom they can relate and with facilitators who support them without infantilizing them.

### Temporary Closure ...

Many autistics and other neuro-diverse people live in the social world without the taken-for-granted, already established, invisible functioning of sensory integration and perceptual-motor development. These lived experiences of ongoing disruptions compel a deeper examination of normalizing narratives of social control, commodity exchange, and bodily/behavioral codes and the expectations of ongoing concordance and "normal functioning." Indeed, the sensory profiles of many "Auties" reveal that the tactile dimensions may be particularly idiosyncratic (e.g., tactile defensiveness / tactile aggressiveness as well as synesthesia). Visual, auditory, and kinesthetic experiences also provide unconventional processing and expressing manifestations, many of which, if not most, could be characterized as intense, terrifying, traumatic and injurious and/or as delightful, absorbing, beautiful and pleasurable.

We claim that ASD and other neuro-diversities are an intersected and complex stressed embodiment frequently in contested relation with master narratives of medical—and psychopathology and minority narratives of the social model of disability. Straus (2011/2013) describes medical culture as having defining attributes. First, medical culture treats neuro-diversity as pathology, either a deficit or an excess with respect to some normative standard. Second, the pathology resides in a determinate, concrete location. Third, the goals of the enterprise are diagnosis and cure. Straus also explains how a social model of disability, in which disability is socially constructed rather than biologically given, but whose activist roots are located within social justice for persons with physical impairments, has been historically less concerned with cognitive impairments, mental or "psychiatrically designated" illness, and developmental diversities like autism. Straus further conceptualizes a reimagining of the "triad of impairments" (popularized in and by the various iterations of the Diagnostic and Statistical Manual into an array of attributes that that can be recast as idiosyncratic differences or features rather than deficits. Straus's recasting is completely resonant with Merleau-Ponty's (1962) conceptualization of embodiment. Embodiment alludes to interconnectedness, the body being a "grouping of lived through meanings which moves toward its equilibrium" (p. 153). Furthermore, this interconnectedness with/in the world "extends an object" so that it literally becomes a part of the body. Using von Uexkull's (1957) term, the object is absorbed into one's *umvelt*, one's "felt sense" of the world, including the limits and plasticity of multiple and transforming horizons. Our best activity stations are instances of an imaginative free variation "solution" to movement challenges that our participants identified as relevant and meaningful. In addressing these challenges using a station that is literally formed out of their embodied preferences, their engagement with the world can be negotiated via their felt sense of the world, which then transforms the limits and plasticity of their multiple and transforming horizons. In other words, those that were previously unattainable and perhaps unimaginable can now be seen as possible movement outcomes.

We began this chapter with a statement of our contentions: that it is humanely and pedagogically necessary—and responsible—to reimagine autism and other neuro-diversities from a non-medical model perspective, that attunement can be taught, practiced, and improved and that this attunement, far from being esoteric or abstract, is deeply practical and embodied and allows for authentic pedagogic engagement between teachers and complex learners. We conclude here by restating these contentions with a resounding exclamation point and by paying tribute to the participants in our programs who are the legacy keepers of embodied wisdom, who have been our foxes and who continue, in ways that are remarkable, discerning, ingenious, and mundane, to teach us to be better teachers and better humans.

## References

Azar, N. R., McKeen, P., Carr, K., Sutherland, C. A., & Horton, S. (2016) Impact of motor skills training in adults with autism spectrum disorder and an intellectual disability. *Journal on Developmental Disabilities, 22*(1), 40–51.

Capio, C. M., Ploolton, J. M., Sit, C. H. P., Equia, K. F., & Masters, R. S. W. (2013) Reduction of errors during practice facilitates fundamental movement skill learning in children with intellectual disabilities. *Journal of Intellectual Disability Research, 57*(4), 295–305. https://doi.org/10.1111/j.1365-2788.2012.01535.x

Chiviacowsky, S., Wulf, G., & Avila, L. T. G. (2013) An external of attention enhances motor learning in children with intellectual disabilities. *Journal of Intellectual Disability Research, 57*(7), 627–634. https://doi.org/10.1111/j.1365-2788.2012.01569.x

Connolly, M. (2008). The remarkable logic of autism: Developing and describing an embedded curriculum based in semiotic phenomenology. *Sport, Ethics & Philosophy, 2*(2), 234–256. https://doi.org/10.1080/17511320802223824

Dott, L. P. (1995) Aesthetic listening: contributions of dance/movement therapy to the psychic understanding of motor stereotypes and distortions in autism and psychosis in childhood and adolescence. *Arts Psychotherapy, 22,* 241–247. https://doi.org/10.1016/1097-4556(95)00033-2

Foroud, A., & Whishaw, I. Q. (2006) Changes in the kinematic structure and non-kinematic features of movements during skilled reaching after a stroke: a Laban movement analysis in two case studies. *Journal of Neuroscience Methods, 158,* 137–149. https://doi.org/10.1016/j.neumeth.2006.05.007

Foroud, A, & Whishaw, I. Q. (2012) The consummatory origins of visually guided reaching in human infants: a dynamic integration of whole body and upper limb movements. *Behavioral Brain Research, 231,* 343–355. https://doi.org/10.1016/j.bbr.2012.01.045

Herzfeld, D., & Shadmehr, R. (2014). Motor variability is not noise, but grist for the learning mill. *Nature Neuroscience, 17*(2), 149–150. https://doi.org/10.1038/nn.3633

Merleau-Ponty, M. (1962). *Phenomenology of perception*. The Humanities Press.
Merleau-Ponty, M. (1963). *The structure of behavior*. Duquesne University Press.
Merleau-Ponty, M. (1968). *The visible and the invisible*. Evanston, IL: Northwestern University Press.
Merleau-Ponty, M. (2002). *Phenomenology of perception*. Routledge.
Poolton, J. M., Masters, R. S. W., & Maxwell, J. P. (2005). The relationship between initial errorless training programs and subsequent performance. *Human Movement Science*, 24(3), 362–378. https://doi.org/10.1016/j.humov.2005.06.006
Shafer, T., Tsacher, R., & Welch, K. (2016) Emotion regulation through movement: unique sets of movement characteristics are associated with and enhance basic emotions. *Frontiers in Psychology*, 6, 1–15. https://doi.org/10.3389/fpsyg.2015.02030
Standal, O. F. (2015). *Phenomenology and pedagogy in physical education*. Routledge.
Stolz, S. (2014) *The philosophy of physical education: A new perspective*. Routledge.
Todes, S. (2001). *Body and world*. MIT Press.
von Uexküll, J. (1957). A stroll through the worlds of animals and men: A picture book of invisible worlds. In C. H. Schiller (Ed.) *Instinctive Behavior: The Development of a Modern Concept* (pp. 5–80). International Universities Press, Inc.

# 5
# Leadership

# 5.1 "It's Not You—It's Us"

Fostering a Behavior Analytical Supervisor/Supervisee Relationship

*Céline Bourbonnais-MacDonald and Alexandra White*

**Introduction**

When reflecting on the contents of the course work prior to becoming a behavior analyst, subject material centered around how to be a supervisor in the field of applied behavior analysis (ABA) is sparse. In today's reality, clinicians need to have the skills to balance their professional knowledge and effective practices with the needs and wishes of clients and other interdisciplinary professionals (Branson, 2010). After becoming certified and beginning to work in the field as a supervisor, it quickly becomes apparent that the supervisor/supervisee relationship is one of utmost importance for the success of the clinic-based setting. Without specific training in this area, behavior analysts rely heavily on previous interactions with past supervisors to act as a model on how to uphold their title as a supervisor. People quickly learn this tactic won't suffice as the only learning experience to rely on. These people skills or, "bed-side" manners, are comprised of key elements such as empathy, respect, listening skills, building rapport, or relationships with clients. As explained in Brower et al. (2000), relational leadership is based on interpersonal exchanges that have benevolence and integrity at their core.

The complex nature of being knowledgeable as well as able to relate to others is clearly communicated in the core principles of the Code of Ethics: (1) benefit others through doing no harm; (2) treat others with compassion, dignity, and respect; (3) behave with integrity; and (4) ensure competence in their profession (Behavior Analyst Certification Board, 2020, p. 4). The behavior analyst needs to be able to demonstrate leadership in both decision-making processes and within professional relationships with supervisees.

The knowledge and practice of the behavior analyst exist within a context of varied relationships, hence, the need for an understanding of the role that relationships have in one's leadership position as a health care and educational professional (Fulop & Mark, 2013; Guerin, 2014). Be it with a client, a colleague within the profession, another profession, or a community member, the behavior analyst must be able to build relationships that reflect the current reality. This is an ongoing social process that involves the interplay of various perspectives, led by the behavior analyst to support the client's needs and goals. As explained by Uhl-Bien (2006), an expert researcher and academic in relational leadership theory, the relational leader is self-aware of their own knowledge, that of others, and how these relate to the current context within their roles.

The relational leader is also aware that relationships are based on communication, which over time evolves through the process of the practice. Behavior analysts often call this process "pairing," a process of conditioned reinforcement whereby the relationship is developed through pairing oneself as a supervisor (neutral stimulus) with already established reinforcers (i.e., praise, supervisor traits that the supervisee finds reinforcing, and so forth.) (Dozier et al., 2012). The established reinforcers would be individual to each supervisee. Behavior analysts can ask supervisees how they prefer to be supervised at the onset of the supervisee/supervisor relationship. This will not only individualize supervision techniques for each person but will help to form that relationship. Although it may seem daunting and time-consuming at the onset of the role as a supervisor, this step needs to be viewed as equally important as pairing with a client. Much like a teacher in a classroom of individuals who differ in preferred learning styles, supervisors must ensure they have a toolbox of supervisory techniques to draw from for each supervisee. Not only are the techniques applicable for each learning style required for the task, but they are also going to pair the supervisor with the supervisee because the supervisee will find that attention, individualization, and presentation in their preferred style of reinforcing. Relational leadership is key to the behavior analyst's work.

Ensuring professional and communicative relationships amongst staff will facilitate all of the above recommendations. Without communicative relationships between staff members, unaddressed conflict that arises can cause breaks in relationships and diminishing respect across coworkers. Being met with criticism or work practices that one does not find reinforcing can act as punishment and thus decrease the occurrence of those communicative behaviors in the future. Evidently, a supervisor's role within an organization is not only to give feedback to the staff members they supervise, but to also act as a mentor and model for those staff members. Staff members will often look to their supervisors to determine the most appropriate ways to interact with each other. Krumhus and Malott (1980) were able to analyze the effectiveness of the components in a commonly used teaching process by behavior analysts entitled Behavior Skills Training (BST). They were able to determine that the modelling component of the

Table 5.1 How to Be a Behavior Analyst Supervisor and a Relational Leader

| Standard 4: Responsibilities to Supervisees and Trainees | Context/Proposed Actions | Relational Leadership | Practical Example |
|---|---|---|---|
| **Expectation 4.02:** Supervisory Competence<br>**Link to BCBA/BCaBA Task List:** Section 2–Applications E-5; I-1 to I-8 | • Context: The Behavior Analyst is aware of the scope of competence by the practitioners and has the skill set to effectively supervise and train others<br>• Action: Continuous building of one's own knowledge of sound supervisory and training competences to support trainees via professional learning/development | • Ability to understand and relate lessons learned through continuous learning, their own practice, current reality, and regulations / policies that support the profession<br>• Ability to construct supervisory and training plans that mesh with the current reality and trainee's skills. | • Looking beyond what was initially taught as supervision techniques.<br>• Frequently revisiting new behavior analytic research pertaining to supervision techniques and strategies.<br>• Attempting new supervision techniques while evaluating their effectiveness. |
| **Expectation 4.07:** Incorporating and Addressing Diversity<br>**Link to BCBA/BCaBA Task List:** Section 2–Applications E-2, E-6, E-7, E-10; H-3, H-6 | • Context: Diversity components that reflect the current reality of the community and may impact the practice are acknowledged (i.e., age, disability, ethnicity, gender expression / identity, immigration status, race, religion, sexual orientation, socioeconomic status)<br>• Action: Intentional discussion of topics related to diversity to help grow the trainee's knowledge and understanding of topics as it relates to the diversity within the current context and its impact on the service plan | • Awareness of the relations between the diverse nature of clients live and the construction of an effective service plan that *respects and values* diversity<br>• Intentional communication of the social realities through the lenses of diversity is part of the leadership role of training and supervision future Behavior Analysts | • Facilitating direct communication with supervisees.<br>• Discussing with supervisees how they prefer to be supervised. For example, how and when they prefer receiving feedback.<br>• Providing alternative supervision methods for those who require it. For example, virtual, written, or in person communication, visual aids, etc. |

| **Expectation** 4.10: Evaluating Effects of Supervision and Training  **Link to BCBA/BCaBA Task List:** Section 2–Applications I-1, 1-2, I-6, I-8 | • Context: Supervision and training of prospective Behavior Analysts must be reviewed to ensure maximum knowledge, competences, and skills to support the practice • Action: There is an ongoing process to review the impact of the supervision and training practices within the profession to ensure the effectiveness of the service plans and the overall wellbeing and of the clients | • Able to *relate* expectations and standards of the profession to the evaluation of the effects of supervision and training • Able to *organize and lead* an effective process for supervision and training that reflects the latest knowledge, practice in the profession | • Performance-based checklists • Using BST by giving instructions, providing models, role-playing, and giving feedback. |
|---|---|---|---|

process resulted in a higher level of improvement in comparison to the other components. Thus, having a supervisor who regularly engages in communicative behaviors with staff, helps to promote these behaviors throughout the workplace (Maurer & Hooper, 2011). Supervisors can promote this communicative environment by engaging in behaviors such as directly bringing up concerns with staff members when warranted, admitting fault, following up on concerns, taking direction from staff members when applicable, and seeking out regular feedback from supervisees (see Table 5.1).

## Conclusion

The behavior analyst demonstrates their relational leadership through specific actions that are directly linked to the Code of Ethics for the profession. They must be able to: (1) relate their overall knowledge, competences, and skills of the practice to varied social realities; (2) co-construct service plans with clients, peers/colleagues, and other key stakeholders; (3) communicate clearly, and on an ongoing basis, through the evolution of the practice as it relates to the clients' ever-changing needs and goals; and, (4) lead the overall service plan through an effective process of organization (Ulh-Bien, 2006). To be able to support their clients, their professional colleagues and stakeholders within a community, behavior analysts must be knowledgeable as well as able to build strong, trusting relationships with those they serve. People skills (Guerin, 2014) are complex and must be a key part to the processes within the practice, when making ethical decisions. As a relational leader within the profession, the behavior analyst must have at the center of their practice a "do no harm" mindset that supports the overall well-being of the clients, the colleagues within the profession as well as other key stakeholders. They need to be able to balance an efficient service process with trustworthy relationships within various contexts and individuals to maximize their overall impact with clients and within the communities they serve.

## References

Behavior Analyst Certification Board. (2020). *Ethics code for behavior analysts.* https://bacb.com/wp-content/ethics-code-for-behavior-analysts/

Branson, C. (2010). Ethical decision-making: Is personal moral integrity the missing link? *Journal of Authentic Leadership in Education, 1*(1), 1–8.

Brower, H. H., Schoorman, F. D., & Tan, H. H. (2000). A model of relational leadership: The integration of trust and leader-member exchange. *Leadership Quarterly, 11*(2), 227–250.

Dozier, C. L., Iwata, B. A., Thomason-Sassi, J., Worsdell, A. S., & Wilson, D. M. (2012). A comparison of two pairing procedures to establish praise as a reinforcer. *Journal of Applied Behavior Analysis, 45*(4), 721–735. https://doi.org/10.1901/jaba.2012.45-721

Fulop, L., Mark, A. (2013). Relational leadership, decision-making and the messiness of context in health care. *Leadership, 9*(2), 254–277. https://doi.org/10.1177/1742715012468785

Guerin, T. (2014). Relationships matter: The role for social-emotional learning in an interprofessional global health education. *Journal of Law, Medicine & Ethics, 42*(S2), 38–44. https://doi.org/10.1111/jlme.12186Krumhus, K. M., & Malott, R. W. (1980). The effects of modeling and immediate and delayed feedback in staff training. *Journal of Organizational Behavior Management, 2*(4), 279–293. https://doi.org/10.1300/J075v02n04_05

Maurer, R., & Hooper, N. (2011). *Feedback toolkit: 16 tools for better communication in the workplace.* Productivity Press.

Uhl-Bien, M. (2006). Relational leadership theory: Exploring the social processes of leadership and organizing. *The Leadership Quarterly, 17*(6), 654–676. https://doi.org/10.1016/j.leaqua.2066.10.007

# 5.2 People Skills for Behavior Analysis
## Putting the *Super* in Your Supervisory Relationships

*Kelly Alves*

### Introduction

The supervisor is a foundational component of the behavior analytic field. As a supervisor, you will play an instrumental role in shaping a supervisee's skill set. This chapter discusses the value of reflective practice, the importance of establishing clear boundaries, and the notion of developing a personal philosophy of supervision. Prospective supervisors are encouraged to consider adding these activities to their practice to potentially enhance the quality of supervisory relationships.

All individuals pursuing certification in applied behavior analysis (ABA) are required to obtain a specified number of supervision hours (Behavior Analyst Certification Board [BACB], 2020a). The field of ABA emphasizes a supervision model to support prospective certificants developing essential skills. Given the reliance the BACB places on supervised practice, it is highly likely that at some point as a behavior analyst, the opportunity to supervise will arise. Prior to committing to take on the role of supervisor, it is worthwhile for a behavior analyst to take time to reflect and appreciate the importance of the supervisory relationship given the amount of time novice practitioners will spend working closely with their supervisors (Helvey et al., 2022; Turner, 2017).

Supervision is a comprehensive term that encompasses teaching, coaching, mentoring, assessing, and evaluating (Corey et al., 2020). More specifically, behavior analytic supervision involves actively and systematically shaping the range of skills required of professionals who formally practice ABA (LeBlanc & Luiselli, 2016; Turner, 2017). As a supervisor, you will be tasked with teaching prospective certificants a variety of competencies that vary in complexity. Therefore, it is understandable that a supervisor is expected to have an established, thorough skill set. Some compulsory skills include competence with developing and explaining contracts, knowledge of regional policies and procedures governing ABA services and supports, refined interpersonal skills, as well as strict adherence to the ethical professional practice guidelines that inform behavior analytic practice (BACB, 2020a; Turner, 2017). Additionally, a supervisor should be proficient in implementing, evaluating, and providing feedback on all competencies listed in the most recent BACB Task List (BACB, 2017). To qualify as a supervisor, you must be an active Board Certified Behavior Analyst (BCBA) in good standing (i.e., no disciplinary actions; BACB, 2021) and have completed an eight-hour competency-based training provided by an Authorized Continuing Education (ACE) provider (BACB, 2020b). Further, behavior analysts' planning to provide supervision must include three hours of continuing education units specific to supervisory practice every recertification cycle (BACB 2020b; Turner, 2017).

The BACB is quite explicit on the procedural requirements of its supervisors (BACB, 2020a). However, there are not as many resources or direction provided regarding tasks and activities supervisors should consider that could enhance the quality of their supervisory relationships. The remainder of this chapter discusses the importance of establishing clear boundaries within supervisory relationships, the value of reflective practice, and the notion of developing a personal supervision philosophy. These practices are widely utilized in counselling, social work, and speech pathology and have the potential to be effective tools for prospective ABA supervisors (Martin & Cannon, 2010; Turner 2017).

**Creating a Safe Supervision Space**

Supervision is an intentional, formal relationship between professionals with clearly outlined roles, responsibilities, and expectations of both the supervisor and supervisee (Corey et al., 2020). Supervising individuals working toward or maintaining a BACB certification is a noteworthy responsibility associated with a prominent level of accountability. The overall quality and long-term integrity of the ABA profession depends on the knowledge and skills professionals acquire through supervision (Sellers et al., 2019). Therefore, it is imperative that the supervisory relationship provides a safe space for novice clinicians to develop the expertise needed to practice effectively as an independent professional.

As a supervisor, a first step to creating a safe supervisory space is to determine that you have time available to provide high-quality supervision (Sellers et al., 2019; Turner, 2017). A behavior analyst should realistically assess their current caseload and work tasks prior to agreeing to supervisory responsibilities. Practical considerations should also be given to logistical factors, such as location (including travel time, if applicable), and scheduling (e.g., similar time zones; Corey et al., 2020). Fostering a safe and supportive relationship with supervisees requires time. A behavior analyst needs to determine they can provide the time and resources necessary to develop a supervisee's skill set. Committing to take on supervision without adequately assessing the time required will likely result in an unsuccessful, unproductive supervision experience for supervisors, supervisees, and potential clients.

Once a behavior analyst determines that time is available to assume supervisory responsibilities, the next task is to involve establishing the framework of the supervisory relationship. This is done by creating a supervision contract as a tool to document the scope and parameters of supervision (BACB, 2020a; Helvey et al., 2022). A supervisor is encouraged to invite supervisees to be actively involved in co-creating this contractual document. Mutual collaboration on contract details may prevent supervisees from perceiving that roles and responsibilities are fixed and pre-determined. Initiating a supervisory relationship that emphasizes open discussion is a precursor to creating a safe supervisory space (Corey et al., 2020; Martin & Cannon, 2010). Results from a 2019 survey by Sellers et al. (2019) support that the vast majority of behavior analysts providing supervision include a formal contract as part of their practice. When questioned, 92 percent of respondents (n =280) reported discussing contract details with supervisees. A well-written contract should be comprehensive, including both general information describing the overall purpose and goals of supervision, as well as pragmatic details such as meeting format, schedule, contact information, and record keeping. If applicable, a supervisor should also plan to discuss compensation and payment schedule at the initial stage of supervision (BACB 2020a; Martin & Cannon, 2010).

### Developing a Supervision Philosophy

A philosophy of supervision is a personal statement that articulates a supervisor's beliefs about the intention, goals, and purposes of high-quality supervision (Martin & Cannon, 2010). Outlining a personal supervision philosophy may seem far removed from day-to-day supervisory responsibilities, but it could enhance the overall quality of the supervisory relationship. A behavior analyst supervisor may even choose to add a statement to a supervision contract about their personal philosophy. Prospective supervisors should reflect and consider the type of supervisor they aspire to be. Drawing from expertise within the American Counseling Association (Martin & Cannon, 2010), and the wisdom of experienced mental health counselors, Dr. F. A. Martin and Dr. W. C. Cannon, below are some reflective questions to help frame a personal philosophy statement.

#### *What Does a Good Supervisor Look Like to You?*

Remember that, as a behavior analyst, at one time you were also a supervisee. Thinking back to past supervisory experiences, both positive and negative, can help inform a personal philosophy statement. It is necessary to remember that a good clinician is not the same as a good supervisor. A good supervisor is a professional who is willing to model clinical skills (as opposed to only discussing them), challenge supervisees to think critically, respect the boundaries of the supervisory relationship, and recognize that client needs supersede the needs of supervisor and supervisee (Martin & Cannon, 2010).

*What Does Good Supervision Look Like to You?*

A behavior analyst has extensive leeway with respect to how to carry out supervision. It is a good learning exercise to think about whether you will aim to conduct formal or informal supervision. Do you envision supervision activities as planned or spontaneous? Answers to these questions can also be included as part of a personal supervision philosophy.

*How Do You Describe the Role of Supervisor?*

A behavior analyst supervisor is required to take on many roles. At varying times, you may serve as a therapist, teacher, mentor, evaluator, and supporter (Helvey et al., 2022; Martin & Canon, 2010; Turner 2017). Summarizing the roles and responsibilities incorporated into your supervision allows potential supervisees to know what they can expect from the outset of the supervisory relationship.

*What Are Your Expectations of a Supervisee?*

Clearly delineating any expectations you may have will allow potential supervisees to determine if priorities and values align. In your personal philosophy statement, you could list a few attributes or personal characteristics you value in others (Martin & Cannon, 2010).

Since a philosophy of supervision is personal, there is no *one size fits all* approach. As a supervisor, you may choose to keep your statement private. You might consider noting your statement in a journal to allow you to see how your values and beliefs pertaining to supervision change over time. Some supervisors may publicize their statement on a professional webpage or marketing materials. Others may present their personal statement visually through an illustration or graphic. The content and clear articulation of personal values and beliefs about supervision should be the focus of the philosophic statement, more so than the format. A personal supervision philosophy is an evolving, dynamic statement. As a supervisor, be prepared to alter your statement. Your personal philosophy of supervision will evolve as new clinical situations are encountered and new candidates are supervised.

### Seek Feedback, Reflect, and Evaluate

Supervision is an iterative process. With each supervisory experience you will refine, improve, and adapt your approach to supervising others. To support fine tuning supervisory practices, it will be necessary to allocate time to obtain feedback, reflect upon, and evaluate supervision activities.

As a behavior analyst you are keenly aware of the behavior-altering effect feedback can have on performance (Miltenberger, 2016). A supervisor should actively attempt to create a trusting and safe supervisory environment where supervisees feel comfortable providing open, honest feedback on their experiences. Supervisors need to be aware of the inherent power dynamic present in all supervisory relationships. It is recommended that, as a supervisor, you are mindful of body language and tone as strategies to mitigate any power imbalances. When a supervisee has a positive rapport with a supervisor, they can practice the important behavior analytic skill of delivering constructive feedback. This feedback is essential to the iterative supervisory process, and in turn, should strengthen the supervisory relationship.

A supervisor may consider providing a supervisee with a rating scale or checklist as a formal tool to gain objective feedback on supervisor competencies. More informal approaches to obtaining feedback can include peer observation or video review (Martin & Cannon, 2010; Turner, 2017). Supervisee feedback is valuable information for a supervisor. Data collected, either formally through assessment tools or anecdotally via discussion, can be used to inform personal learning goals or professional development opportunities (Sellers et al., 2019). Through feedback, a supervisor may uncover that they could benefit from additional training on effective communication, or supervision documentation. A supervisor could then register for supervision continuing education related to these topics.

Supervision is an integral component of the field of ABA. Supervisors are required to take on a myriad of roles as well as monitor and assess a range of clinical skills. The long-term integrity of the ABA depends on the skill set of its supervisors. While there are adequate resources available to behavior analyst supervisors outlining the procedural competencies that need to be overseen, the field of ABA needs to look to other disciplines such as counselling and speech pathology for guidance on auxiliary factors a behavior analyst should consider. In addition to demonstrating competency in procedures listed in the most recent BCBA Task List and Ethics Code for Behavior Analysts (BACB, 2017; BACB, 2020c). A behavior analyst planning to supervise practitioners working toward or maintaining certification should prioritize establishing a safe supervisory relationship, form a personal supervision philosophy and engage in active, ongoing reflective practice.

## A Supervisee's Perspective

The following comments are from a supervisee who recently received supervision from a behavior analyst. Seeking feedback and perspectives from recent supervisees can assist greatly in shaping one's practice.

Hello! My name is Rayanne Roberge, and I am a behavior therapist working in Northern Ontario. I am also in the process of completing the requirements for BCaBA certification. Navigating the processes of the BACB can seem overwhelming at times but the BACB website has a lot of helpful resources and a podcast that breaks the specific aspects of the field down. When considering qualities I look for in potential supervisors, I think it is important to ensure that whoever you choose can provide the feedback and support that you feel you need most. Going to your clinical supervisor(s) when you are unsure about something is always a good idea to ensure you don't miss anything! For me this means asking a lot of questions and being clear with my expectations about what I wanted to gain from them before signing a contract. In my experience, having multiple supervisors has been beneficial as I see different styles of teaching and learning. This is allowing me to discover what my style is and what I want to be able to provide while supervising in the future. I have learned that this field is comprised of the most dedicated and compassionate people and my biggest advice is to ask for the help and the direction you need, the main goal is to continue to grow and learn!

## References

Behavior Analyst Certification Board. (2017). *BCBA/BCaBA task list* (5th ed.). www.bacb.com/wp-content/uploads/2020/05/170113-BCBA-BCaBA-task-list-5th-ed-.pdf

Behavior Analyst Certification Board. (2020a). *Fieldwork standards.* www.bacb.com/wp-content/uploads/2022-BCBAFieldwork-Standards_200501.pdf

Behavior Analyst Certification Board. (2020b). *Supervisor training curriculum outline (2.0)* www.bacb.com/wpcontent/uploads/2020/05/Supervision_Training_Curriculum_190813.pdf

Behavior Analyst Certification Board. (2020c). *Ethics code for behavior analysts.* www.bacb.com/wp-content/uploads/2020/11/Ethics-Code-for-Behavior-Analysts-201228.pdf

Behavior Analyst Certification Board. (2021). *Consultation supervisor requirements and documentation.* www.bacb.com/wp-content/uploads/2020/11/Consultation-Supervisor-Requirements-and-Documentation_211130.pdf

Corey, G., Haynes, R., Moulton, P., & Muratori, M. (2020). *Clinical supervision in the helping professions: A practical guide* (3rd ed.). Wiley.

Helvey, C. I., Thuman, E. & Cariveau, T. (2022). Recommended practices for individual supervision: Considerations for the behavior-analytic trainee. *Behavior Analysis in Practice, 15*(1), 370–381. https://doi.org/10.1007/s40617-021-00557-9

LeBlanc, L. A., & Luiselli, J. (2016). Refining supervision practices in the field of behavior analysis: Introduction to the special section on supervision. *Behavior Analysis in Practice, 9*(4), 271–273. https://doi.org/10.1007/s40617-016-0156-6

Martin, F. A., & Cannon, W. C. (2010). *The necessity of a philosophy of clinical supervision.* Counseling Outfitters. www.counseling.org/resources/library/vistas/2010-v-online/Article_45.pdf

Miltenberger, R. G. (2016). *Behavior modification: Principles and procedures.* (6th edn.). Cengage Learning.

Sellers, T. P., Valentino, A. L., Landon, T. J., & Aiello, S. (2019). Board Certified Behavior Analysts' supervisory practices of trainees: Survey results and recommendations. *Behavior Analysis in Practice, 12*(3), 536–546. https://doi.org/10.1007/s40617-019-00367-0

Turner, L. B. (2017). Behavior analytic supervision. In J. K. Luiselli (Ed.), *Applied behavior analysis advanced guidebook: A manual for professional practice* (pp. 1–10). Elsevier Academic Press. http://dx.doi.org/10.1016/B978-0-12-811122-2.00001-2

# 5.3 Giving Performance Feedback that Makes a Difference

*Michael G. Palmer*

## Introduction

Performance feedback from supervisors is one of those topics that is so ubiquitous that if you were to ask anyone inside or outside of your field whether they know what it is they would likely say, "of course." However, if you were to ask those individuals to define what it is, what parts of it make it more or less successful, or how to best deliver it, you would likely get different answers for each of those questions. Indeed, feedback is studied from various perspectives, across several fields, including behavior analysis, industrial/organizational psychology, and business. What will be presented here is an attempt to bring together research from various fields to help guide practitioners to use feedback to its maximum potential.

Despite feedback being studied extensively within the field of behavior analysis, and a consensus that it is generally effective, there is not a consensus on the *definition* of feedback (Mangiapanello & Hemmes, 2015; Sleiman et al. 2020). Because of this, Mangiapanello and Hemmes (2015) created an operational definition, that seems to capture most instances of feedback. In essence, the definition stated that feedback is the presentation of a stimulus to a person, and the characteristics of that stimulus changes based on the person's previous responding (Mangiapanello & Hemmes, 2015). We also know that, most of the time, doing this results in a change in behavior in the desired direction; however, there isn't a consistent direction of behavior change, which is why a functional definition is difficult to use (Sleiman et al., 2020). While not perfect, the operational definition offered by Mangiapanello and Hemmes (2015) is a great starting point in our discussion about using feedback.

Those in supervisory roles will need to know how to use feedback as a method of behavior change. It is likely because it has been shown to be so effective in changing behavior in organizational settings (Sleiman et al., 2020) that it is included in both the Behavior Analyst Certification Board (BACB) 5th edition Task List and BACB Ethics Code for Behavior

DOI: 10.4324/9781003300465-27

Analysts. Thus, those entering the field will need to learn about, and will be tested on, the use of feedback (see Task List number I-5; BACB, 2017). Additionally, it is expected that those in supervisory roles monitor performance and provide feedback to those they are supervising to improve performance (see Ethics Standard 4.08; BACB, 2020).

Last, before getting to recommendations to make your feedback process better, it is important to acknowledge that while this topic has been studied extensively with behavior analysis, there are large components of the performance feedback process that have yet to be explored at all, or in-depth, within the field. Therefore, what will be presented will be an amalgamation of recommendations, not only from within behavior analysis, but also from other fields, mainly industrial/organizational psychology. What follows is a series of questions someone may pose if they want to learn more about how to improve their use of feedback along with an answer to that question based on research within behavior analysis and, when appropriate, accompanied by research from other fields.

### Who Should Deliver the Feedback?

Feedback can come from multiple sources. For instance, when using *360° feedback* (see Garavan et al., 1997 for a review of *360° feedback*), sources include those the individual reports to, peers, and those who report to that individual. In their review, Sleiman and colleagues (2020) reported supervisory, researcher, self-generated, mechanical, and expert sources. All sources, except for expert-provided feedback, were found to produce large to very large effect sizes (Sleiman et al., 2020). Unfortunately, none of the research reviewed by Sleiman and colleagues (2020) included peer- or subordinate-delivered feedback. Additionally, this tells us very little about the source of the feedback besides their position with regard to the individual receiving the feedback.

An important factor when it comes to who will be providing the feedback is the quality of the relationship between the person providing the feedback and the person receiving the feedback (Gregory & Levy, 2015). This can impact whether the person receiving the feedback will be more or less likely to seek out feedback again in the future (Gregory & Levy, 2015). In a lab study, Williams and colleagues (1999) manipulated how supportive a manager was and measured the percentage of participants who sought feedback. They found those with a supportive manager (or one who shows empathy, is considerate of employee perspectives, builds positive relationships with employees, etc.) sought feedback more often than those without a supportive manager (Williams et al., 1999). Survey research on actual employee-leader dyads corroborated this research, finding that when employees had an unsupportive leader, they actively avoided feedback from that leader (Moss et al., 2009). In general, the more supportive is the person providing the feedback, the more willing the person receiving the feedback will be to listen. If an employee or supervisee is unwilling to listen to the feedback or actively avoids the feedback, it is as if the feedback is not being delivered at all, which could result in no improvement in performance. From a behavior analytic perspective, this makes conceptual sense because supportive supervisors who provide effective feedback help employees and supervisees obtain positive consequences.

### How Should the Feedback Be Delivered?

Feedback can be delivered in multiple ways. In many review articles, this is referred to as "the medium" (Alvero et al., 2001; Sleiman et al., 2020). It is likely no surprise to readers that the most common way behavior analysts like to provide feedback information is in the form of a graph (Sleiman et al., 2020). Other ways of providing feedback include simply telling the person the information, using some form of written communication, or a form of automated feedback (such as when software tracks the number of supervision hours per month and reports it automatically) or self-generated feedback (such as when using self-monitoring; Sleiman et al., 2020). These can also be combined. Sleiman and colleagues found that a combination of graphical display, verbal explanation, and written description produced the largest changes in behavior (2020).

With advances in technology, it can be tempting to turn to it to provide feedback faster. For instance, email and multiple forms of instant messaging allow for immediate, short feedback to be delivered, sometimes at the moment to allow someone to immediately change their behavior. Gregory and Levy (2015) urge caution when using these forms of technology due to the possibility of misinterpretation of messages and quick replies that are more harmful than helpful. An alternative to these mediums could be the use of bug-in-the-ear technologies, which have been identified as empirically supported interventions for teaching practitioners (Schaefer & Ottley, 2018).

While face-to-face feedback has been identified as generally more preferred over other mediums, there is one important thing to consider when deciding whether to provide face-to-face or technology-mediated feedback: whether the feedback will possibly be negative. Matey and colleagues (2019) found that when required to deliver negative feedback following observations, participants were more likely to record inaccurate observations. These researchers hypothesized this occurred because participants wanted to avoid delivering negative feedback (Matey et al., 2019). Gregory and Levy (2015) suggested the use of structured feedback, such as the situation-behavior-impact model, where the feedback describes the context, the behavior, and why the behavior matters. This can help people from talking too much around the issue, which could leave the person receiving the feedback unclear about the situation (Gregory & Levy, 2015). Supervisors may also be tempted to use popular management traditions of pairing negative feedback with positive feedback, such as using the feedback sandwich, to help ease the impact of the negative feedback. Based on the research, it is not recommended that supervisors use these traditions (Henley & DiGennaro-Reed, 2015) and instead stick to using consistent feedback sequencing, such as positive-positive, negative-negative, or deliver one type of feedback at a time (Choi et al., 2018).

**What Information Should Feedback Convey?**

Feedback can convey various pieces of information. The most common way to break down this aspect of feedback is to look at whether it is positive or negative. Positive feedback is indicative of behavior that is improving or has met a performance standard. Negative feedback is indicative of behavior that is not improving or is not at a performance standard. Sleiman and colleagues' (2020) review indicated that positive feedback produced larger effect sizes than negative feedback; however, both were found to improve performance.

Feedback may contain a subjective evaluation component (e.g. "That was a great try") or an objective component (e.g. "You completed 7 out of 10 steps correct."). Johnson (2013) found that a combination of evaluative and objective components produced the largest changes in behavior, though each presented separately also improve behavior over no feedback.

Feedback may also focus on the process (the specific behaviors involved) or an outcome (the result of the behaviors). Gregory and Levy (2015) suggest using process feedback as much as possible as it provides information to the performer on exactly what to change if it needs changing. Additionally, a focus on outcomes could lead to unwanted behaviors by employees or supervisees to get to specified outcomes.

Notice that under all of these circumstances, the feedback focuses on the behaviors involved (behavior-focused) and not on the individual's characteristics (person-focused). This is because behavior-focused feedback has consistently been shown to be more effective. After all, those receiving it are better able to change their behavior than change aspects about themselves (Gregory & Levy, 2015). This is particularly true if you need to present negative feedback (Gregory & Levy, 2015).

### What Should I Know about the Person Receiving the Feedback?

The person receiving the performance feedback is not an empty vessel into which you place your feedback. They will respond to it in different ways, depending on several individual factors. In behavior analysis, we may refer to these as histories of reinforcement or learning. In other fields, these are referred to as individual differences (Gregory & Levy, 2015). One of the most important individual differences you could learn about your employee or supervisee is their feedback orientation. This refers to how receptive the person will be to the feedback, how comfortable they are with receiving feedback, whether they will seek out feedback, and whether they will change their behavior due to the feedback received (Gregory & Levy, 2015).

Luckily you don't have to guess at what someone's feedback orientation is; Linderbaum and Levy (2010) created the Feedback Orientation Scale (FOS) to measure how much an individual finds utility in feedback, how accountable they feel to do something with feedback, how much they understand that others will perceive them, and how much they believe they can do something with the feedback. Someone who does not find utility in feedback, does not feel accountable to change, is not aware of how others perceive them, and does not believe they can do something with the feedback and will need to be approached with feedback very differently than someone who is the opposite. This may indicate they have a history of performance feedback being delivered only when performance is low (i.e., experience with negative feedback) or with poorly delivered feedback.

Fortunately, feedback orientation can be developed or change over time (Gregory & Levy, 2015). Readers are encouraged to review Ehrlich and colleagues (2020) as well as Walker and Sellers (2021) for guidance on how to improve feedback reception skills using behavior change tactics. Due to the behaviors targeted, the type of intervention used by Ehrlich and colleagues (2020) and Walker and Sellers (2021) would likely improve feedback orientation.

### How Often Should Feedback Be Delivered?

Feedback can be delivered at various frequencies. Sleiman and colleagues (2020) reported finding frequencies of daily, weekly, bi-weekly, monthly, quarterly, once, after every observation, and various combinations of all of the previously mentioned. Sleiman and colleagues (2020) found that daily and weekly, weekly and after each observation, and after each observation produced the largest changes in behavior. Generally speaking, the more frequent the better, but this must be balanced with other factors such as how the feedback is being presented and what information is being presented (Gregory and Levy, 2015). For example, if the feedback is being delivered via a graph by a supervisor and includes an evaluative component, it may not be feasible for the supervisor to do this after every observation or even daily. Additionally, if the feedback is presenting information about the count of behaviors across a week, then a weekly update may suffice rather than a daily update of the previous seven days. It may be worthwhile to try out different frequencies to find the ideal one for your employee, while balancing supervisor responsibilities.

**When Should Feedback Be Delivered?**

Feedback can be delivered at various times with respect to the behavior. For instance, it can be delivered right after the behavior occurs, right before the next behavior occurs, or sometime in between. This is somewhat tied to the frequency of feedback delivery. Sleiman and colleagues (2020) found that most feedback is not delivered within 60 seconds of the behavior being observed but found that this was slightly more effective at changing behavior. Research by Aljadeff-Abergel et al. (2017) suggested that providing feedback right before the individual had an opportunity to engage in the behavior again resulted in better performance than performance provided right after the individual engaged in the behavior. However, other research has not corroborated this finding (Wine et al., 2019). It is likely that as long as it is frequent and relevant to the employee or supervisee, it will be helpful for behavior change.

**What Other Contextual Factors Should I Be Considering?**

Other things to consider include the privacy of the feedback, the culture of the organization as it pertains to feedback, and whether the feedback is going to be used with other behavior change tactics. Feedback can be delivered privately, publicly, or as a combination of both. Sleiman and colleagues (2020) found that both private and public forms of feedback produced similar changes in behavior; however, Gregory and Levy (2015) caution against the use of public feedback as people have been found to report feelings of nervousness when provided feedback in public. This will likely depend on the nature of the feedback; for instance, public feedback about group data may not evoke those feelings of nervousness. However, an additional consideration is that, while it is used frequently and found to change behavior, most behavior change tactics in organizational behavior management do not have social validity data collected on them (Nastasi et al., 2021).

Another contextual factor is that the delivery of feedback should be part of the organizational or supervisory relationship culture. In other words, feedback should not be provided only when things need to be improved. If this is done, it is likely pairing feedback with negative consequences and in the long term may cause feedback avoidance behaviors (Gregory & Levy, 2015). Ideally, the organization should value learning and provide opportunities for employees and supervisees to build mastery in a space where feedback is provided on behaviors in a non-judgmental way, consistently (Gregory & Levy, 2015). This can be done by providing a favorable feedback environment where high-quality feedback is provided by a trusted source, in an effective delivery method, who can provide both positive and negative feedback when needed, and is readily available, and where feedback-seeking is valued (Gregory and Levy, 2015).

The last contextual factor to consider is that feedback can be delivered alone or in combination with other behavior change tactics. For instance, feedback can be used with antecedents, goal setting, behavioral consequences, or various combinations of these (Sleiman et al., 2020). Sleiman and colleagues (2020) found that feedback along with antecedents and behavioral consequences produced the largest changes in behavior. However, feedback paired with goal setting, and feedback paired with goal setting and behavioral consequences produced similar changes in behavior. That being said, feedback when used alone also produced large changes in behavior (Sleiman et al., 2020).

### Putting It All Together

A summary of all the features of feedback that help make a difference can be found in Figure 5.1. When used together, Gregory and Levy (2015) suggested that these lead to a favorable feedback environment. In turn, a favorable feedback environment can lead to better job satisfaction, more instances of employees helping each other, higher performance, better

| Features of Feedback That Makes a Difference |
|---|
| ☐ Provided by an individual that has a good relationship with the employee or supervisee |
| ☐ Provided face-to-face using graphical, written, and verbal explanation of the feedback, when possible |
| ☐ Negative feedback is presented in isolation from positive feedback (i.e., no feedback sandwich) |
| ☐ Provides a subjective and objective evaluation of the process behaviours |
| ☐ Provided to an individual who is trained on how to use the feedback |
| ☐ Provided frequently, ideally around times that are relevant to the employee and the behaviour the feedback pertains to |
| ☐ Provided in a private space |
| ☐ Provided in an organizational culture that values learning and feedback |
| ☐ Provided with other antecedent or consequential interventions |

*Figure 5.1* Features of feedback that makes a difference.

organizational relationships, and lower intentions of turnover (Gregory & Levy, 2015). An intervention so simple, so effective, and that typically costs little to nothing to implement can lead to ripples of positive effects throughout an organization.

**References**

Aljadeff-Abergel, E., Peterson, S. M., Wiskerchen, R. R., Hagen, K. K., & Cole, M. L. (2017). Evaluating the temporal location of feedback: Providing feedback following performance vs. prior to performance. *Journal of Organizational Behavior Management, 37*(2), 171–195. https://doi.org/10.1080/01608 061.2017.1309332

Alvero, A. M., Bucklin, B. R., & Austin, J. (2001). An objective review of the effectiveness and essential characteristics of performance feedback in organizational settings (1985–1998). *Journal of Organizational Behavior Management, 21*, 3–29. http://dx.doi.org/10.1300/J075v21n01_02

Behavior Analyst Certification Board. (2017). *BCBA task list* (5th ed.). Littleton, CO.

Behavior Analyst Certification Board. (2020). *Ethics code for behavior analysts.* Littleton, CO.

Choi, E., Johnson, D. E., Moon, K., Oah, S. (2018). Effects of positive and negative feedback sequence on work performance and emotional responses. *Journal of Organizational Behavior Management, 38*(2–3), 97–115. https://doi.org/10.1080/01608061.2017.1423151

Ehrlich, R. J., Nosik, M. R., Carr, J. E., & Wine, B. (2020). Teaching employees how to receive feedback: A preliminary investigation. *Journal of Organizational Behavior Management, 40*(1–2), 19–29. https://doi.org/10.1080/01608 061.2020.1746470

Garavan, T. N., Morley, M., & Flynn, M. (1997). 360 degree feedback: Its role in employee development. *Journal of Management Development, 16*(2), 134–147. https://doi.org/10.1108/02621719710164300

Gregory, J. B. & Levy, P. E. (2015). *Using feedback in organizational consulting.* American Psychological Association. http://dx.doi.org/10.1037/14619-000

Henley, A. J. & DiGennaro-Reed, F. D. (2015). Should you order the feedback sandwich? Efficacy of feedback sequence and timing. *Journal of Organizational Behavior Management, 35*(3–4), 321–335. https://doi.org/10.1080/01608 061.2015.1093057

Johnson, D. A. (2013). A component analysis of the impact of evaluative and objective feedback on performance. *Journal of Organizational Behavior Management, 33*(2), 89–103. https://doi.org/10.1080/01608061.2013.785879

Linderbaum, B. G., & Levy, P. E. (2010). The development and validation of the Feedback Orientation Scale (FOS). *Journal of Management, 36*, 1372–1405. http://dx.doi.org/10.1177/0149206310373145

Mangiapanello, K. A. & Hemmes, N. S. (2015). An analysis of feedback from a behavior analytic perspective. *The Behavior Analyst, 38*, 51–75. https://doi.org/10.1007/s40614-014-0026-x

Matey, N., Gravina, N. Rajagopal, S., & Betz, A. (2019). Effects of feedback delivery requirements on accuracy of observations. *Journal of Organizational Behavior Management, 39*(3–4), 247–256. https://doi.org/10.1080/01608 061.2019.1666773

Moss, S. E., Sanchez, J. I., Brumbaugh, A. M., & Borkowski, N. (2009). The mediating role of feedback avoidance behavior in the LMX-performance relationship. *Group & Organization Management, 34*, 645–664. http://dx.doi.org/10.1177/1059601109350986

Nastasi, J., Simmons, D., & Gravina, N. (2021). Has OBM found its heart? An assessment of procedural acceptability trends in the Journal of Organizational Behavior Management. *Journal of Organizational Behavior Management, 41*(1), 64–82. https://doi.org/10.1080/01608061.2020.1853000

Schaefer, J. M. & Ottley, J. R. (2018). Evaluating immediate feedback via bug-in-ear as an evidence-based practice for professional development. *Journal of Special Education Technology, 33*(4). https://doi.org/10.1177/0162643418766870

Sleiman, A. A., Sigurjonsdottir, S., Elnes, A., Gage, N. A., & Gravina, N. E. (2020). A quantitative review of performance feedback in organizational settings (1998–2018). *Journal of Organizational Behavior Management, 40*(3–4), 303–332. https://doi.org/10.1080/01608061.2020.1823300

Walker, S. & Sellers, T. (2021). Teaching appropriate feedback reception skills using computer-based instruction: A systematic replication. *Journal of Organizational Behavior Management, 41*(3), 236–254. https://doi.org/10.1080/01608 061.2021.1903647

Williams, J. R., Miller, C., Steelman, L. A., & Levy, P. E. (1999). Increasing feedback seeking in public contexts: It takes two (or more) to tango. *Journal of Applied Psychology, 84*, 969–976. http://dx.doi.org/10.1037/0021-9010.84.6.969

Wine, B., Lewis, K., Newcomb, E. T., Gambin, J. G., Chen, T. Liesfeld, J. E., Matthews, K. M., Morgan, C. A., & Newcomb, B. B. (2019). The effects of temporal placement of feedback on performance with and without goals. *Journal of Organizational Behavior Management, 39*(3–4), 308–316. https://doi.org/10.1080/01608061.2019.1632244

## 5.4 Interviewing Strategies

Integrating Practices from Industrial/
Organizational Psychology
into Behavior Analysis
for Successful Hiring

*Michael G. Palmer*

Introduction

Howard Schulz, CEO of Starbucks, stated, "Hiring people is an art, not a science, and resumes can't tell you whether someone will fit into a company's culture." Those pursuing certification at the master's or doctoral level spend years completing coursework and accruing experience hours in preparation for a certification exam. This is often done in the hopes of becoming a supervisor: completing assessments, making clinical decisions about clients' programs, and overseeing work done by front-line clinicians. However, what often comes with this new supervisory role in an organization are other managerial tasks. Some of these tasks the new supervisor has been trained to do, such as track employee performance and deliver performance feedback. Other tasks, such as employee selection, they may have never had the training to complete. What follows is a summary of various ways someone can use the interviewing component of hiring to select a new employee.

    Before getting to this, however, I must throw out one cautionary note: if you are going to be involved in the hiring process and are going to deviate from a traditional interview in any way, it is recommended that you work closely with someone familiar with employment laws within the jurisdiction where you are located. This could be a human resource professional, a lawyer with specialization in labor and employment law, or an industrial/organizational (I/O) psychologist who specializes in employee recruitment and selection. Because this is a scope of practice on its own, those trained in applied behavior analysis usually do not have this skill within their scope of practice. Practicing this without the help of someone who has competence in this area could lead to accusations of improper hiring practices, bias, or discrimination. Readers are encouraged to review chapter 13 of Dipboye (2018) or chapter 20 of Goldstein et al. (2017) for a summary of related

DOI: 10.4324/9781003300465-28

issues. Therefore, it is best to consult with someone who understands the legal requirements for employee selection to ensure whatever hiring practices you are using are legally and professionally sound.

A new behavior analyst supervisor may attempt to turn to the field's journals or textbooks to learn more about the employee hiring and selection process. However, this search often turns up little to no helpful guidance about how to go about doing this (Biagi, 2020). Instead, what will be found are articles about how to teach interview skills (Barker et al., 2019; Edgemon et al., 2020) that, while being helpful in other contexts, are not helpful in this situation as they focus on the behavior of the interviewee rather than the interviewer. Additionally, some of the behaviors they focus on are fairly superficial, such as smiling, eye contact, and posture and, as we will discuss later, may not predict whether someone will be a good fit for the position. Because there is generally a lack of research regarding employee selection in behavior analysis, we must look to related fields for guidance. What will be presented is a summary of findings from I/O psychology and human resource management to help provide direction to supervisors in behavior analysis. When possible, these findings will be translated to behavioral terms to be more conceptually systematic, and examples of how to use them in a behavioral organization will be provided.

**What's the Function?**

Let's first start this discussion by exploring the purpose(s) behind hiring interviews. Contrary to Howard Shultz's quote at the opening, employee selection can be a science, and we can turn to the science to help us understand more about the employee selection process. The main purposes of an interview in the hiring process are to predict the candidate's abilities to do the job and determine whether the candidate will fit in with the organization's culture (Goldstein et al., 2017). I would be remiss to note that interviewing is but one strategy used to predict the candidate's abilities to do the job; for instance, cognitive tests and personality tests can also be used in the selection process to make predictions (Dipboye, 2018). However, these strategies are beyond the scope of this discussion. Moreover, some argue that if an interview is set up correctly, and with the right questions, they can provide similar pictures of the applicant in terms of their personality, motivations, person-to-job fit, willingness to help, and so on (Goldstein et al., 2017). Additionally, with the right setup, interviews can be just as predictive as other hiring strategies (Dipboye, 2018). The interviewing strategies that will be reviewed below include unstructured interviews, structured interviews, and behavioral interviews.

*Unstructured Interviews*

Contrary to their label, most unstructured interviews are not completed unstructured (i.e., a situation where the interviewer makes up questions as they go). What characterizes these interviews is the lack of script while asking questions (Goldstein et al., 2017). Therefore, you may have a series of questions selected beforehand, but following an unstructured interview format allows you to ask the questions in any order and ask follow-up or supplementary questions (Goldstein et al., 2017). Many small businesses rely solely on unstructured interviews; I remember my first job interview being an unstructured one, and when I became a manager in that same restaurant, I used unstructured interviews. However, unstructured interviews are subject to many biases, such as primacy effects or similar-to-me effects, among others (Dipboye, 2018).

That being said, unstructured interviews may have a place in the selection process. Some I/O psychologists have found that personality judgments based on unstructured interviews can be just as accurate as structured personality measures, even when they don't contain personality-related questions (Blackman, 2002; Goldstein et al., 2017). If you are uncomfortable with the use of the word *personality*, you can translate this to *consistent patterns of covert and overt behavior*. Unstructured interviews can be conducted over beverages or snacks, while providing a tour of the organization, or when moving from one space to another. The key for conducting unstructured interviews to glean information about an individual's personality is to get the applicant comfortable with the interviewer such that they become more relaxed, show more of their nonverbal behavior, and engage in more natural conversation (Goldstein et al., 2017). When this occurs, an interviewer can better predict how much someone wants to do their job well (conscientiousness); how likely they will develop and maintain prosocial relationships (agreeableness); whether they are reliable (dependability); whether they will engage in behaviors above and beyond what is required (organizational citizenship); and whether the person will fit the job (Goldstein et al., 2017). Taken together, unstructured interviews may help determine whether someone will fit into the organization's culture (i.e., will you get along with them?). However, unstructured interviews do a poor job at predicting whether someone will do the job well (Dipboye, 2018; see Table 5.2).

*Structured Interviews*

Unlike unstructured interviews, structured interviews are made up of carefully selected questions that are asked using a specific script and are scored against specific criteria (Goldstein et al., 2017). In I/O psychology and

*Table 5.2* Examples of Unstructured Interview Questions

*Examples of Unstructured Interview Questions*

- What got you interested in the field?
- What would your ideal boss be like?
- Where do you see yourself in 5 years?
- What are your thoughts about (recently published journal or newspaper article)?

human resources management, these are sometimes referred to as behavioral interviews (Dipboye, 2018); however, I am reserving that label for interviews that may align more with the field of behavior analysis. The questions that are used are selected based on a job analysis of the position you are hiring for; in other words, the questions are designed specifically to evoke responses related to skills that the position requires to be successful (Goldstein et al., 2017). Every applicant is asked the same questions, in the same order, with little to no follow-up questioning (Goldstein et al., 2017). Because of how they are designed, these questions do a superior job of being able to predict which candidate will be better at the job than unstructured interviews (Dipboye, 2018).

Questions in a structured interview should be based on a job analysis (see chapter 10 of Dipboye, 2018 for a thorough description of job analysis procedures); however, in the absence of a job analysis, the Society of Human Resource Management (SHRM) suggests creating questions centered around job competencies that will get the applicant to talk about how they demonstrate the competencies through their behaviors (2016). SHRM recommends using the situation, task, action, results (STAR) model to create questions (2016). Applicants are given a situational frame ("Tell me about a time…" or "What would you do if…") along with a job-related competency ("Where a client presented with a behavior that…"). Then, applicants are asked what they did or would do to complete the task ("What did you do and…" or "What would you do and…") and what the results were or what they think they would be by taking those actions ("How did it turn out?" or "What do you think could happen next?").

Once questions are created, a scoring system must also be created to ensure consistency across applicants (Diboye, 2018). SHRM (2016) outlines two methods for scoring structured interview questions: Likert-type rating scales that require interviewers to rate the applicants' answers from "below requirements" to "far exceeding requirements" or behaviorally anchored rating scales (BARS), which align specific behaviors that would suggest they would do the job well to the applicant's answers

(SHRM, 2016). For example, in a question asking about when a client presented with a behavior that might suggest something was physically or medically wrong, a superior answer may be that the applicant indicated to the client's family that they would feel more comfortable moving forward with behavioral services after the client has visited with their primary care physician to rule out any medical causes, and an unsatisfactory response may be one in which the applicant ignores this detail and suggests moving forward with behavioral services.

Dipboye (2018) and Goldstein et al. (2017) suggest that at least two people be involved during structured interviews so that both may rate the responses and an average be obtained. This will help reduce some forms of bias, such as attractiveness, age, sex, race, disability status, or sexual orientation (Dipboye, 2018; Goldstein et al., 2017). Additionally, Goldstein and colleagues recommend conducting the structured interview first, followed by the unstructured interview (2017). Research from I/O psychology has demonstrated that structured interviews are predictive of on-the-job skills (Dipboye, 2018); however, from a behavioral perspective, they seem to tell us more about the applicant's verbal behavior than actual on-the-job skills (Table 5.3).

*(More) Behavioral Interviews*

While the traditional interview has historically been based on the verbal exchanges of individuals, there are other selection strategies that can be brought into the interview space (Goldstein et al., 2017). What would make an interview more behavioral (i.e., consistent with behavior analysis) would be to incorporate overt applicant behavior into the interview process. In I/O psychology, these types of activities are referred to as simulation exercises and are one of the best predictors of job performance (Goldstein et al., 2017).

Based on the research about structured interviews and guidance from SHRM, Biagi (2020) suggested how behavior analysts could incorporate simulation exercises into interviews. As with structured interviews, one must start with a job analysis or with a list of critical behaviors that would make an individual successful in that position (Biagi, 2020). Next, identify behaviors that can be observed quickly in an interview setting. You will not be able to observe all job-related behaviors, thus you will need to choose those that are most representative of that position. This is where you can get creative about the different ways you could see the applicant engaging in that skill. For example, if you are hiring for a front-line technician, perhaps you are interested in whether they will seek out clarification. You could indicate that you are going to engage in role-play and instruct them to implement a specific procedure with you as the client. This would allow you to assess whether they seek out clarification and other skills as

*Table 5.3* Examples of Structured Interview Questions

*Examples of Structured Interview Questions*

- Think of a time when you were new to an organization, or maybe you will be new to this organization. What steps did you, or will you, take to establish yourself as credible and trustworthy to your clients and coworkers?
- Imagine you are working with a client and their parents become very angry with you over the procedures you are implementing. What would you do and what would your next steps be?
- Tell me about a time when you needed to keep information confidential. What actions did you take and how did it turn out?

*Table 5.4* Examples of Behavioral Interview Scenarios

*Examples of Behavioral Interview Scenarios*

- Role-play with the applicant.
- Provide a hypothetical graph of data and ask the applicant to decide what to do next.
- Create a hypothetical behavior plan and ask the applicant to identify errors or suggest ways to improve the behavior plan.

well, such as how well they implement that procedure or other inter-/intra-personal. Or let's say you are hiring for an assistant behavior analyst and you want to see how well they deliver performance feedback. You could create a video of a front-line technician implementing a routine procedure with a few intentional errors in it. This could test whether they identified the errors, whether they felt comfortable making suggestions for the front-line technician to improve, and how well they deliver feedback.

As with structured interviews, this form of behavioral interviewing should be evaluated using some form of standardized evaluation measure (Goldstein et al., 2017). To reduce bias, every candidate should be asked to complete the same activities in the same order and be evaluated against the same standards. You can create an evaluation scoring system in the same way as outlined by SHRM for structured interviews. You can also pilot the scoring system using performers who are already on the job (see Table 5.4).

## Putting It All Together

While there may not be research within the field of behavior analysis to help guide supervisors through the hiring process, there is ample research and guidance from I/O psychology and human resource management

Table 5.5 Various Interview Formats and Their Usage

| Interview Format | Description | Predictive Criteria | Behavioral Translation of Predictive Criteria | Optimal Use |
|---|---|---|---|---|
| Unstructured Interview | Small-talk, unstructured questions, follow-up questions, answers not rated by interviewer(s) | Personality characteristics, integrity, organizational citizenship, dependability | Consistent patterns of day-to-day work and social behaviors | Assessing whether the applicant will be a good fit in the organization |
| Structured Interview | Standardized questions, no follow-up questions, answers rated by interviewer(s) | Skill sets, job performance | Verbal repertoire | Indirect assessment of a wide range of skills, comparing between applicants |
| Behavioral Interview | Standardized scenarios or simulations, no follow-up questions, behavior rated by interviewer(s) | Skill sets, job performance | Overt repertoire | Direct assessment of specific skills, comparing between applicants |

(Dipboye, 2018; Goldstein et al., 2017; SHRM, 2016). Until this is research from within the field of behavior analysis, supervisors will need to learn to translate approaches from other fields to fit into our conceptual framework. An example of this is in Table 5.5. Each of the interviewing methods reviewed provides a snapshot of slightly different behavioral repertoires. Unstructured interviews provide insight into what it would be like to interact with that person on a day-to-day basis, and whether they would be a good fit for the organization. Structured interviews provide insight into the applicant's verbal repertoire and can be thought of as equivalent to an indirect assessment of the applicant's overt repertoire. Behavioral interviews provide insight about specific skills in the applicant's overt repertoire and can be thought of as equivalent to a direct assessment of an applicant's skills. Information from each type of interview should be taken into consideration together and compared against all other applicants when deciding whether someone should be offered the position.

## References

Barker, L. K., Moore, J. W., Olmi, J., & Rowsey, K. (2019). A comparison of immediate and post-session feedback with behavioral skills training to improve interview skills in college students. *Journal of Organizational Behavior Management, 39*(3–4), 145–163. https://doi.org/10.1080/01608061.2019.1632240

Biagi, S. (2020, May 23). *Behavioral interviewing: Strategies for successful hiring in ABA organizations.* [Paper session]. Association for Behavior Analysis International Annual Conference, Online.

Blackman, M. C. (2002). Personality judgment and the utility of the unstructured employment interview. *Basic and Applied Social Psychology, 24*(3), 240–249. https://doi.org/10.1207/S15324834BASP2403_6

Dipboye, R. L. (2018). *The emerald review of industrial and organizational psychology.* Emerald Publishing. https://doi.org/10.1108/9781787437852

Edgemon, A. K., Rapp. J. T., Brogan, K. M., Richling, S. M., Hamrick, S. A., Peters, R. J., & O'Rourke, S. A. (2020). Behavioral skills training to increase interview skills of adolescent males in a juvenile residential treatment facility. *Journal of Applied Behavior Analysis, 53*(4), 2303–2318. https://doi.org/10.1002/jaba.707

Goldstein, H. W., Pulakos, E. D., Semedo, C., & Passmore, J. (2017). *The Wiley Blackwell handbook of the psychology of recruitment, selection and employee retention.* John Wiley.

Society for Human Resource Management (2016). *A guide to conducting behavioral interviews with early career job candidates.* www.shrm.org/LearningAndCareer/learning/Documents/Behavioral%20Interviewing%20Guide%20for%20Early%20Career%20Candidates.pdf

# 6
# Innovative, Creative Ways to Use ABA

# 6.1 Making Applied Behavior Analysis Accessible to Consumers

*Nicole Neil and Anastasia Klimova*

### Legislation and Standards Related to Accessibility

Legislative requirements and standards provide people with disabilities legal protection of their rights and seek to decrease or eliminate barriers of access, while protecting both public and private sectors against undue burden or hardship. Jurisdictions take different approaches to accessibility. At the global level, the United Nations Convention on the Rights of Persons with Disabilities (UNCRPD, 2006) establishes the rights to education and to the highest attainable standard of health without discrimination based on disability (Articles 45 and 25). In Article 26, this information is expanded, requiring state parties to "take effective and appropriate measures, including through peer support, to enable persons with disabilities to attain and maintain maximum independence, full physical, mental, social, and vocational ability, and full inclusion and participation in all aspects of life" (UNCRPD, 2006, para. 1).

In the United States, the Americans with Disabilities Act (ADA, 1990) seeks to provide equal access to opportunities for people with disabilities in all aspects of society, including social, vocational, occupational, economic, and educational. It ensures legal protections for people with disabilities and legal obligations for public and private service sectors. In Canada, the Accessible Canada Act (ACA, 2019) seeks to identify, remove, and prevent barriers to access that are experienced by people with disabilities, and the Accessibility for Ontarians with Disabilities Act (AODA, 2005) is a provincial parallel of the ACA. With the AODA, Ontario became one of the first places in the world to establish a law with a goal and timeframe for accessibility. Standards under the act are laws that businesses and organizations must follow to identify, remove, and prevent barriers. AODA also sets out core principles (independence, dignity, integration, and equality of opportunity for people with disabilities) that should be upheld as organizations seek to meet the requirements. Other disability rights references include the Equality Act (Equality Act, 2010) in the

DOI: 10.4324/9781003300465-30

United Kingdom, the European Social Charter (Revised) in the European Union, and the Disability Discrimination Act, 1992, in Australia, to name only a few.

The Behavior Analyst Certification Board (BACB) Ethics Code for Behavior Analysts (BACB, 2020) outlines the ethical standards for practicing behavior analysts. However, to date, there are no explicit references to accessibility in the Code. The Code requires behavior analysts to use understandable language and to ensure comprehension of all communications with stakeholders. Further, behavior analysts are required to address the environmental conditions that may interfere with or prevent the delivery of ABA services. Despite an overall global commitment to accessibility, expressed in the UNCRPD, as well as national, state, provincial, and territorial level standards governing accessibility, challenges remain, and concrete actions to improve access to applied behavior analysis are needed.

**Barriers to Accessibility in ABA**

In this section, we provide an overview of factors influencing the accessibility of ABA services. Given local variation, it is important to develop community-based models in which community stakeholders shape the identification of relevant factors that influence access in their community.

*Individual and Caregiver Factors*

Child and parent factors, including age, diagnosis, severity of symptoms, parent education level, beliefs about diagnosis, ethnicity and income are linked to ABA service use for people with disabilities, including those diagnosed with autism (Gibson & O'Connor, 2010; Wilson et al., 2018). Intersections among these personal factors have the potential to delay or prevent access to ABA. For example, an autism diagnosis is often a key factor in access to ABA services. Most ABA practitioners report providing services only to autistic consumers (BACB, 2019). Insurance coverage for ABA services is only available in some jurisdictions. Where available, access to reimbursement often requires justification of medical necessity, usually via diagnosis. Similarly, to qualify for most government funding in Canada, a valid diagnosis is required—for example, the Ontario Autism Program (OAP; Ministry of Children Community and Social Services, 2022). Diagnosis then interacts with other personal factors resulting in access or barriers to ABA services. Insurance mandates limit the age at which companies must reimburse ABA services, ranging from 9 to 21 years (Kennedy-Hendricks et al., 2018; National Conference of State Legislators, 2021). Children who are non-white, lower socioeconomic status, do not have intellectual disability, or require fewer supports, are less likely to have a diagnosis despite meeting autism surveillance definitions (Mandell et al., 2002, 2009; Mazurek et al., 2014; Miller et al., 2016; Rosenberg et al., 2011; Valicenti-Mcdermott et al., 2012; Wiggins et al., 2020). That is, they meet diagnostic criteria according to population surveillance criteria but have not been identified by the health care or education systems as having autism. Race-ethnicity and neighborhood affluence are associated with the use of early intensive behavioral intervention (EIBI) services. EIBI is a comprehensive ABA intervention and typically includes 20–40 hours per week of individualized instruction for autistic children who begin treatment at the age of four years or younger, and who usually continue for 2–3 years (ONTABA, 2017). Yingling and Bell, (2019) found racialized children were less likely to enroll in behavioral intervention and that black children who enrolled in EIBI used fewer hours.

*Financial Barriers*

Financial barriers are one of the most significant obstacles to ABA services. Costs associated with intervention are a key determinant of intervention choice by parents of children with autism (Hebert, 2014; Serpentine et al., 2010). High-income earning parents access ABA more frequently than do low-income earning parents (Irvin et al., 2012), and children who live in neighborhoods with higher poverty levels are less likely to enroll in EIBI (Yingling et al., 2019). Finances also intersect with other individual level factors. In Canada, for example, the household income trajectories of families of children living with neurodevelopmental disabilities are consistently lower than those of families whose children do not (Salvino et al., 2022).

Structural barriers also directly affect access to funding, including the lack of centralization of information and the process of disability verification (Salvino et al., 2022). In Canada, families of children with neurodevelopmentals express concerns regarding the accessibility of financial supports. For example, verification of disability under the Disability Tax Credit acts as a gateway to other financial programs (such as the Child Disability Benefit), yet it is undergoing a government review due to a reported lack of accessibility (Colbert, 2020; Salvino et al., 2022). The lack of accessible information on financial support programs and geographic discrepancies also pose a barrier (Salvino et al., 2022).

*Health Literacy*

Health literacy is the degree to which people can access, understand, interpret, and use health information to make informed decisions (Sørensen et al., 2012). It is an important predictor of intervention use because health literacy is required to trust and make treatment decisions based on reliable sources. Approximately 28–50 percent (Payakachat et al., 2018;

Zuckerman et al., 2017) of young children with autism use evidence-based services, such as ABA, while 45 percent (Owen-Smith et al., 2015; Perrin et al., 2012) use approaches with limited evidence supporting their efficacy and/or safety (Hyman et al., 2020). The decision by parents to use ABA is influenced by ABA literacy, with higher parent health literacy facilitating behavior service use for young children with autism (Green et al., 2006; Lindly et al., 2022; Tzanakaki et al., 2012). Further, literacy may play the greatest role when deciding between behavioral services versus other services, for example, prescription medication (Lindly et al., 2022). Parents report pursuing complementary approaches because of their perceived effectiveness and beliefs they are "natural," while decisions to use ABA are often influenced by both empirical and anecdotal evidence (Christon et al., 2010; Lindly et al., 2018). Recommendations from others, including therapists, service providers, teachers, doctors, other parents and friends are commonly reported to be important to treatment choices (Wilson et al., 2018).

*Communication*

People with disabilities may experience challenges in two-way communication with ABA service providers, which can negatively impact the quality of care (Health Care Standards Development Committee, 2022). Consumers of ABA may require accommodations to fully participate in their care. Rights may be compromised when policies and practices about consent and capacity do not take into consideration the needs of the person with the disability (Health Care Standards Development Committee, 2022). The BACB ethics code (BACB, 2020) advocates for involving clients and stakeholders in therapeutic decisions (Code 2.09). Although choice-making has long been endorsed in related literature (e.g., Bannerman et al., 1990) critiques continue against ABA about the lack of consumers' agency and voice in intervention and treatment (Gore et al., 2021; Wilkenfeld & McCarthy, 2020). Choice is seldom incorporated into research on behavioral interventions (Ferguson et al., 2019) and when choice is offered, it may be trivial (Hanley, 2010) compared to options such as choosing not to participate at all (Rajaraman et al. 2021).

The communication of behavior analysts may be experienced as inaccessible. Several studies suggest that the non-experts find the communication of behavior analysts hard to understand and unpleasant (Becirevic et al., 2016; Critchfield, Doepke, et al., 2017; Critchfield & Doepke, 2018; Lindsley, 1991; Neuman, 2018). Technical jargon is less preferred by the public and ABA terms have been referred to as harsh, abrasive, difficult to understand, and awkward sounding (Becirevic et al., 2016; Critchfield,

Becirevic, et al., 2017; Critchfield, Doepke, et al., 2017; Critchfield & Doepke, 2018).

*Lack of Service Providers and Waitlists*

Several studies indicate that families have difficulty accessing providers, resulting in delayed or no treatment (Yingling et al., 2017). More than half of all counties in the United States have no Board Certified Behavior Analysts (Yingling et al., 2021). Families continue to experience long waitlists due to provider shortages in the United States and Canada (Yingling et al., 2021). In Ontario, one survey found the majority of caregivers of autistic children (73.6%) reported long waitlists as the largest barrier to services provided under the OAP (Autism Ontario, 2018). Waitlists to access autism services in Ontario, primarily ABA, grew from 1,600 in 2011–2012 to 27,600 in 2019–2022 (Financial Accountability Office of Ontario, 2020).

*Education and Training*

ABA providers may lack education on how to provide accommodations during the delivery of services. The Ethics Code for Behavior Analysts supports addressing diversity, including disability, in behavior analytic practice and requires that certificants obtain training in this area (Behavior Analyst Certification Board, 2020). However, deficits in addressing accessibility in the field of behavior analysis continue to be present. For example, although there are a growing number of articles focused on diversity, equity, and inclusion as it relates to race in the field (Deochand & Costello, 2022; Hilton et al., 2021; Jimenez-Gomez & Beaulieu, 2022; Levy et al., 2021; Mathur & Rodriguez, 2021), we found no studies focusing on accessibility and inclusion of people with disabilities in ABA. Further, the BACB has only introduced requirements for verified course sequences to integrate content related to diversity, equity, and inclusion in 2022 (BACB, 2022). ABAI currently has no requirements related to equity, diversity, and inclusion or accessibility in their accreditation standards (Association for Behavior Analysis International, 2022). It is clear that the formal curriculum and training for behavior analysts do not adequately address accessibility (Veneziano & Shea, 2022).

*Stigma*

Responses to people with disabilities in the general public will influence the likely success or failure of policies and practices related to accessibility. The largest consumers of ABA are people with autism and/or intellectual disability. Research has shown that people with intellectual disabilities face more prejudice and stigma than other forms of disability (Mitter et al., 2019; Nowicki & Sandieson, 2010; Scior et al., 2020; Werner & Shulman, 2015) and autistic people report being stereotyped, judged, and discriminated against by others (Han et al., 2021).

Stigma toward disabilities also comes from service providers. Studies examining care providers' attitudes toward people with intellectual disabilities find providers tend to be ambivalent about whether people with intellectual disabilities should be protected or empowered (Pelleboer-Gunnink et al., 2021). People with intellectual disabilities also report overprotection, lack of recognition, and dependence on support as expressions of stigmatizing treatment (e.g., Jahoda & Markova, 2004, Jahoda et al. 2010, Giesbers et al. 2019). Stigma surrounding homosexuality, parenthood, the priority of sexuality, privacy, and self-determination concerning sexuality and disability may result in these areas being deprioritized in service provision (Pelleboer-Gunnink et al., 2021).

Stigma may also play a role in the progress, or lack thereof, the field of ABA has made in addressing criticisms from autistic self-advocates. Critical feedback from recipients of behavior analytic programming include claims that therapy can be abusive, cause trauma, and has the overarching goal of making autistic people indistinguishable from non-autistic people (Devita-Raeburn, 2016; Li, 2018; Lynch, 2019; Ne'eman, 2016; Perry, 2018; Ward, 2015). Dismissing the criticism of service recipients based on their disability could be considered an intentional form of ableism (Veneziano & Shea, 2022).

*Policy and Administration*

People with disabilities experience multiple barriers within ABA service delivery related to policy and administration. There is often a lack of engagement of persons with disabilities in planning and decision-making (Health Care Standards Development Committee, 2022). This occurs at the highest levels; an analysis of Western European autism policies found that most countries consider autistic people to lack capacity, expertise, and value. As a result, autistic people rarely participate in determining policy (Precious, 2021). It follows that people with disabilities are unlikely to be represented in policy determination at the level of ABA service provision (Yingling & Bell, 2019b). Examples of collaborative decision-making with autistic individuals and their families are less common (Canadian Academy of Health Sciences, 2022), but are promising models for future engagement.

Lack of oversight and adequate feedback systems (particularly where clients may fear repercussions to their care) also presents an accessibility barrier (Health Care Standards Development Committee, 2022). Licensure allows consumers to easily identify practitioners with training, and those without, and consumer protection (Dorsey et al., 2009), and provides legal authority to respond to complaints from the general public. At this time, no Canadian provinces have a mechanism for licensing behavior analysts, and only 36 U.S. states have licensure laws (Behavior Analyst Certification Board, 2022). Currently, Ontario is actively working toward including behavior analysis as a registered health profession under the yet to be proclaimed Psychology and Applied Behavior Analysis Act, 2021 (College of Psychologists of Ontario, 2022).

*Other Barriers*

Other obstacles to accessibility identified in the literature include financial barriers, where additional costs are required for accommodations (Health Care Standards Development Committee, 2022), developing and maintaining well-trained service delivery teams (Johnson & Hastings, 2002), barriers to employment in the ABA field (Health Care Standards Development Committee, 2022), and the nature of the ABA services themselves—for example, perceived inflexibility, disruption to family life (Johnson & Hastings, 2002).

**Toward Accessibility in Applied Behavior Analysis**

Removing systematic barriers (e.g., physical, communicative) to ABA services will require a culture of accessibility. The Health Care Standards Development Committee of Ontario put forward guiding principles recognized as essential to developing a culture of accessibility in care: (1) Equity, diversity, and inclusion; (2) Independence and dignity; (3) Respect for an individual's abilities; (4) Person first and identity first language; (5) Human rights; (6) Dimensions of patient-centered care; (7) Intersectionality. The process of building a culture of accessibility is iterative; a continuing cycle of working to make ABA practitioners and their practice more accessible and inclusive. We present a cycle modeled on the UBC Equity and Inclusion Office's (2020) Activating Inclusion Toolkit to support practitioners in accessibility efforts.

The first phase in this model (Figure 6.1) is building a team. Planning and implementing accessibility initiatives are most effective when guided by a group of committed individuals. Behavior analysts should identify an individual within their organization to function as the accessibility lead. Depending on the size of the organization, creating a committee or working group to support accessibility efforts may result in broader support and ensure a diversity of viewpoints is considered. Behavior analysts should ensure there are formal mechanisms to meaningfully secure

# Making Applied Behavior Analysis Accessible to Consumers 269

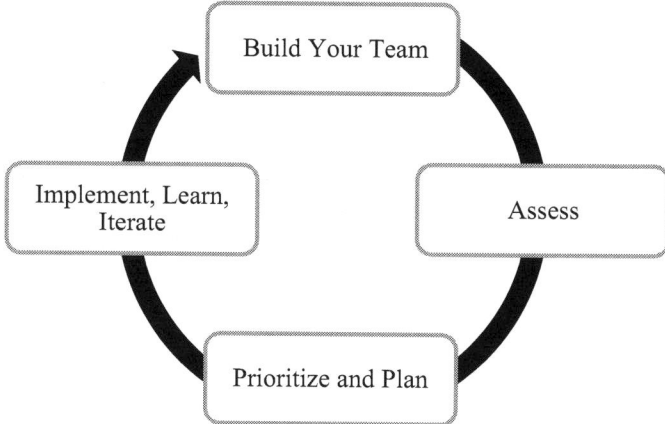

*Note.* This cycle was modeled from the UBC Equity & Inclusion Office's (2020) Activating Inclusion Toolkit.

*Figure 6.1* Cycle to support ABA practitioners in accessibility efforts.

effective representation of persons with lived experience of disability and diverse identities to participate in accessibility planning. Behavior analysts should actively recruit persons with disabilities and/or lived experiences for advisory committees and appointments at all organizational levels. The next phase in the cycle is to assess the current state and understand accessibility in your practice. This might include the use of accessibility self-evaluation tools (e.g. Checklist for Web Content Accessibility Guidelines 2.0, World Wide Web Consortium, 2006; ADA Checklist for Existing Facilities, Institute for Human Centered Design, 2016; UNICEF Toolkit on Accessibility UNICEF Disability Team, 2022).

The next stage is to identify and prioritize concrete changes and make a clear plan for achieving those changes. Following this state, the team implements their plans for change. As they do so, they learn and re-evaluate what is working, adjust, and try again. They may return to other parts of the cycle and build again from that point. The cycle shown in Figure 6.1. might be used to address some of the following recommendations:

1 Appointing an accessibility lead in your organization.
2 Engaging with persons with disabilities in planning and decision-making.
3 Providing access to accessibility equipment (e.g., assistive devices, communication devices, materials).
4 Documenting and sharing accessibility accommodations for clients.

5 Developing policies and practices to advance an accessible and inclusive care philosophy and ensure persons with disabilities can be full participants in the design and delivery of their care.
6 Ensuring policies and practices provide persons with disabilities accommodations for effective communication and provision of informed consent.
7 Developing and implementing mandatory education for staff regarding accessibility that includes topics such as the understanding of legislative responsibilities and the duty to accommodate.
8 Developing policies and practices for an accessible work environment for all employees.
9 Completing an accessibility self-evaluation.
10 Reviewing and declaring values, including a commitment statement to accessibility and inclusion.
11 Developing and providing mechanisms for consumers to provide feedback to your organization.

## Specific Considerations for Accessibility

### Environments (Clinics, Classrooms)

Several are available to support the implementation of physical accessibility. For example, Building for Everyone: A Universal Design Approach (Centre for Excellence in Universal Design CEUD], 2012) guides how to design, build, and manage buildings and spaces to be accessed and used by everyone. The International Organization for Standardization brings experts from all over the world together to develop and publish International Standards—for example, ISO Standard 21542 on Accessibility and Usability of the Built Environment. In general, the built environment, such as a clinic or a school, should be in easy reach of the community and close to public transport, have accessible approach routes and a site layout that is accessible including parking areas and pedestrian paths. Similarly, buildings must follow accessibility standards, which might include ramps, lifts, or elevators whenever there are stairs, automatic doors and width doorways at entrances, accessible public washrooms, barrier-free paths of travel into and through buildings, and visual fire alarms. Older buildings may be closed or unwelcoming to people with disabilities, and planning space in newer buildings may provide opportunities to increase access to services.

### Written Materials and Visuals

Organizations should create, provide, and receive information and communications (e.g., written, visual), including feedback systems, in ways that are accessible and inclusive to people with disabilities. There are many guides for creating clear and understandable content guidelines (*Simply Put* (U.S. Centers for Disease Control and Prevention, 2010); *Clear Communication Index* (Centers for Disease Control and Prevention Office of the Associate Director for Communication, 2019); *Toolkit for Making Written Material Clear and Effective*, (Centers for Medicare & Medicaid Services, 2012); as well as for considering bias and disabling language (*Health Equity Guiding Principles for Unbiased, Inclusive Communication* (Centers for Disease Control and Prevention, 2022); *Advancing Health Equity: A Guide to Language and Concepts* (American Medical Association and Association of American Medical Colleges, 2021). Written materials and visuals must be provided in ways that are accessible for people with disabilities, such as accessible electronic formats, Braille, closed captions, alternative communication devices, text transcripts of audio and visual information, and opportunities to have information repeated, restated, or clarified. Accessible formats must be available and provided upon the request of the individual promptly and at no extra cost.

*Websites*

The Web Content Accessibility Guidelines (WCAG) 2.1 provide the technical standards required to create accessible websites recognized worldwide (The World Wide Web Consortium (W3C, 2018). Information about ABA found on the web may not be accessible because it does not comply with the WCAG 2.1 standards (Acosta-Vargas et al., 2021). Standards include providing text alternatives for non-text content (e.g., image descriptions), alternatives for time-based media (e.g., recordings, options to pause/rewind/fast forward), and adaptable content (e.g., simpler layout) and distinguishable (e.g., high contrast). Overall, organizations should create websites that are functional and adaptable, to increase the readability of the information and maximize compatibility with assistive technologies and mobile applications.

## Conclusion

Creating more universally accessible ABA services, including environments, materials, and websites benefits all consumers, including those with disabilities. Providing accessible behavior, analytic services begins with consideration of accessibility standards and legislation in addition to barriers to accessibility of ABA services at the individual, community, and policy levels. Going beyond requires that behavior analysts foster a culture of accessibility in partnership with persons with disabilities. The process of building a culture of accessibility is iterative; a continuing cycle of working to make ABA practitioners and their practice more accessible and inclusive. In doing this, behavior analysts can meet the intent of relevant legislation and foster an environment that welcomes everyone.

## References

Accessible Canada Act, no. SC 2019, c. 10 (2019).

Acosta-Vargas, G., Acosta-Vargas, P., Jadán-Guerrero, J., Salvador-Ullauri, L., & Gonzalez, M. (2021). Improvement of accessibility in medical and healthcare websites. *Lecture Notes in Networks and Systems*, 265, 266–273. https://doi.org/10.1007/978-3-030-79816-1_33

American Medical Association and Association of American Medical Colleges. (2021). *Advancing Health equity: A guide to language, narrative, and concepts.* www.ama-assn.org/equity-guide

Americans with Disabilities Act, Pub. L. No. 12101 (1990).

Association for Behavior Analysis International. (2022). *Accreditation standards.* https://accreditation.abainternational.org/apply/accreditation-standards.aspx.

Autism Ontario. (2018). *Provincially speaking survey results.* www.autismontario.com/provincially-speaking-survey-results-2018

Bannerman, D. J., Sheldon, J. B., Sherman, J. A., & Harchik, A. E. (1990). Balancing the right to habilitation with the right to personal liberties: The rights of people with developmental disabilities to eat too many doughnuts and take a nap. *Journal of Applied Behavior Analysis*, 23(1), 1286212. https://doi.org/10.1901/jaba.1990.23-79

Becirevic, A., Critchfield, T. S., & Reed, D. D. (2016). On the social acceptability of behavior-analytic terms: Crowdsourced comparisons of lay and technical language. *Behavior Analyst*, 39(2), 305–317. https://doi.org/10.1007/s40614-016-0067-4

Behavior Analyst Certification Board. (2019). *BACB certificant data.* www.bacb.com/bacb-certificant-data/

Behavior Analyst Certification Board. (2020). *Ethics code for behavior analysts.* www.bacb.com/wp-content/uploads/2022/01/Ethics-Code-for-Behavior-Analysts-220316-2.pdf

Behavior Analyst Certification Board. (2022, March). *BACB newsletter: March 2022.* www.bacb.com/wp-content/uploads/2022/01/BACB_March2022_Newsletter-220713.pdf

Behavior Analyst Certification Board. (2022). *U.S. licensure of behavior analysts.* www.bacb.com/u-s-licensure-of-behavior-analysts/

Canadian Academy of Health Sciences. (2022). *Autism in Canada: Considerations for future public policy development: Weaving together evidence and lived experience.* The Oversight Panel on the Assessment on Autism, CAHS. www.cahs-acss.ca

Centers for Disease Control and Prevention. (2022). *Health Equity guiding principles for inclusive communication.* www.cdc.gov/healthcommunication/Health_Equity.html

Centers for Disease Control and Prevention Office of the Associate Director for Communication. (2019). *Clear communication index: A tool for developing and assessing CDC public communication products (User Guide)*. www.cdc.gov/ccindex/pdf/clear-communication-user-guide.pdf

Centers for Medicare & Medicaid Services. (2012). *Toolkit for making written material clear and effective*. www.cms.gov/Outreach-and-Education/Outreach/WrittenMaterialsToolkit

Centre for Excellence in Universal Design (CEUD). (2012). *Centre for excellence in universal design– Home*. https://universaldesign.ie/home/

Christon, L. M., Mackintosh, V. H., & Myers, B. J. (2010). Use of complementary and alternative medicine (CAM) treatments by parents of children with autism spectrum disorders. *Research in Autism Spectrum Disorders, 4*(2), 249–259. https://doi.org/10.1016/J.RASD.2009.09.013

Colbert Y. (2020, June 12). *COVID-19 puts disability tax credit reform on back burner. CBC News Nova Scotia*. www.cbc.ca/news/canada/nova-scotia/disabilities-tax-credits-high-fees-private-companies-1.5606214

College of Psychologists of Ontario. (2022). *Preparing for Applied Behavior Analysis (ABA) Regulation–CPO Public*. https://cpo.on.ca/preparing-for-applied-behavior-analysis-aba-regulation-january-2022/

Critchfield, T. S., Becirevic, A., & Reed, D. D. (2017). On the social validity of behavior-analytic communication: A call for research and description of one method. *The Analysis of Verbal Behavior, 33*(1), 1–23. https://doi.org/10.1007/s40616-017-0077-7

Critchfield, T. S., & Doepke, K. J. (2018). Emotional overtones of behavior analysis terms in English and five other languages. *Behavior Analysis in Practice, 11*(2), 97–105. https://doi.org/10.1007/s40617-018-0222-3

Critchfield, T. S., Doepke, K. J., Kimberly Epting, L., Becirevic, A., Reed, D. D., Fienup, D. M., Kremsreiter, J. L., & Ecott, C. L. (2017). Normative emotional responses to behavior analysis jargon or how not to use words to win friends and influence people. *Behavior Analysis in Practice, 10*(2), 97–106. https://doi.org/10.1007/s40617-016-0161-9

Deochand, N., & Costello, M. S. (2022). Building a social justice framework for cultural and linguistic diversity in ABA. *Behavior Analysis in Practice*, 1–16. https://doi.org/10.1007/s40617-021-00659-4

Devita-Raeburn, E. (2016). The controversy over autism's most common therapy. *Spectrum*. https://doi.org/10.53053/rlll6075

Disability Discrimination Act, No. 135. (1992).

Dorsey, M. F., Weinberg, M., Zane, T., & Guidi, M. M. (2009). The case for licensure of applied behavior analysts. *Behavior Analysis in Practice, 2*(1), 53–58. https://doi.org/10.1007/bf03391738

*Equality Act*, (2010) Testimony of her majesty's stationery office.

Ferguson, J. L., Cihon, J. H., Leaf, J. B., Van Meter, S. M., McEachin, J., & Leaf, R. (2019). Assessment of social validity trends in the Journal of Applied Behavior Analysis. *European Journal of Behavior Analysis, 20*(1), 146–157.

Financial Accountability Office of Ontario. (2020). *Autism Services: A financial review of autism services and program design considerations for the new Ontario Autism Program*. www.fao-on.org/en/Blog/Publications/autism-services-2020

Gibson, J. C., & O'Connor, R. J. (2010). Access to health care for disabled people: A systematic review. *Social Care and Neurodisability*, 1(3), 21–31. https://doi.org/10.5042/SCN.2010.0599/FULL/XML

Giesbers, S. A., Hendriks, L., Jahoda, A., Hastings, R. P., & Embregts, P. J. (2019). Living with support: Experiences of people with mild intellectual disability. *Journal of Applied Research in Intellectual Disabilities*, 32(2), 446–456.

Gore, N. J., McGill, P. & Hastings, R. P. (2021). Personalized goals for positive behavioral support: Engaging directly with children who have intellectual and developmental disabilities. *Journal of Child and Family Studies*, 30, 375–387. https://doi.org/10.1007/s10826-020-01867-2

Green, V. A., Pituch, K. A., Itchon, J., Choi, A., O'Reilly, M., Sigafoos, J., O'Reilly, M., & Sigafoos, J. (2006). Internet survey of treatments used by parents of children with autism. *Research in Developmental Disabilities*, 27(1), 70–84. https://doi.org/10.1016/j.ridd.2004.12.002

Han, E., Scior, K., Avramides, K., & Crane, L. (2022). A systematic review on autistic people's experiences of stigma and coping strategies. *Autism Research*, 15(1), 12–26.

Hanley, G. P. (2010). Toward effective and preferred programming: A case for the objective measurement of social validity with recipients of behavior-change programs. *Behavior Analysis in Practice*, 3, 13–21.

Health Care Standards Development Committee. (2022). *Development of health care Sstandards– Final recommendations report 2022*.

Hebert, E. B. (2014). Factors affecting parental decision-making regarding interventions for their child with autism: *Focus on Autism and Other Developmental Disabilities*, 29(2), 111–124. https://doi.org/10.1177/1088357614522291

Hilton, J., Syed, N., Weiss, M. J., Tereshko, L., Marya, V., Marshall, K., Gatzunis, K., Russell, C., & Driscoll, N. (2021). Initiatives to address diversity, equity, and inclusion within a higher education ABA department. *Behavior and Social Issues*, 30(1), 58–81. https://doi.org/10.1007/s42822-021-00082-y

Hyman, S. L., Levy, S. E., Myers, S. M., Kuo, D. Z., Apkon, C. S., Davidson, L. F., Ellerbeck, K. A., Foster, J. E. A., Noritz, G. H., O'Connor Leppert, M., Saunders, B. S., Stille, C., Yin, L., Brei, T., Davis, B. E., Lipkin, P. H., Norwood, K., Coleman, C., Mann, M., ... Paul, L. (2020). Identification, evaluation, and management of children with autism spectrum disorder. *Pediatrics*, 145(1), e20193447. https://doi.org/10.1542/peds.2019-3447

Institute for Human Centered Design. (2016). *ADA checklist for existing facilities based on the 2010 ADA standards for accessible design*. www.adachecklist.org/doc/fullchecklist/ada-checklist.pdf

Irvin, D. W., McBee, M., Boyd, B. A., Hume, K., & Odom, S. L. (2012). Child and family factors associated with the use of services for preschoolers with autism spectrum disorder. *Research in Autism Spectrum Disorders*, 6(1), 565–572. https://doi.org/10.1016/J.RASD.2011.07.018

Jahoda, A., & Markova, I. (2004). Coping with social stigma: People with intellectual disabilities moving from institutions and family home. *Journal of Intellectual Disability Research*, 48(8), 719–729.

Jahoda, A., Wilson, A., Stalker, K., & Cairney, A. (2010). Living with stigma and the self-perceptions of people with mild intellectual disabilities. *Journal of Social Issues*, 66(3), 521–534.

Jimenez-Gomez, C., & Beaulieu, L. (2022). Cultural responsiveness in applied behavior analysis: Research and practice. *Journal of Applied Behavior Analysis*, 55(3), 650–673. https://doi.org/10.1002/JABA.920

Johnson, E., & Hastings, R. P. (2002). Facilitating factors and barriers to the implementation of intensive home-based behavioral intervention for young children with autism. *Child: Care, Health and Development*, 28(2), 123–129. https://doi.org/10.1046/j.1365-2214.2002.00251.x

Kennedy-Hendricks, A., Epstein, A. J., Mandell, D. S., Candon, M. K., Marcus, S. C., Xie, M., & Barry, C. L. (2018). Effects of state autism mandate age caps on health service use and spending among adolescents. *Journal of the American Academy of Child & Adolescent Psychiatry*, 57(2), 125–131. https://doi.org/10.1016/j.jaac.2017.10.019

Levy, S., Siebold, A., Vaidya, J., Truchon, M.-M., Dettmering, J., & Mittelman, C. (2021). A look in the mirror: How the field of nehavior analysis can become anti-racist. *Behavior Analysis in Practice*, 1–14. https://doi.org/10.1007/s40617-021-00630-3

Li, E. (2018, April 20). *"Treating" autism: The controversy of applied behavior analysis*. https://intr100neurodsp18burk.sites.wm.edu/2018/04/20/treating-autism-the-controversy-of-applied-behavior-analysis

Lindly, O. J., Cabral, J., Mohammed, R., Garber, I., Mistry, K. B., & Kuhlthau, K. A. (2022). "I don't do much without researching things myself": A mixed methods study exploring the role of parent health literacy in autism services use for young children. *Journal of Autism and Developmental Disorders*, 52(8), 3598–3611. https://doi.org/10.1007/s10803-021-05240-0

Lindly, O. J., Thorburn, S., Heisler, K., Reyes, N. M., & Zuckerman, K. E. (2018). Parents' use of complementary health approaches for young children with Autism Spectrum Disorder. *Journal of Autism and Developmental Disorders*, 48(5), 1803–1818. https://doi.org/10.1007/s10803-017-3432-6

Lindsley, O. R. (1991). From technical jargon to plain English for application. *Journal of Applied Behavior Analysis*, 24(3), 449–458. https://doi.org/10.1901/jaba.1991.24-449

Lynch, C. L. (2019, March 28). *Invisible abuse: ABA and the things only autistic people can see*. NeuroClastic Change. https://neuroclastic.com/invisible-abuse-aba-and-the-things-only-autistic-people-can-see/

Mandell, D. S., Listerud, J., Levy, S. E., & Pinto-Martin, J. A. (2002). Race differences in the age at diagnosis among Medicaid-eligible children with autism. *Journal of the American Academy of Child and Adolescent Psychiatry*, 41(12), 1447–1453. https://doi.org/10.1097/00004583-200212000-00016

Mandell, D. S., Wiggins, L. D., Carpenter, L. A., Daniels, J., DiGuiseppi, C., Durkin, M. S., Giarelli, E., Morrier, M. J., Nicholas, J. S., Pinto-Martin, J. A., Shattuck, P. T., Thomas, K. C., Yeargin-Allsopp, M., & Kirby, R. S. (2009). Racial/ethnic disparities in the identification of children with autism spectrum disorders. *American Journal of Public Health*, 99(3), 493–498. https://doi.org/10.2105/AJPH.2007.131243

Mathur, S. K., & Rodriguez, K. A. (2021). Cultural responsiveness curriculum for behavior analysts: A meaningful step toward social justice. *Behavior Analysis in Practice*, 1–9. https://doi.org/10.1007/s40617-021-00579-3

Mazurek, M. O., Handen, B. L., Wodka, E. L., Nowinski, L., Butter, E., & Engelhardt, C. R. (2014). Age at first autism spectrum disorder diagnosis: The role of birth cohort, demographic factors, and clinical features. *Journal of Developmental and Behavioral Pediatrics*, 35(9), 561–569. https://doi.org/10.1097/DBP.0000000000000097

Miller, A., Shen, J., & Mâsse, L. C. (2016). Child functional characteristics explain child and family outcomes better than diagnosis: Population-based study of children with autism or other neurodevelopmental disorders/disabilities. *Health Reports*, 27(6), 9–18.

Ministry of Children Community and Social Services. (2022). *Ontario autism program*. www.ontario.ca/page/ontario-autism-program

Mitter, N., Ali, A., & Scior, K. (2019). Stigma experienced by families of individuals with intellectual disabilities and autism: A systematic review. *Research in Developmental Disabilities*, 89, 10–21. https://doi.org/10.1016/J.RIDD.2019.03.001

National Conference of State Legislators. (2021). *Autism and insurance coverage state laws*. www.ncsl.org/research/health/autism-and-insurance-coverage-state-laws.aspx

Ne'eman, A. (2016, January 21). *The errors—and revelations—in two major books about autism*. Vox.

Neuman, P. (2018). Vernacular selection: What to say and when to say it. *The Analysis of Verbal Behavior*, 34(1), 62–78. https://doi.org/10.1007/S40616-018-0097-Y

Nowicki, E. A., & Sandieson, R. (2010). A meta-analysis of school-age children's attitudes towards persons with physical or intellectual disabilities, *International Journal of Disability, Development and Education*, 21(1), 243–265. https://doi.org/10.1080/1034912022000007270

ONTABA. (2017). *Evidence-based practices for individuals with autism spectrum disorder: Recommendations for caregivers, practitioners, and policy makers*. ONTABA.

Owen-Smith, A. A., Bent, S., Lynch, F. L., Coleman, K. J., Yau, V. M., Pearson, K. A., Massolo, M. L., Quinn, V., & Croen, L. A. (2015). Prevalence and predictors of complementary and alternative medicine use in a large insured sample of children with Autism Spectrum Disorders. *Research in Autism Spectrum Disorders*, 17, 40–51. https://doi.org/10.1016/j.rasd.2015.05.002

Payakachat, N., Tilford, J. M., & Kuhlthau, K. A. (2018). Parent-reported use of interventions by toddlers and preschoolers with autism spectrum disorder. *Psychiatric Services*, 69(2), 186–194. https://doi.org/10.1176/appi.ps.201600524

Pelleboer-Gunnink, H. A., W J van Oorsouw, W. M., van Weeghel, J., & C M Embregts, P. J. (2021). Stigma research in the field of intellectual disabilities: A scoping review on the perspective of care providers. *International Journal of Developmental Disabilities*, 67(3), 168–187. https://doi.org/10.1080/20473 869.2019.1616990

Perrin, J. M., Coury, D. L., Hyman, S. L., Cole, L., Reynolds, A. M., & Clemons, T. (2012). Complementary and alternative medicine use in a large pediatric autism sample. *Pediatrics*, 130 Suppl 2, S77–S82. https://doi.org/10.1542/peds.2012-0900E

Perry, David. M. (2018). *art of stimming*. Pacific Standard. https://psmag.com/education/the-art-of-stimming

Precious, K. (2021). Informed, involved, or empowered? Three ideal types of autism policy design in Western Europe. *European Policy Analysis*, 7(1), 185–206. https://doi.org/10.1002/EPA2.1092

Rajaraman, A., Hanley, G. P., Gover, H. C., Staubitz, J. L., Staubitz, J. E., Simcoe, K., & Metras, R. L. (2021). Minimizing escalation by treating dangerous problem behavior within an enhanced choice model. *Behavior Analysis in Practice*. Advance online publication. https://doi.org/10.1007/ s40617-020-00548-2

Rosenberg, R. E., Landa, R., Law, J. K., Stuart, E. A., & Law, P. A. (2011). Factors affecting age at initial autism spectrum disorder diagnosis in a national survey. *Autism Research and Treatment*, 2011, 1–11. https://doi.org/10.1155/2011/874619

Salvino, C., Spencer, C., Filipe, A. M., & Lach, L. M. (2022). Mapping of financial support programs for children with neurodisabilities across Canada: Barriers and discrepancies within a patchwork system. *Journal of Disability Policy Studies*. Advanced online publication. https://doi.org/10.1177/10442073211066776

Scior, K., Hamid, A., Hastings, R., Werner, S., Belton, C., Laniyan, A., Patel, M., & Kett, M. (2020). Intellectual disability stigma and initiatives to challenge it and promote inclusion around the globe. *Journal of Policy and Practice in Intellectual Disabilities*, 17(2), 165–175. https://doi.org/10.1111/JPPI.12330

Serpentine, E. C., Tarnai, B., Drager, K. D. R., & Finke, E. H. (2010). Decision making of parents of children with autism spectrum disorder concerning augmentative and alternative communication in Hungary, *Communication Disorders Quarterly*, 32(4), 221–231. https://doi.org/10.1177/1525740109353938

Sørensen, K., Van den Broucke, S., Fullam, J., Doyle, G., Pelikan, J., Slonska, Z., Brand, H., & (HLS-EU) Consortium Health Literacy Project European (2012). Health literacy and public health: a systematic review and integration of definitions and models. *BMC Public Health*, 12, 80. https://doi.org/10.1186/1471-2458-12-80

Tzanakaki, P., Grindle, C., Hastings, R. P., Hughes, J. C., Kovshoff, H., & Remington, B. (2012). How and why do parents choose early intensive behavioral intervention for their young child with autism? *Education and Training in Autism and Developmental Disabilities*, 47(1), 58–71.

UNICEF Disability Team. (2022). *Toolkit on accessibility*. Section G: Accessibility checklists.

United Nations Convention on the Rights of Persons with Disabilities. (2006). www.ohchr.org/en/hrbodies/crpd/pages/conventionrightspersonswithdisabilities.aspx

United Nations Department of Economic and Social Affairs. (2016). *Identifying Social Inclusion and Exclusion.* https://doi.org/10.18356/5890648C-EN

University of British Columbia. (2020). *Activating inclusion toolkit.* https://equity.ubc.ca/resources/activating-inclusion-toolkit.

U.S. Centers for Disease Control and Prevention. (2010). *Simply Put: A guide for creating easy-to-understand materials* (3rd ed.). www.cdc.gov/healthliteracy/pdf/simply_put.pdf

Valicenti-Mcdermott, M., Hottinger, K., Seijo, R., & Shulman, L. (2012). Age at diagnosis of autism spectrum disorders. *The Journal of Pediatrics, 161*(3), 554–556. https://doi.org/10.1016/J.JPEDS.2012.05.012

Veneziano, J., & Shea, S. (2022). They have a voice; are we listening? *Behavior Analysis in Practice,* 1–18. https://doi.org/10.1007/s40617-022-00690-z

Ward, T. A. (2015, April 10). *10 myths of applied behavior analysis.* https://aba-easysteps.co.uk/wp/10-myths-of-applied-behavior-analysis/

Werner, S., & Shulman, C. (2015). Does type of disability make a difference in affiliate stigma among family caregivers of individuals with autism, intellectual disability or physical disability? *Journal of Intellectual Disability Research, 59*(3), 272–283. https://doi.org/10.1111/jir.12136

Wiggins, L. D., Durkin, M., Esler, A., Lee, L. C., Zahorodny, W., Rice, C., Yeargin-Allsopp, M., Dowling, N. F., Hall-Lande, J., Morrier, M. J., Christensen, D., Shenouda, J., & Baio, J. (2020). Disparities in documented diagnoses of autism spectrum disorder based on demographic, individual, and service factors. *Autism Research, 13*(3), 464–473. https://doi.org/10.1002/AUR.2255

Wilkenfeld, D. A., & McCarthy, A. M. (2020). Ethical concerns with applied behavior analysis for autism spectrum "Disorder." *Kennedy Institute of Ethics Journal* 30(1), 31–69. doi:10.1353/ken.2020.0000.

Wilson, M., Hamilton, D., Whelan, T., & Pilkington, P. (2018). A systematic review of factors related to parents' treatment decisions for their children with autism spectrum disorders. *Research in Autism Spectrum Disorders, 48,* 17–35. https://doi.org/10.1016/J.RASD.2018.01.004

World Wide Web Consortium. (2006). *Checklist for web content accessibility guidelines 2.0.* www.w3.org/TR/2006/WD-WCAG20-20060427/appendixB.html

World Wide Web Consortium (W3C). (2018). *Web content accessibility guidelines (WCAG) 2.1.* www.w3.org/TR/WCAG21/#adaptable

Yingling, M. E., & Bell, B. A. (2019a). Racial-ethnic and neighborhood inequities in age of treatment receipt among a national sample of children with autism spectrum disorder. *Autism, 23*(4), 963–970. https://doi.org/10.1177/1362361318791816

Yingling, M. E., & Bell, B. A. (2019b). Underutilization of early intensive behavioral intervention among 3-year-old children with autism spectrum disorder. *Journal of Autism and Developmental Disorders, 49*(7), 2956–2964. https://doi.org/10.1007/s10803-019-04005-0

Yingling, M. E., Bell, B. A., & Hock, R. M. (2019). Treatment utilization trajectories among children with autism spectrum disorder: Differences by race-ethnicity and neighborhood. *Journal of Autism and Developmental Disorders, 49*(5), 2173–2183. https://doi.org/10.1007/S10803-019-03896-3/TABLES/2

Yingling, M. E., Hock, R. M., Cohen, A. P., & McCaslin, E. M. (2017). Parent perceived challenges to treatment utilization in a publicly funded early intensive behavioral intervention program for children with autism spectrum disorder., *64*(4–5), 272–282. https://doi.org/10.1080/20473869.2017.1324352

Yingling, M. E., Ruther, M. H., Dubuque, E. M., & Mandell, D. S. (2021). County-level variation in geographic access to Board Certified Behavior Analysts among children with Autism Spectrum Disorder in the United States. *Autism, 25*(6), 1734–1745. https://doi.org/10.1177/13623613211002051

Zuckerman, K. E., Lindly, O. J., Reyes, N. M., Chavez, A. E., Macias, K., Smith, K. N., & Reynolds, A. (2017). Disparities in diagnosis and treatment of autism in Latino and Non-Latino White families. *Pediatrics, 139*(5), e20163010. https://doi.org/10.1542/peds.2016-3010

# 6.2 Telehealth to Implement Applied Behavior Analysis Treatment

*Natalie Paquet Croteau and Erica Franco*

**Applied Behavior Analysis (ABA) Treatment via Telehealth**

Applied Behavior Analysis (ABA) has, until recently, been implemented primarily in person. In the traditional model, a clinical supervisor oversees the implementation of treatment delivered by a behavior therapist who works with the client directly. The pandemic, COVID-19, that began in March 2020 wreaked global havoc, with the impact felt from industries to individuals. This sudden global crisis caused lockdowns and meant an immediate pivot and departure from what we all knew. Given the length of this crisis, industries and individuals had to adapt. With technological advances over the last decade, many health services turned to the use of *telehealth* as an alternative to provide services. According to the Health Resources Services Administration, telehealth is "the use of electronic information and telecommunications technologies to support long-distance clinical health care, patient and professional health-related education, public health and health administration" (Health Information Technology, n.d.).

Telehealth has been a service modality familiar to ABA for many years. While it has been used effectively in the past with individuals with an autism spectrum disorder (ASD) diagnosis, it has mainly been used in the capacity of supervising professionals and for parent coaching and training (Pollard et al., 2021). In this format, service has been delivered with the aim of teaching caregivers and other professionals to conduct assessments and implement strategies such as functional behavioral assessments (Baretto et al., 2006; Benson et al., 2018), preference assessments (Higgins et al., 2017), increasing communication and language (Akemoglu et al., 2020; Barkaia et al., 2017; Ingersoll et al., 2016), social (Wainer & Ingersoll, 2015), and daily living skills (McLay et al., 2020). More recently, telehealth services in ABA have expanded to include delivering direct service to a client by a behavior therapist. COVID-19 saw restrictions imposed by governing bodies across the world that were unprecedented

DOI: 10.4324/9781003300465-31

and prolonged. Because the containment was so lengthy and it was unclear when the restrictions would be lifted, ABA service providers expanded treatment to include direct treatment to the client via telehealth as a means to overcome these barriers.

Research in the use of telehealth to provide direct ABA treatment to clients is emerging. Pollard et al. (2021) provided 17 cases where clients moved from in-person services to telehealth services and showed improved performance. As restrictions have eased and life is returning to "normal," research in this area will continue, given Pollard et al.'s (2021) preliminary results showing effective change. There is a need for further research in the use of telehealth as its use has been demonstrated to facilitate accessibility in extreme circumstances, but this may have also facilitated an opportunity to broaden its application with consideration to client adaptation, economic status, limited capacity of clinicians available, and geographic boundaries (Meadan & Daczewitz, 2015; Zoder-Martell et al., 2020). There are many circumstances that lend themselves for this type of service to be applied via a digital platform. As we continue to explore this area, there are various considerations relevant to its application, from the technology itself, to aspects of the intervention and all the ethical considerations that are attached.

## ABA Considerations and Telehealth

The use of telehealth for the direct delivery of ABA treatment to children with ASD is relatively new, though many more service providers have had some experience with it at the height of the global pandemic. As a clinician, regardless of how the service is delivered, in-person or via a digital platform, for learning to occur there are prescribed treatment parameters that must be assessed and defined. In this section we will explore the clinical considerations that are necessary for the use of this modality. Some of these prescribed treatment parameters include: modality, intensity of treatment, client goals, how to teach the goals, how to measure and monitor the client progress, and defining the environmental variables where the learning needs to take place.

## Practical Considerations

Before delving too deep into considerations, as there are many, it is important to establish the modality of the treatment service. The potential to use a digital platform needs to be established based on the family's availability of technology and their capacity and comfort in using it. It is necessary to determine if the family has consistent access to a computer with a web-cam (built in or otherwise), the Internet, the required Internet speeds to support the digital platform, and that the Internet connection is reliable. Technological difficulties can be disruptive to sessions so it is imperative to ensure that the technology itself will be reliable for the most part.

Once it is determined that the technology is available and reliable, it is important to check in with caregivers about their skill level and comfort in setting up and navigating the technology. When something goes wrong, behavior technicians will work with caregivers to troubleshoot. It is recommended that, if telehealth is chosen as the modality to deliver service, that a practice session is scheduled to review the use of the technology and navigate through the digital platform to familiarize the caregiver. Also established at the practice session is a plan of action, should the digital platform be faulty, during the session. For example, log off and try to log on again. Establishing these processes in advance will help to have a more seamless service and reduce the potential for miscommunications. While we recommend that this brief practice session does not come at a cost to the client and is considered a part of providing the overall service, however a practitioner builds their fee structure, this must be outlined in the service agreement and explicitly explained to the caregiver so that they are informed.

**Intensity of Treatment**

Intensity of treatment is often determined by the severity of the diagnosis and assessment outcomes (Linstead et al., 2017). There are multiple facets to consider specific to the client's age, strengths, needs, and current integration in other community services (e.g., speech and language pathology, occupational therapy). There are additional factors to consider when exploring services provided by telehealth. It is important to have all of this information to assist the clinician's recommendation to use telehealth solely for treatment, as one modality in the child's ABA treatment or to determine that telehealth is not an appropriate modality for direct service. It is important to note that while direct treatment may not be determined to be the modality of direct service, the use of a digital platform can still be considered for other aspects of service delivery (e.g., parent meetings, report reviews, parent training, parent-mediated service). For considering direct treatment to be delivered through telehealth, the clinician must have all the information related to the client's portfolio to determine the client's behavior plan. Some factors that inform the prescribed treatment include the skill set of the client at the onset of treatment, joint attention, sitting duration, imitation skills, and if there are any maladaptive behaviors to consider.

While it is ideal to consider the aforementioned in determining if telehealth is an appropriate modality for a client, there are circumstances in which one may consider this modality despite it not being the most appropriate for the client. This modality may be used as a last resort in situations where treatment might not otherwise be accessible. Some of the factors that could influence this may be due to the family's geographic location and limited qualified service providers. In situations like this, alternative solutions are considered, such as whether the best approach is parent mediation or starting services in small increments for the child to acquire the skills necessary to participate in this treatment modality.

## Client Goals

Parent interviews and assessment of skills help to determine what skills to teach the client and, as in-person services, goals must be clearly defined, measured, and monitored. Additional considerations when teaching skills through telehealth include the use of reinforcement, how it is delivered, and generalization of skills. These are factors that require creativity and planning. Given that the client is expected to interact with a therapist who is not present in the same physical room as the client, there are novel approaches that must be part of the client's behavior plan. For example, if they like specific entertainment characters (i.e., Paw Patrol) we may embed some of that preference in the stimuli used to teach the skill or play a game that would include those characters, similar to the approaches used for in-person services.

## Client Environment for Learning

There are various environmental factors that also need to be considered when providing treatment via telehealth, from learning how to apply the technology, to using the digital hardware, to the actual learning space of the client. All of these factors may impact the ability to successfully use telehealth.

Since the behavior technician and the client are not in the same physical space, some strategies that would typically be used in person may not be feasible in the usual way and other means need to be considered. Reinforcement is an essential component for learning to take place; however, when using a digital platform, it may not be a physical exchange (e.g., providing a slinky), and as such it is important to expand the preference assessment to include various activities such as videos, digital books, games, and pictures that may be provided as reinforcement. Some of these preferences may also be used to keep the learner's attention during sessions.

In the initial stage of telehealth treatment, a caregiver may assist the learner through the session and be the primary person who delivers tangible reinforcement. In this time, where caregiver support is not required, behavior technicians may work to expand preference repertoires to include reinforcement delivered digitally. Once further expanded, collaboration with caregivers may help support the delivery of delayed reinforcement using a token economy system that includes mixed reinforcement modalities (e.g., tokens are accrued and exchanged with caregivers for preferred activities such as going to the park, having a special treat).

Similar to in-person treatment, when using a digital platform it is important to minimize distractions and sustain the client's attention. There is a greater limitation when using telehealth as we are not able to physically redirect or limit access to things in the client's environment that may motivate or distract the learner in the immediacy. Service agreements and ongoing communication with caregivers is essential to limit these barriers to effective and efficient treatment. Environmental variables to consider when delivering services using telehealth include: location where the treatment takes place, the learning space (e.g., desk and chair versus couch), distractors (people, items in the room, noises, windows or other scenery, lighting), and seating arrangements. Similar to in-person services, it is important to include a desk and chair in the learning space. The learner is not recommended to be seated on the couch, that is not the ideal learning environment. Consideration to the distractions available in the space where the telehealth takes place should be considered based on the child's attending skills. Focus on the child's level of attending skills will help make these determinations to increase the effectiveness of the intervention. In review of these elements, this may be one of the reasons research shows that the average age of clients receiving telehealth services is 10 years and 9 months (Pollard et al., 2021).

The treatment plan for telehealth services has similar components and variables as in-person services. However, how some of the components are administered may vary. For example, the duration of sessions is a variable in the treatment plan that may need to be systematically increased as the learner's attending skills improve and develop. There are also occasions where telehealth treatment sessions are delivered with the caregiver also present in the session, as a physical prompt to the learner to increase attending skills. Of course, the aim in this situation is to systematically fade out the presence of the caregiver, as the learner's skills increase.

### Ethical and Cultural Considerations

While behavior analysts are held to professional standards guided by the Ethics Code for Behavior Analysts (BACB, 2020) and are trained to have competence in areas outlined by the BACB Task List (BACB, 2017), there are specific areas that need special attention when providing services via telehealth. The ethical considerations are greater when considering telehealth, as this modality poses more unknowns for the client as the traditional modality for ABA service delivery has predominantly been in person. The pandemic opened the possibilities of this modality being more widely used and, as this has been a relatively new approach to service delivery, we recommend clinical practitioners reference these resources to guide their practice.

The Ethics Code for Behavior Analysts (BACB, 2020) highlights four core principles to guide the work behavior analysts partake in: that behavior analysts work in a manner that benefit others; that they treat others with compassion, respect, and dignity; that they behave with integrity; and that they ensure they are competent in their work. While these are core principles guiding a behavior analyst's practice, when considering the use of technology such as telehealth, there may be extra emphasis on particular aspects of the activities that take place when considering the modality and when service provisions are defined: defining the service and outlining the risks and benefits of the service in addition to the risks and benefits of this modality compared to other more traditional modalities; informed consent; confidentiality; and having parameters in place should the technology break down or stop working in the middle of a session. Worth discussion as well are the "people skills" required to develop a collaborative rapport with families, who are also guided by the BACB Task List and the Ethics Code for Behavior Analysts (BACB, 2020). Both these tools are excellent sources, recommended for reference and actively used, in particular when making decisions about the potential to use telehealth as a service modality for a client.

The first core principle outlines that it is the behavior analyst's responsibility to do no harm and behave in a manner that benefits the client (Ethics Code 2.01; BACB, 2020). This guiding principle encompasses putting the client's best interest first and protecting their wellbeing and rights. Services are assessed for both short- and long-term effects and shared ongoingly with the client as to obtain informed consent and assent (Ethics Code 2.11; BACB, 2020). Regardless of the modality or platform used to deliver the treatment, the client is entitled to be informed of what the treatment entails, what their active role is, and what are the risks and benefits.

The second core principle emphasizes treating others with compassion, dignity, and respect. Considerations to ensure that the behavior analyst sets out practices that promote independence, and respect privacy and confidentiality, are central to this principle. How technology is going to be used in programming in a manner that promotes independence is a serious consideration. Additionally, the use of technology also increases risks to privacy and confidentiality and needs to be reviewed with families so they can make an informed decision. For example, the use of a digital platform may increase the potential risk of a stranger logging on to the session, so protocols indicating how these situations are handled need to be considered and shared prior to treatment implementation. It is important that clients are aware of the risks and what is being done to mitigate and minimize those risks.

The third core principle of behaving with integrity highlights the importance of being honest and trustworthy, following through on what is said and representing oneself honestly. This could entail informing a family about the behavior analyst's level of experience delivering service using telehealth, as this is relevant to the information a family may require to make a decision about proceeding with service with the behavior analyst or delivery using telehealth at all. Informing the family on the risks and benefits of the treatment plan and answering all their questions help to form a collaborative rapport.

The fourth core principle relates to the behavior analyst working within their scope of practice and ensuring that they have the necessary competencies to execute the work or, if they do not, that they consult or get support to ensure effective delivery of the service. Some of the many components that fall within the scope of the behavior analysts' practice include assessment of the behavior in question, defining the intervention based on the results of the assessment, and defining the treatment plan to ultimately change the behavior. These areas must be clearly defined to the client (Ethics Code 2.08, 2.13 and 2.14; BACB, 2020). Working within one's scope of practice primarily means that the behavior plan being prescribed is not based on what one "should do," but on what one has the knowledge to implement.

**Capacity Building**

In many places, the capacity to provide ABA services continues to be limited. Professionals with Behavior Analyst Certification Board (BACB) credentials continue to be in high demand, despite evidence showing an increase in the total number of professionals in the last four years (BACB, 2022). Applied Behavior Analysis is a prevalent treatment recommendation when there is a diagnosis of ASD. Though the field of ABA has expanded, and more professionals are certified as Board Certified Behavior Analysts (BCBA), the field has not been able to keep up with the demand. One of the many advantages of telehealth is that it provides opportunities for people in more remote areas to access mentorship and supervision by a qualified behavior analyst. Telehealth has afforded candidates seeking supervision to access it without geographical limitations. While North America has a number of behavior analysts, in other continents the field is emerging to meet the demands of the population. Behavior analysts can provide supervision to people in remote areas and, in turn, as individuals are certified, they too, can provide supervision and mentorship of others in their areas.

Through telehealth, a behavior analyst may mentor, observe sessions, provide feedback, and train students in the necessary competencies. Telehealth also offers those in remote areas to have a wider pool of professionals to gain mentorship from, which is particularly important when developing skills. Importantly, building capacity also opens the door to accessibility.

## Accessibility

There are many potential barriers to people accessing the behavior analyst's services. For some, transportation may be an issue, for others, it may be living in a remote area where behavior analysts may not be available, or distance would increase the cost of accessing the service. Telehealth provides people, who may not otherwise have access, an opportunity to receive the service (Meadan & Daczewitz, 2015). It also allows a caregiver an option for how they might like to access the service, if it is feasible and recommended. Cutting travel costs, whether it be for the caregiver or the service provider could mean access to more service or providing service to more people. This issue is especially meaningful in more rural and remote areas. For some this may even be the difference between being able to access service versus not accessing service at all.

The pandemic caused the world to pause. During that pause it seems that people contemplated the meaning of many things. That pause expanded our views of what is possible, had us evaluate what we care about, and let us see what we are capable of. Many industries continue to struggle with finding people to fill vacancies. Time is certainly a commodity that the pandemic had us weigh, and many discovered that working remotely, using technology, was beneficial as a cost-saving measure but also gave us time to do other things necessary within the household. In the field of behavior analysis, this is something that stands out as well and, if service remains effective then telehealth may be a service option that families can confidently turn to.

## References

Akemoglu, Y., Muharib, R., & Meadan, H. (2020). A systematic and quality review of parent-implemented language and communication interventions conducted via telepractice. *Journal of Behavioral Education, 29*, 282–316. https://doi.org/10.1007/s10864-019-09356-3

Barkaia, A., Stokes, T. F., & Mikiashvili, T. (2017). Intercontinental telehealth coaching of therapists to improve verbalizations by children with autism. *Journal of Applied Behavior Analysis, 50*(3), 582–589. https://doi.org/10.1002/jaba.391

Barretto, A., Wacker, D. P., Harding, J., Lee, J., & Berg, W. K. (2006). Using telemedicine to conduct behavioral assessments. *Journal of Applied Behavior Analysis, 39*(3), 333–340. https://doi.org/10.1901/jaba.2006.173-04

Behavior Analyst Certification Board. (2017). *BCBA task list* (5th ed.). Author.

Behavior Analyst Certification Board. (2020). *Ethics code for behavior analysts.* https://bacb.com/wp-content/ethics-code-for-behavior-analysts/

Behavior Analyst Certification Board. (2022). *BACB certificant data.* www.bacb.com/BACB-certificant-data

Benson, S. S., Dimian, A. F., Elmquist, M., Simacek, J., McComas, J. J., & Symons, F. J. (2018). Coaching parents to assess and treat self-injurious behaviour via telehealth. *Journal of Intellectual Disability Research, 62*(12), 1114–1123.

Health Information Technology. (n.d., June 2). *Telemedicine and Telehealth.* HealthIT. www.healthit.gov/topic/health-it-initiatives/telemedicine-and-telehealth

Higgins, W. J., Luczynski, K. C., Carroll, R. A., Fisher, W. W., & Mudford, O. C. (2017). Evaluation of a telehealth training package to remotely train staff to conduct a preference assessment. *Journal of Applied Behavior Analysis, 50*(2), 238–251. https://doi.org/10.1002/jaba.370

Ingersoll, B., Wainer, A. L., Berger, N. I., Pickard, K. E., & Bonter, N. (2016). Comparison of a self-directed and therapist-assisted telehealth parent-mediated intervention for children with ASD: A pilot RCT. *Journal of Autism and Developmental Disorders, 46*(7), 2275–2284. https://doi.org/10.1007/s10803-016-2755-z

Linstead, E., Dixon D. R., Hong, E., Burns, C. O., French, R., Novack, M. N., & Granpeesheh, D. (2017). An evaluation of the effects of intensity and duration on outcomes across treatment domains for children with autism spectrum disorder. *Translational Psychiatry, 7*, e1234. https://doi.org/10.1038/tp.2017.207

Meadan, H., & Daczewitz, M. E. (2015). Internet-based intervention training for parents of young children with disabilities: A promising service-delivery model. *Early Child Development and Care, 85*, 155–169. https://doi.org/10.1080/03004430.2014.908866

Pollard, J. S., LeBlanc, L. A., Griffin, C. A., & Baker, J. M. (2021). The effects of transition to technician-delivered telehealth ABA treatment during the COVID-19 crisis: A preliminary analysis. *Journal of Applied Behavior Analysis, 54*(1), 87–102. https://doi./10.1002/jaba.803

Wainer, A. L., & Ingersoll, B. R. (2015). Increasing access to an ASD imitation intervention via a telehealth parent training program. *Journal of Autism and Developmental Disorders, 45*(12), 3877–3890. https://doi.org/10.1007/s10803-014-2186-7

Zoder-Martell, K. A., Markelz, A. M., Floress, M. T., Skribal, H. A., & Sayyah, L. E. N. (2020). Technology to facilitate telehealth in Applied Behavior Analysis. *Behavior Analysis in Practice, 13*, 596–603. https://doi.org/10.1007/s40617-020-00449-4

# 6.3 "You're on Mute!" People Skills Behind the Screens and Behind the Scenes

*Anders Lunde and Kimberly Maich*

## Introduction

The COVID-19 pandemic has seen an increase in the use of service provision through remote, online means in order to maintain access to in-person services that met with imposed restrictions. One of the main challenges with online service delivery is that such a method requires an additional skillset beyond that of in-person services. For behavioral services specifically, these challenges are most evident when it comes to technical skills and people skills, and how these skills can be transferred from in-person to virtual settings. Whether the services are provided in-person or virtually, the importance of interpersonal skills of behavioral practitioners remain paramount.

The increased of use of smartphones—in essence, devices that function as miniature computers with Internet access—has (in theory) eased access to such technological tools (e.g., video-based communication) that can support the delivery of telehealth. Eighty-five percent of Americans use a smartphone (Pew Research Center, 2022) and the same percent of Canadians are expected to be using one of these tools by 2024 (Statista, 2022). Remote access to services, including routine medical services (e.g., a prescription renewal), can also be eased with the simple use of a phone appointment. One author of this monograph experienced such a drug prescription renewal event. The prescription renewal took 28 seconds to complete with the general practitioner. While it is suspected the medical doctor in question made the call from his office, the call could just as easily have been completed from home, reducing the need for an office space. The phone appointment similarly saved the patient the need to take time off work, find a replacement to cover their shift, *and* travel to the appointment, arguably providing cost savings across multiple dimensions, financial as well as environmental given the reduced need to travel by either personal or public transportation. It is not without reason, then,

DOI: 10.4324/9781003300465-32

*Figure 6.2* Working at a computer.

that efforts have been made to enhance virtual medical care (Lawrence et al., 2020), including behavioral analytic services (see Figure 6.2).

Additionally, it has been argued that health care services delivered virtually are equally efficient, if not *more* efficient, than more traditional, in-person service delivery (Henry et al., 2021; Hilty et al., 2019; Hilty et al., 2020). While such an argument may be valid in some cases, especially considering the prescription renewal example included above, not all services, practitioners, and clients will result in speed, ease, and efficiency simply due to the use of telehealth. While Hilty and colleagues (2019) outlined overcoming geographical barriers, geography still presents a challenge for many potential clients, including those of behavioral therapists. Access to phones—including home phones (landlines), cellphones, and smartphones—is not a significant challenge for many people across the United States as reportedly nearly all Americans have access to a phone of some kind (Pew Research Centre, 2022). Should service demands be expanded outside of a simple phone call, however, challenges are more evident. Nearly a quarter of Americans in rural areas do not have access to broadband Internet, with slightly less access for those living on Native American lands (Federal Communications Commission, 2020). In Canada's north, 5 to 22 percent of the population still do not have access to cellphones, the Internet, or landline services (Environics Research, 2020). Including rural Canada as a whole, just over 55 percent of the population have access to the Internet (Zarum, 2022), severely limiting the services available and deliverable through more advanced means than the telephone. Geography or limited access to the required technology would make clients unable to receive services, and the argument that accessing telehealth services is more efficient and cost effective is moot. When the necessary technological networks are in place, however, the potential for telehealth seems *almost* limitless, even being described as being able to extend traditional care in large part due to the ability to provide access remotely and reach clients without their needing to present themselves in a physical facility away from their home (Hility et al., 2020). It is also worth considering that the current generation grew up with technologies such as the Internet and smartphones; therefore, both service providers *and* clients in the population born during and after the 1980s both "want *and* expect a digital health care experience" (Hilty et al., p. 9).

This not to say that *all* technology *at all times* will be suitable when we consider access to complex behavioral services. Professional decisions to use "the *right* technology at the *right* time" remains because some clients dislike phone communications and others do not enjoy the "being watched" aspect of video conversations (Hilty et al., 2020, p. 4, see also Henry et al., 2021). For example, there may be a time to consider asynchronous delivery of services, through playback of sessions (e.g., modelling of particular behavioral techniques; Murphey et al., 2019), while other times and clients call for, or even demand, synchronous services to discuss complex situations. Both clients and professionals may need to manage a variety of ethical and legislative considerations—including risks—associated with online therapeutic services, such as boundaries, privacy, and confidential client information, depending on the platform chosen for service delivery (Hilty et al.; Blakemore & Agllias, 2020).

**Interpersonal Skills for Use Via Online Service Delivery Must Be *Taught***

Online service delivery of applied behavior analytic (ABA) services, as with other types of health care service delivery, requires different skills than traditional, person-to-person service delivery (Henry et al., 2021; Hilty et al., 2019; Hilty et al., 2020; Lawrence et al., 2020; Michaud, 2019). It is also worth noting that there is limited research published on behavioral therapy delivered through online means (Ferguson et al., 2019). One of the more challenging aspects of online service delivery is to be able to demonstrate the necessary inter- and intra-personal skills so crucial in any successful therapy, but done remotely, often through a screen. Behavior therapists not only rely on the cooperation or optimal functioning of the technology to be used, but a successful session would necessitate optimal camera placement, lighting, and so on, let alone the absence of distractions (see, for example, Hilty et al., 2019). The absence of distractions could be of significant importance. We have all experienced children, in particular, being more than eager to share their pets up on the screen when instructions quickly switched to online platforms as the COVID-19 pandemic hit in the spring of 2020, not to mention incessant background noise (including parents swearing) and as indicated with the title of this monograph, the ongoing issue of people remaining on mute far longer than necessary or desirable. The ABA therapist functions as both therapist and technical support rolled into one, with a required skillset not typically covered in current graduate programs (Henry et al., 2021; Hilty et al., 2019; Hilty et al., 2020; Michaud, 2019). Please note, however, that the above requirements apply equally well to both analysts and clients as it is unlikely clients will fully benefit from sessions if practitioners are attempting to deliver services from a cluttered and distracting space as well.

*Specific Screen Skills*

Applied behavior analysis is considered by some as an art form, going as far as describing ABA as "behavioral artistry" and, consequently, ABA is very closely associated with "care, attentiveness, and creativity" (Callahan et al., 2019, p. 3557), as well as requiring the strictly technical skills of the ABA approach. An effective ABA therapist thus demonstrates flexibility through the people skills outlined above, as well as a potential adherence to scripts and manuals when delivering services. Consequently, effective delivery of behavioral therapy arguably relies on two distinctly separate sets of skills, namely those who rely on technical skills alone on the one hand and on the other those who demonstrate exceptional interpersonal skills—skills which include "warmth, emotional stability, liveliness, social boldness, self-assurance, openness to change, self-reliance, and perfectionism" (Callahan et al., 2019, p. 3561). Furthermore, interacting with and delivering services to clients virtually requires an ability to use technology that "promotes interactive experiences by conveying verbal or nonverbal cues" through non-traditional means—yet another layer of competence (Henry et al., 2021, p. 1). Interpersonal skills, however, remain at the forefront, as demonstrated through social presence theory (Henry et al., 2021). Without the ability to make and maintain eye contact, smiling, and open and inviting body language, it is challenging to even begin imagining any type of professional intimacy required for advancements with therapy. Relationships, or at least the *illusion* of relationships, through a screen, is key.

*Are These Skills Obtainable?*

Demonstrating the above-noted skills with a virtual presence may present a challenge—but perhaps not an insurmountable one. In fact, the development and necessity of people skills is being noted as important in a variety of fields not related to healthcare (online or otherwise), such as engineering (Munir, 2022) and information technology (Dubey et al., 2022. Henry and colleagues (2022) outlined 12 specific skills for delivery of virtual health services, including clarity, active listening, focusing on the individual being served, tone of voice, eye contact, body language, rapport, respect, therapeutic alliance, and environment (p. 5). A significant overlap of desirable skills has also been identified elsewhere (Hilty et al., 2020; Michaud, 2019). While these skills are not necessarily directly taught, they likely make a great difference in terms of the quality of service and support the ABA therapist is able to deliver (Callahan et al., 2019).

One way these interpersonal skills can be taught is, perhaps ironically, through technology and machines, building onto a more traditional text and discussion-based case-study approach (Keenan et al., 2020; Maich et al., 2016). Goldberg et al. (2021). This approach examined machine-learning evaluations of video-recorded responses by university students who responded to recorded psychotherapy sessions. The authors found that while it was promising in many aspects, and likely more so in the future, the approach, especially the technology used, was not currently all that helpful or beneficial. Another promising approach involves the use of a virtual learning environment and students learning through interactions with a virtual patient (Bánszki et al., 2018). The benefit of virtual clients is that they can be programed to change behavior based on a students' strengths and weaknesses as they continue to learn. These changes can be made rather quickly (in this particular study, changes to the virtual client were immediate, controlled by a clinician from a laptop), and certainly far more expeditiously than relying on video recordings of ABA therapists-in-training interacting with real-life clients. While Bánszki's team (2018) studied speech-language pathology students, there is little doubt that the training approach can be extrapolated to other disciplines, such as ABA. A significant benefit would be the opportunity to quickly re-do an interaction should it be less than ideal and, furthermore, there is no possibility to cause any harm to an actual client if a misstep were to occur. There is also safety in place for the students in the sense that, should they be unsuccessful in reaching the client, the session, unlike in real life, can pause until students can recompose themselves.

Keenan and their team (2020) suggested yet another approach through which technology can be beneficial in expanding the reaches of ABA, especially within Europe where behavior analysis is less widely applied than in

North America. Keenan and colleagues' list of a number of technologies which may be beneficial in distributing and teaching ABA skills, such as QR codes to quickly access information, animation to recreate behaviors, Smartrooms to recreate social settings and situations, and social robots to simulate social situations. All of these can help provide access to the skills ABA therapists require. A teaching approach and information-spreading approach in regard to ABA and the skills it requires intuitively makes sense given the propensity for technology often found among those diagnosed with autism, especially children (Costantin et al., 2017). Given the inclination toward technology, therapy through technology may also rather quickly serve the purpose of establishing rapport between the clinician and client.

**Problem Solving with Online Service Delivery**

Certainly, the burgeoning use of telehealth during the pandemic taught us that we not only need to proactively prepare in order to effectively meld what we already know about the provision of ABA services and the provision of remote health care, but also that we are learning as we go. Perhaps the well-despised pandemic term *pivot* applies here. No matter how well-prepared we were—we are—problems will occur that we need to competently solve while keeping our people skills well intact or even leveraging them to defuse situations. For example, it might not be the best time to think you are safe and muted with the camera off to yell downstairs about your husband bringing home some fish to grill while you are in a collegial meeting with colleagues, but of course this is the ONE TIME when you know you were going to end up with NOT using the mute button at the right time—instead of over-using it! Or, you would never think that your young, clothing-optional grandson would throw himself at your locked door so many times during a live webinar that you would have to quickly let him ... and then to have that young child flatulate very loudly under the view of the camera ... but not out of range of the computer audio—like what happened with one author. Does anyone ever believe that term "It wasn't me?" In the field of ABA, Keenan et al. noted that, "In any discussion of teaching, it would be at best naïve, and at worst irresponsible, to ignore the technological requirements needed" (2020, p. 48). The same is true for how we apply our people skills to problem-solving in online service delivery: a different kind of problem-solving.

Cham et al. mentioned that in the optometry field, it appears that what they term *soft* skills (truly a misnomer) tend to be informal, hidden, and learned through everyday professional opportunities such as expectations, values—and teaching and learning interactions—a contrast to the technical competencies required and its overt curriculum-based planned activities (Cham et al., 2020; Michaud, 2019). The field of behavior analysis has no doubt encountered this same contrast. Taking this issue one step further, as above-noted, when using virtual services, therapists have to not only engage the technical competencies of their clinical professional but also problem-solve on the fly. When clinicians are isolated—behind screens and behind the scenes—from one another and their clients, these more subtle sets of complex skills are more difficult to develop without the modelling that comes with time, experience, and of course watching others (e.g., professors, supervisors) learn to problem-solve. It is difficult to develop a cognitive hierarchy, flow chart, or confidence in knowing multiple ways to solve issues that may come up—and not the technical skills of the professional—but ones that relate to the people skills of it. What happens during contentious, confrontational, or otherwise anxiety-inducing moments when behavior analysts might be asked or told something like:

What kind of credentials do you have to do this with my child?
It seems like all of this behavior work is just mumbo jumbo to me.
All autistic adults hate ABA.
I really don't like the home therapist that comes to our house for our ABA weekend hours.

Keenan (2020) argued that "existing misrepresentations [about behavior analysis] are confronted directly" (p. 48), and education to clear up misconceptions can be done through a multitude of technological teaching tools. However, it is nevertheless difficult to know what to do next when these situations have been abstract ones that turned into concrete reality (Cham et al., 2020) all while attempting to convey empathy, concern, and care with your face, voice, and body language—to which listeners may not even be attending. What is left? Your carefully chosen words, but in a situation where you might not know what to say due to a lack of experience, specific training, or in-person mentoring. Where appropriate, it's okay to say, "I will get back to you with this information tomorrow." This is where a trusted colleague, educator, or supervisor comes in handy; where you can now spend some non-time-pressured moments investigating the right next step in how to respond using the best people skills possible in a solution-driven approach (Michaud, 2019). It's okay to turn a contentious issue back to your client or parent and ask, "In an ideal world, what would you like to see happen?" It's okay to give them some time to rant, or vent, if that is what they need to do before getting back to more concrete problem-solving. In Michaud's metaphor: "Before you fix the equipment, you need to 'fix' the customer" (2019, p. 439) with the caveat that, of course, our customers are clients, and that our equipment is our programming—and that our customers and clients of course don't need fixing.

In all of these considerations, we can also plan proactively to help develop complex, flexible problem-solving skills, perhaps especially so in situations where telehealth requirements and necessities make that in-person service a less preferred or possible option. Where in-person experiences or even virtual clients (Bánszki et al, 2018) are not possible, ethical, or healthy, consider building case-study options into teaching and learning, not only during initial course sequences or degree programs, but for ongoing professional learning throughout career pathways. Take the opportunity to build professional learning partnerships, groups, teams and build your capacities for professional problem-solving through dialogue (even if on-screen), leveraging those people skills already in place or in development.

**Conclusion**

ABA skills still apply to virtual delivery—when services are behind the screens and everyone involved might feel more behind the scenes. These

*Table 6.1* Task List and Ethics Code Links When Behind the Screen

| Fifth Edition Task List: | Ethics Code: |
|---|---|
| Section 2: Applications | Ethics Standards |
| E1: Responsible conduct of behavior analysts | Section 1—Responsibility as a Professional |
| | 1.06 Maintaining Competence |
| E2: Behavior analysts' responsibility to clients | 1.07 Cultural Responsiveness and Diversity |
| | 1.15 Responding to Requests |
| E6: Behavior analysts' ethical responsibility to the profession of behavior analysis | Section 2—Responsibility in Practice |
| | 2.03 Protecting Confidential Information |
| | 2.05 Documentation Protection and Retention *(especially electronic documentation)* |
| Behavior Analyst Certification Board. (2017). BCBA/BCaBA task list (5th ed.). Author. www.bacb.com/wp-content/uploads/2020/05/170113-BCBA-BCaBA-task-list-5th-ed-.pdf | 2.08 Communicating About Services |
| | 2.11 Obtaining Informed Consent |
| | 2.19 Addressing Conditions Interfering with Service Delivery |
| | Section 5—Responsibility in Public Statements |
| | 5.10 Social Media Channels and Websites (see 1.02, 2.03, 2.04, 2.11, 3.01, 3.10) |
| | 5.11 Using Digital Content in Public Statements (see 1.02, 1.03, 2.03, 2.04, 2.11, 3.01, 3.10) |
| | Behavior Analyst Certification Board. (2020). Ethics code for behavior analysts. https://bacb.com/wp-content/ethics-code-for-behavior-analysts/ |

skills must be transferred from in-person services to online services, placing a significant emphasis on strong communication skills as part of so-called *people skills* in order to convey, for example, verbal and nonverbal clues and personal connection to virtual conversations. In order for successful service delivery to occur, adaptability is important for both the providers and the client, with the onus being on practitioners to ensure that the needs of clients can indeed be met by learning and using those people skills, if they are not already part of the therapist's or analyst's repertoire (see Table 6.1).

## References

Bánszki, F., Beilby, J., Quail, M., Allen, P. J., Brundage, S. B., & Spitalnick, J. (2018). A clinical educator's experience using a virtual patient to teach communication and interpersonal skills. *Australasian Journal of Educational Technology*, 34(3), 60–73. https://doi.org/10.14742/ajet.3296

Blakemore, T., and Agllias, K. (2020). Social media, empathy and interpersonal skills: Social work students' reflections in the digital era. *Social Work Education*, 39(2), 200–213. https://doi.org/10.1080/02615479.2019.1619683

Callahan, Foxx, R. M., Swierczynski, A., Aerts, X., Mehta, S., McComb, M.-E., Nichols, S. M., Segal, G., Donald, A., & Sharma, R. (2019). Behavioral artistry: Examining the relationship between the interpersonal skills and effective practice repertoires of applied behavior analysis practitioners. *Journal of Autism and Developmental Disorders*, 49(9), 3557–3570. https://doi.org/10.1007/s10803-019-04082-1

Cham, K. M. C., Gaunt, H., Delany, C. (2020). Pilot study: Thinking outside the square in cultivating "soft skills"—going beyond the standard optometric curriculum. *Optometry & Vision Sciences*, 97(11), 962–969. https://doi.org/10.1097/OPX.0000000000001594

Costantin, A., Johnson, H., Smith, E., Lengyel, D., & Brosnan, M. (2017). Designing computer-based rewards with and for children with autism spectrum disorder and/or intellectual disability. *Computers in Human Behavior*, 75, 404–414. http://dx.doi.org/10.1016/j.chb.2017.05.030

Dubey, R. S., Paul, J. & Tewari. V. (2022) The soft skills gap: A bottleneck in the talent supply in emerging economies. *The International Journal of Human Resource Management*, 33(13), 2630–2661. https://doi.org/10.1080/09585192.2020.1871399

Environics Research (2020). *Research on telecommunications services in northern Canada—Final report prepared for the Canadian radio-television and telecommunications commission (CRTC)*. https://epe.lac-bac.gc.ca/100/200/301/pwgsc-tpsgc/por-ef/crtc/2021/023-20-e/POR023-20-Final-Report.html#a0301

Federal Communications Commission (2020). *2020 Broadband Deployment Report*. https://docs.fcc.gov/public/attachments/FCC-20-50A1.pdf

Ferguson, J., Craig, E. A., & Dounavi, K. (2019). Telehealth as a model for providing behavior analytic interventions to individuals with autism spectrum disorder: A systematic review. *Journal of Autism and Developmental Disorders, 49*(2), 582–616. https://doi.org/10.1007/s10803-018-3724-5

Goldberg, S. B., Tanana M., Imel, Z. E., Atkins, D. C., Hil, C. E. l. & Anderson, T. (2021) Can a computer detect interpersonal skills? Using machine learning to scale up the facilitative interpersonal skills tasks. *Psychotherapy Research, 31*(3), 281–288. https://doi.org/10.1080/10503307.2020.1741047

Henry, B. W., Billingsly, D., Block, D. E., & Ehrmann, J. (2021). Development of the teaching interpersonal skills for telehealth checklist. *Evaluation & the Health Professions.* https://doi.org/10.1177/0163278721992831

Hilty, D.M., Chan, S., Torous, J., Lou, J., & Boland, R. J. (2019). A telehealth framework for mobile health, smartphones, and apps: Competencies, training, and faculty development. *Journal of Technology in Behavioral Science, 4,* 106–123. https://doi.org/10.1007/s41347-019-00091-0

Hilty, D. M., Chan, S., Torous, J., Lou, J., and Boland, R. J. (2020). A framework for competencies for the use of mobile technologies in psychiatry and medicine: Scoping review. *JMIR mHealth and uHealth, 8*(2), 1–14. https://doi.org/10.2196/12229

Keenan, M., Presti, G., & Dillenburger, K. (2020). Technology and behavior analysis in higher education. *European Journal of Behavior Analysis, 21*(1), 26–54. https://doi.org/10.1080/15021149.2019.1651569

Lawrence, K., Hanley, K., Adams, J., Sartori, D. J., Greene, R., & Zabar, S. (2020). Building telemedicine capacity for trainees during the novel coronavirus outbreak: A case study and lessons learned. *Journal of General Internal Medicine, 35*(9), 2675–2679. https://doi.org/10.1007/s11606-020-05979-9

Maich, K., Levine, D., & Hall, C. (2016). *Applied behavior analysis: 50 cases studies in home, school, and community settings.* Springer Science & Business Media. https://doi.org/10.1007/978-3-319-44794-0_1

Michaud, S. (2019). Feature: As HTM evolves, soft skills become more important. *Biomedical Instrumentation & Technology, 53*(6), 438–442.

Munir, F. (2022). More than technical experts: Engineering professionals' perspectives on the role of soft skills in their practice. *Industry and Higher Education, 36*(3), 294–305. https://doi.org/10.1177/09504222211034725

Murphy, D., Slovak, P., Thieme, A., Jackson, D., Olivier, P. and Fitzpatrick, G. (2019). Developing technology to enhance learning interpersonal skills in counsellor education. *British Journal of Guidance and Counselling, 47(3),* 328–341. https://doi.org/10.1080/03069885.2017.1377337

Statista (2022). *Number of smartphone users in Canada from 2018 to 2024 (in millions).* www.statista.com/statistics/467190/forecast-of-smartphone-users-in-canada/

Pew Research Centre (2022). *Mobile fact sheet.* www.pewresearch.org/internet/fact-sheet/mobile/

Zarum, D. (2022). *Rural internet could hold the key to Canada's economic future.* www.cpacanada.ca/en/news/pivot-magazine/2021-02-14-rural-internet

# 6.4 Making a Document People Will Actually Read

*Carmen Hall*

Introduction

A primary role of a behavior analyst is coaching and training others (e.g., instructor therapists, parents, educators, community members). Having effective print materials (e.g., handouts) is important for the behavior analyst to ensure the information is passed onto others. In addition, it needs to be easy to read, appealing, and reinforcing. An effective way to provide handouts is embedded in the model of Behavior Skills Training (BST; Anderson et al., 2023; Kirkpatrick et al., 2019; Schaefer & Andzik, 2001). BST is an evidence-based training methodology that includes four components: instruction, modeling, role-play, and feedback (Schaefer & Andzik, 2001). During the instruction component, behavior analysts may pair verbal directions with visual handouts to present the skills they are teaching.

Many behavior analysts are trained in academic writing, such as graduate level essay composition using American Psychological Association (APA) style. While this skill may not be taught directly, it is shaped through instructor feedback and preparing articles for publication. However, behavior analysts may be required to create handouts for mediators that use less technical terminology and more accessible language. Handouts developed without the intended audience in mind run the risk of reducing a mediators' ability to comprehend the material and their general willingness to read it. In other words, handouts that contain too much information are visually unappealing and/or are not relevant, are less likely to be reinforcing for the mediator and, therefore, are less likely to be read.

Why People Don't Read Documents?

*Too Much Information*

Since the creation of the Internet, the task for humans has not been to find information, but to find *pertinent* information from the vast amount

DOI: 10.4324/9781003300465-33

available (Badi et al., 2006). People spend more time triaging relevant information than actually reading it (Badi et al., 2006). Moreover, the rise of social media and other technologies introduce new competing reinforcers that divide one's attention and bypass critical-thinking skills (Ward, 2022). The purpose of this chapter is to provide strategies to pick important information to include on handouts and how to design it to get the most out of the text put on the page.

When creating documents, assume that no one will read the text (Harvard Business Review Press, 2016). Although behavior analytic terminology promotes clear understanding, others may perceive the language as harsh and having pre-existing negative connotations (e.g., consequences meaning punishment) (Critchfield et al., 2017). Technical information for the layperson is designed differently than novels, newspapers, and academic papers with readers requiring the assistance of charts, lists, different font sizes and typefaces, and headings to understand the technical, behavior analytic information (Lannon & Klepp, 2009). Unlike viewing pictures, reading text requires a higher cognitive load, more eye fixations, greater effort, and a longer processing time (Garcia-Madariaga et al., 2019). Furthermore, even the amount of text on a page can impact comprehension, with too much text being associated with reduced retention and requiring too much response effort for the reinforcement (i.e., Matching Law) (Jayes et al., 2022). Although topographical differences (e.g., headings, position, sentence length) act as cues for reading comprehension, including too much information can take away from the reader's ability to discern the importance of various parts of the text, especially when skim reading (Jayes et al., 2022).

In addition, people may be distracted when reading technical documents, reading it only because they have to—thus the harder and more complex a document is, the harder they have to work (Lannon & Klepp, 2009). If the response effort is high, the mediator is less likely to read the behavior analysts' work. Ginns et al. (2013) found that technical information that was presented in a conversational tone (e.g., praising your child) rather than a formal tone (e.g., reinforcement for children) led to increased learning in both retention and transfer. With the quantity of information available, readers want methods to get them the information they really need (Lannon & Keep, 2009).

*Layouts*

Spatial layouts, the interaction between words and pictures presented in documents, are significant elements that aid in attracting and sustaining the reader's attention (Lannon & Klepp, 2009). Integrated formatting that uses the spatial contiguity principle (words and pictures near each other on a page) and the dual scripting principle (establishing a specific reading path

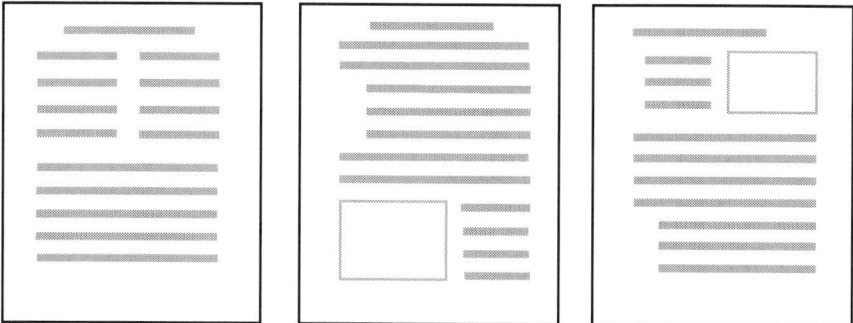

*Figure 6.3* Sample page layouts to assist the reader.

following a top-down guidance) will not only initially attract the reader's attention, but can help to sustain the reader's interest, thereby resulting in a deeper engagement with the material (Holsanova et al., 2009). See Figure 6.3 for examples of page layouts that allow the text to be chunked and be easily digestible for the reader.

*What's in It for Me?*

The key to developing a handout that will meet and keep a mediator's attention is to answer the simple question: "What's in it for me?" The mediator requires you to highlight what the reinforcing value is for them. Readers who had greater interest in the content showed significantly higher retention of what they had read (Freeman et al., 2021).

The motivation to engage and interact with a technical document is also primarily influenced by the individual's self-interest. This means that one should also have that sense of personal gain to make that connection to the written words and motivate the individual. One effective strategy is storytelling. This is where analysis and synthesis are bridged together as a process to become an interesting story. Stories are significant methods whereby knowledge can be stored and experiences can be given meaning (Parrish, 2006). In fact, this potential has influenced the landscape in scientific writing. It is now slowly embracing scientific storytelling so science can be conveyed through an individual's emotions and imagination, where mediators can relate (Martinez–Conde et al., 2019). Hence, creating an interesting story with complete parts (beginning with exposition, middle with climax, and ending with resolution) while using an integrated format should be the goal of the handout or document so the reader is reinforced with interesting information.

## Ways to Make a Document People WILL Read

### White Space

White space takes text and divides it into smaller, easily understandable components (Lannon & Keep, 2009). When used skillfully, white space calls attention to important information by defining the subjects and bringing balance to visual scenes (Amare & Alan, 2013). Organizing text in columns, using indents, and maintaining ample margins provides clarity, keeps related elements together, and emphasizes important points—ensuring all margins are at least 1-inch wide (Lannon & Keep, 2009).

In a study done by Chaparro et al. (2004), reading performance and overall user experience are altered when changing white space layout. The use of margins also affected both reading speed and comprehension. In terms of interactive content, white space helps to build focal points and direct the user's attention to specific layout parts, which makes reading comfortable and essential in making interface content legible (Soegaard, 2014).

### Line Length

It is important to have a line length that is not too long or too short. A line length that is too long makes it hard for the eye to find its place and too short makes the eyes move back and forth rapidly, disrupting the normal rhythm of reading (Kinross, 1984; Lannon & Keep, 2009). Dyson (2004) highlighted that long lines of words are normally perceived by readers as overwhelming and intimidating, which acts as an abolishing operation for reading. It is suggested that lines be 70–80 characters in length (12 to 14 words) for 8.5 x 11 inch pages and single line text is suggested for technical information rather than columns, as used in newspapers (Lannon & Keep, 2009; See Figure 6.4).

### Lists

Lists serve to break up the text, indent and add more white space, and are easier to read than continuous paragraphs (Lannon & Keep, 2009). Vijgen (2014) states lists are powerful as they have the capacity to establish structure, simplicity, content recognition, originality, and can appeal to a diverse audience. On the contrary, lists that are too long result in greater frustration and can negatively impact these outcomes (Freeman et al., 2021).

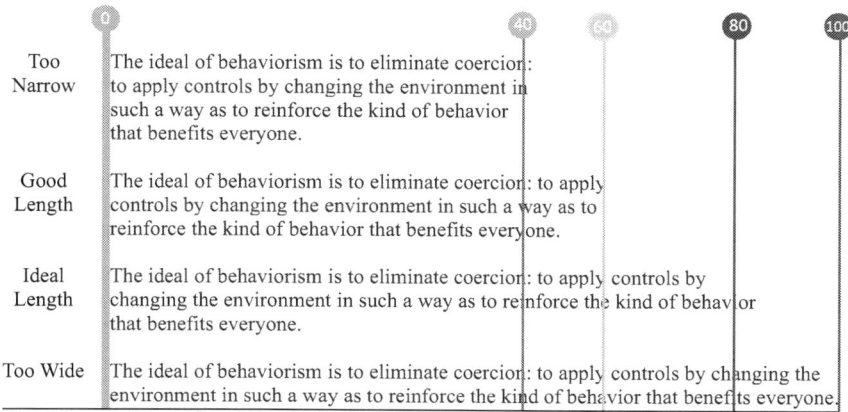

*Figure 6.4* Ideal line lengths for readability.

An effective list has:

- A clear title or introduction (as shown here);
- Proper punctuation;
- Consistent tense and grammatical structure; and
- Additional white space before and after the list.

*Typeface*

Selecting a typeface is vital as it can help make a technical document accessible and increase the mediator's readability. Typeface, or font, is referred to as the "personality" of a document. Typeface has a significant influence on readers' responses to the information presented and can even increase reading speed up to 30 percent (Lannon & Keep, 2009). Readability of text is also necessary in ensuring effective engagement with the material (Hojjati & Munjandy, 2014). A typeface that also looks familiar may influence reading performance and reading preferences (Beier & Larson, 2013).

There are two primary groups of fonts: Serif fonts, which include small ticks (serifs) on the edges of the letters to help the eye connect one letter to the next, and sans serif fonts, which are purely vertical and have no serifs. Most North American texts use sans serif fonts for headings and serif fonts for large blocks of texts; however, different cultures use different

| | |
|---|---|
| Serif Fonts | Times New Roman, Georgia, etc (for blocks of text) |
| Sanserif Fonts | Arial, Calibri, Helvetica, etc. (for titles and PowerPoint) |

| | |
|---|---|
| HEADLINES | Fail to provide height for visual tracking |
| Signature | Provide visual differences in height of letters |

*Figure 6.5* Caps and typefaces.

font patterns (Lannon & Keep, 2009). Hojjati and Munjandy (2014) demonstrated that the sans serif is a better font type in displaying long text for on-screen display and for accessibility.

Emphasizing text is important and can be done using spacing and headings. All caps headings can be difficult to read because the spacing is flush on both the top and the bottom of the text, rather than when capital and lowercase letters are used together (see Figure 6.5; Lannon & Keep, 2009). Italics and underlining are not considered accessible fonts and are discouraged in many accessibility documents. In a study by Rello and Baeza-Yates (2013), people with dyslexia demonstrated better reading performances of sans serif fonts over italics and serif fonts. Many accessibility guides discourage use of italics or underlining. Bold fonts can be effective in emphasizing specific words, or short phrases or sentences, when used sparingly (Lannon & Keep, 2009). Large font sizes and combinations of fonts should be avoided unless a specific statement is needed (Lannon & Keep, 2009).

*Headings*

Headings are an important feature of documents as they help guide the reader, facilitate comprehension and retention, and organize information into accessible blocks (Hyönä & Lorch, 2004; Lannon & Keep, 2009). In addition, they can help a reader decide whether or not a section is worth reading (Lannon & Keep, 2009). These headings are also indicators to those who at first will only scan the document and will later appreciate that the topics are relevant, interesting and are worth their time. Headings should be informative but not too wordy (e.g., Benefits of Reinforcement), and visually and grammatically consistent (Lannon & Keep, 2009). To help with transitions, subheadings can be used to carry the flow of the text forward.

## Conclusion

As the field of behavior analysis continues to grow, the reliance on training mediators (parents, educators, etc.) grows as well. In addition to print documents, the accessibility of handouts to share information about the science is critical. Designing a layout is important, not only for the behavior analysts to be recognized as professional and credible, but also to enhance the comprehension of the reader. Ultimately, the design and layout of a document can have a powerful evocative (or abative!) effect on reading behavior.

## References

Amare, N., & Manning, A. (2013). *A unified theory of information design: Visuals, text and ethics*. Routledge.

Anderson, B. M., Kozluk, A., Morgan, M.-C., MacDonald, M., Friedel, J., & Cox, A. D. (2023). Exploring characteristics of implementer training associated with improved learner outcomes: A meta-analysis. *Journal on Behavioral Education*, 1–30. https://doi.org/10.1007/s10864-022-09504-2.

Badi, R., Bae, S., Moore, J. M., Meintanis, K. A., Zacchi, A., Hsieh, H., Shipman, F. M., & Marshall, C. C. (2006). Recognizing user interest and document value from reading and organizing activities in document triage. *International Conference on Intelligent User Interfaces*, 218–225. https://doi.org/10.1145/1111449.1111496.

Beier, S., & Larson, K. (2013). How does typeface familiarity affect reading performance and reader preference? *Information Design Journal*, 20(1), 16–31. https://doi.org/10.1075/idj.20.1.02bei.

Chaparro, B., Baker, J. R., Shaikh, A. D., Hull, S., & Brady, L. (2004). Reading online text: A comparison of four white space layouts. *Usability News*, 6(2), 1–7.

Critchfield, T. S., Doepke, K. J., Kimberly Epting, L., Becirevic, A., Reed, D. D., Fienup, D. M., Kremsreiter, J. L., & Ecott, C. L. (2017). Normative emotional responses to behavior analysis jargon or how not to use words to win friends and influence people. *Behavior Analysis in Practice*, 10(2), 97–106. https://doi.org/10.1007/s40617-016-0161-9.

Dyson, M. C. (2004). How physical text layout affects reading from screen. *Behavior & Information Technology*, 23(6), 377–393. https://doi.org/10.1080/01449290410001715714.

Freeman, J. R. (2017). *The rise of the Listicle: Using eye-tracking and signal detection theory to measure this growing phenomenon* (ISSN: 2572-4479). [Doctoral dissertation, Brigham Young University]. Scholars Archive.

Freeman, J., Buckley, C., Triptow, C., & Chai, Y. (2021). For the love of lists: Identifying the effects of listicle type and length. *New Review of*

Hypermedia and Multimedia, 27(4), 301–323. https://doi.org/10.1080/13614 568.2021.2001581

García-Madariaga, J., Blasco López, M. F., Burgos, I. M., & Virto, N. R. (2019). Do isolated packaging variables influence consumers' attention and preferences?. *Physiology & Behavior, 200,* 96–103. https://doi.org/10.1016/j.physbeh.2018.04.030

Ginns, Martin, A. J., & Marsh, H. W. (2013). Designing instructional text in a conversational style: A meta-analysis. *Educational Psychology Review, 25*(4), 445–472. https://doi.org/10.1007/s10648-013-9228-0

Harvard Business Review Press. (2016). *HBR guide to delivering effective feedback.* Author.

Hojjati, N., & Muniandy, B. (2014). The effects of font type and spacing of text for online readability and performance. *Contemporary Educational Technology, 5*(2), 161–174. https://doi.org/10.30935/cedtech/6122

Holsanova, J., Holmberg, N., & Holmqvist, K. (2009). Reading information graphics: The role of spatial contiguity and dual attentional guidance. *Applied Cognitive Psychology, 23*(9), 1215–1226. https://doi.org/10.1002/acp.1525

Hyönä, J., & Lorch, R. F. (2004). Effects of topic headings on text processing: Evidence from adult readers' eye fixation patterns. *Learning and Instruction, 14*(2), 131–152. https://doi.org/10.1016/j.learninstruc.2004.01.001

Jayes, L. T., Fitzsimmons, G., Weal, M. J., Kaakinen, J. K., & Drieghe, D. (2022). The impact of hyperlinks, skim reading and perceived importance when reading on the Web. *PLoS ONE, 17*(2), Article e0263669 https://doi.org/10.1371/journal.pone.0263669

Kinross, R. (1984). Emil Ruder's typography and 'Swiss typography'. *Information Design Journal, 4*(2), 147–153. https://doi.org/10.1075/idj.4.2.04kin

Kirkpatrick, M., Akers, J., & Rivera, G. (2019). Use of behavioral skills training with teachers: A systematic review. *Journal of Behavioral Education, 28,* 344–361. https://doi.org/10.1007/s10864-019-09322-z

Lannon, J. M., & Klepp, D. (2009). *Technical communication.* (4th Canadian ed.). Pearson.

Martinez-Conde, S., Alexander, R. G., Blum, D., Britton, N., Lipska, B. K., Quirk, G. J., & Macknik, S. L. (2019). The storytelling brain: How neuroscience stories help bridge the gap between research and society. *Journal of Neuroscience, 39*(42), 8285–8290. https://doi.org/10.1523/JNEUROSCI.1180-19.2019

Parrish, P. (2006). Design as storytelling. *TechTrends, 50*(4), 72–82. https://doi.org/10.1007/s11528-006-0072-7

Rello, L., & Baeza-Yates, R. (2013). Good fonts for dyslexia. *Proceedings of the 15th International ACM SIGACCESS Conference on Computers and Accessibility, 14,* 1–8. https://doi.org/10.1145/2513383.2513447

Schaefer, J. M., & Andzik, N. R. (2021). Evaluating behavioral skills training as an evidence-based practice when training parents to intervene with their children. *Behavior Modification, 45*(6), 887–910. https://doi.org/10.1177/01454 45520923996

Soegaard, M. (2014). *The power of white space in design.* Medium.

Vijgen, B. (2014). The Listicle: An exploring research on an interesting shareable new media phenomenon. *Studia Universitatis Babes-Bolyai-Ephemerides*, *59*(1), 103–122.

Ward, S. M. (2022). Goldfish, horns on skulls, and attention spans: Using the SIFT Method to determine the credibility of online sources. *The Journal of Innovation, Teaching, and Digital Learning Excellence (2)*, 49–53.

# 6.5 A Trauma-Informed Approach to Applied Behavior Analysis

*Brianna M. Anderson and Dana Kalil*

## Introduction

Trauma is a pervasive mental health condition that can affect anyone, regardless of age, race, gender, or ability. Trauma has been defined as a "subjective experience in which an individual experiences an overwhelming threat to their own or a loved one's physical or psychological safety that compromises their ability to cope" (Rich et al., 2021, p. 604). In other words, it is the emotional or psychological response to a distressing event (Goodman et al., 2016). Trauma consists of three components—the event, the individual's experience of the event, and the resulting effects—and can stem from a single adverse event (e.g., a violent incident, natural disaster, death), chronic condition (e.g., ongoing abuse, war, illness), discrimination, or racism (Butler, 2011; Rich et al., 2021).

## The Effects of Trauma

Trauma can lead to significant distress and interference with everyday functioning (Rajaraman et al., 2022). The effects of adverse events can be long-lasting and pervasive, impacting an individual's mental, emotional, and physical well-being long after the adverse event has occurred. Trauma can cause fear, agitation, confusion, diminished affect, or similar initial responses, and may lead to more severe, long-lasting effects, such as the development of post-traumatic stress disorder (PTSD), sleep disorder, or depression (Center for Substance Abuse Treatment [CSAT], 2014).

How an individual experiences an adverse event is shaped by a number of factors, including learning history, biological stress response, and availability of social supports. The behavioral responses that arise to cope with traumatic events can also impact the ways individuals interact with others and their communities (Kirmayer et al., 2008). Further, individuals can become conditioned to certain stimuli (e.g., sounds, smells) that are related to their trauma and which can elicit negative physiological reactions (Van

DOI: 10.4324/9781003300465-34

der Kolk, 1994). For example, an unexpected loud noise may trigger a violent startle response in a refugee with PTSD. Because individuals can have impaired memory of traumatic events, trauma responses can be unexpected and unpredictable (Harris & Fallot, 2001).

## The Prevalence of Trauma

In a survey by Statistics Canada (2022), approximately 64 percent of adults reported being exposed to at least one traumatic event in their lives. Of those individuals, 8 percent met the criteria for PTSD, 21 percent of whom indicated their worst traumatic event had occurred within the past two years. Some of the populations at greatest risk for developing significant adverse effects as a result of trauma include individuals with a history of victimization (i.e., members of marginalized communities), limited access to social support (i.e., people experiencing homelessness), and genetic predisposition for trauma (i.e., individuals with existing mental health conditions; Sayed et al., 2015).

Given the widespread prevalence of trauma, it is important to be aware of potential trauma when working with others, particularly individuals at greater risk of experiencing trauma (Rajaraman et al., 2022). As it is not always possible to know who has experienced trauma and who has not, it can be helpful to assume trauma and approach treatment from a trauma-informed perspective (Goodman et al., 2016).

## Trauma-Informed Care

Trauma-informed care is a universal approach to preventing or minimizing retraumatization for those who may have a history of trauma (Goodman et al., 2016). It involves recognizing an individual's history (or potential history) of trauma and how trauma can shape behavior (Harris & Fallot, 2001). While an agreed upon definition of trauma-informed care has yet to be developed, common themes include understanding trauma and its effects, establishing a safe relationship with the individual, taking proactive steps to prevent retraumatization, making accommodations where needed, and providing services that support the individual in engaging in steps that promote healing and facilitate participation (Centers for Disease Control and Prevention, 2020).

Adopting a trauma-informed approach can lead to a number of benefits for individuals experiencing trauma.[1] Such an approach can promote a safe environment by reducing the probability of individuals encountering trauma-related triggers, and therefore, lessening potentially harmful behavioral responses (Reeves, 2015). For example, a behavior analyst may avoid bringing their client in contact with an antecedent that has historically elicited an aggressive startle response requiring the use of physical restraint. Trauma-informed care can also foster trust and collaboration within therapeutic relationships, which has been shown to lead to increases in both client and staff satisfaction (Fallot & Harris, 2009). Finally, the proactive, strengths-based nature of trauma-informed care can empower individuals experiencing trauma compared to approaches that only focus on deficits.

A trauma-informed approach can also benefit behavior analysts working with individuals experiencing trauma. Screening for trauma during intake and/or assessment can provide behavior analysts with important information for selecting evidence-based treatment strategies that are less likely to cause retraumatization (CSAT, 2014). Training in trauma-informed care can also assist behavior analysts with their ability to cope with secondary traumatic stress (also known as vicarious trauma) when working with individuals experiencing trauma (Deblinger et al., 2020).

## Trauma-Informed Care and Applied Behavior Analysis

Although research explicitly examining a trauma-informed approach to behavior analytic treatment is limited, many studies have examined trauma or incorporated elements of trauma-informed care into their treatment methodologies with promising results (Fiorillo et al., 2017; Friman et al., 1998; Prather & Golden, 2009). The Substance Abuse and Mental Health Services Administration (SAMHSA, 2014) has developed four key assumptions of trauma-informed care, all of which can be directly applied to the field of behavior analysis. Further, Rajaraman et al. (2022) has proposed a theoretical framework for incorporating trauma-informed care into behavior analysis, which has been integrated below.

### *Realization*

*Realization* requires behavior analysts to develop a basic understanding of trauma and its effects, not only for individuals experiencing trauma but for those working with these individuals (SAMHSA, 2014). For example, individuals affected by trauma may have behavioral reactions to specific antecedents (e.g., sudden aggressive outbursts) or turn to unhealthy coping mechanisms (e.g., substance abuse). These reactions can negatively affect their ability to develop therapeutic rapport, to follow through with treatment recommendations, or to access much-needed services. Trauma can even go so far as to cause secondary traumatic stress for those close to them, such as family members, friends, and direct care professionals. For example, the mother of an adolescent accessing treatment for challenging behavior may be experiencing depression as a result of her child's history of trauma and subsequent effects.

*Recognition*

Behavior analysts are able to *recognize* the risk factors for and signs of trauma, and incorporate trauma screening into intake and/or behavioral assessments (SAMHSA, 2014). Identifying risk factors can be done through a comprehensive screening process that includes questions about trauma history, symptoms, sudden changes in behavior, mental health concerns, coping methods, and availability of resources (CSAT, 2014). Because not everyone will willingly disclose their trauma, is aware of their trauma, or has the capability of communicating their trauma, it can be helpful to use a screening or assessment tool that is sensitive to potential effects of trauma even if the trauma is not disclosed outright. For example, direct questions (e.g., "Have you ever experienced a traumatic event?") could be supplemented with indirect questions (e.g., "Have you had a sudden negative change in behavior as a result of a significant life event?").[2]

*Responding*

*Responding* involves the application of a trauma-informed approach, at both an individual and organizational level (SAMHSA, 2014). Anyone who may interact with a client has the potential to engage in behaviors that retraumatize the individual without knowledge of trauma-informed care. For this reason, trauma awareness should occur at all organizational levels, from the administrative team and instructor therapists to behavior analysts and management. Implementing a trauma-informed approach can be accomplished through professional development and staff training, as well as ongoing support to maintain trauma-informed care skills. Trauma-informed care can also be written directly into behavioral contracts and organizational policies. Finally, resources should be available for individuals accessing behavioral services, such as the contact information for local mental health organizations.

*Resistance*

Lastly, behavior analysts should strive to *resist* established systems that exacerbate the effects of trauma (SAMHSA, 2014). This process involves ongoing monitoring of behaviors and practices that may cause retraumatization and actively working toward eliminating them. As described by Rajaraman et al. (2022), trauma-informed behavior analysts should strive to:

*Table 6.2* Case Illustration

Charlie is a behavior analyst working at an agency that provides services to the community. Olive, a 55-year-old woman with a developmental delay, was recently added to Charlie's caseload. Olive can communicate using simple words and phrases via an augmentative communication device. She has strong receptive language skills and requires little support completing basic daily routines. She has been living in her current group home since her mother passed away fifteen years ago and is generally healthy. During the initial meeting, her care provider informs Charlie that his primary goal is teaching Olive to respond to the fire alarm at the group home. Olive's support staff have tried teaching her this skill but have been unsuccessful thus far.

Charlie asks the care provider to describe how Olive reacts to the fire alarm. Despite being given verbal prompts by her housemates or other staff members, Olive must be guided to the evacuation site. During this time, she will scream, bite her hand, and bite others, which has caused injury to several staff members in the past. Her provider says this is the only time Olive ever engages in these behaviors. With the care provider's consent and Olive's assent, Charlie collects baseline data on Olive's ability to follow familiar multi-step routines. Charlie observes her independently making her bed, retrieving the mail, and helping her housemate make dinner. During this time, Olive is exposed to multiple loud noises, none of which appear to cause Olive any distress.

Because Charlie is trained in trauma-informed care and *realizes* the widespread prevalence of trauma and its effects, he *recognizes* the possibility that Olive's behavior is a trauma response. Charlie chooses to conduct an indirect assessment of Olive's behavior, determining the risks of conducting an in vivo assessment (e.g., ABC descriptive assessment, functional analysis) outweigh the benefits at this time. He also asks the support staff to complete the SLE Screening tool on Olive's behalf (CSAT, 2014).

Olive's support staff have little information on Olive's life before the group home; therefore, they can only complete the SLE Screening tool based on her experiences over the past 15 years, which has not included any significant traumatic events. The indirect assessment indicates that Olive's behavior may be multiply controlled, serving both escape demand and automatic functions.

Given this information, Charlie *responds* by consulting with a psychologist who specializes in trauma in people with developmental disabilities. Together, they develop a strength-based treatment plan with the goal of increasing Olive's functional communication skills and *very* gradually desensitizing Olive to scenarios related to fire safety without causing her undue distress.

Lastly, Charlie *resisted* what he would have done in the past, which was immediately conducting a standard functional analysis to determine the function of Olive's behavior. While functional analyses are incredibly beneficial in many cases, Charlie used his clinical judgment and knowledge of trauma to find another way of assessing and treating Olive's behavior. In the end, Charlie's trauma informed approach lessened the likelihood of retraumatizing Olive and putting both her and her support staff at risk of further harm.

- develop an environment that prioritizes client safety;
- build and maintain trust and rapport;
- provide opportunities for choice-making and shared governance;
- use alternatives to intrusive procedures (i.e., physical restraint, punishment) when warranted; and
- use a strength-based approach to teach functional skills.

## Summary

Overall, there are many benefits to implementing a trauma-informed approach to applied behavior analysis, as clients, members of the behavior analytic team, and members of the community are all susceptible to trauma. This approach proactively reduces the likelihood of an individual experiencing further physical and/or psychological harm. By realizing the effects of trauma, recognizing trauma in those we work with, responding to said effects, and resisting systems that have historically perpetuated the effects of trauma, behavior analysts can better serve their clients and their communities (see Table 6.2).

## Notes

1 For the purpose of this chapter, the phrase "individuals experiencing trauma" is used to describe trauma victims, survivors, and traumatized individuals. While there is no universally-accepted language used to describe individuals experiencing trauma, clinicians can take a trauma-sensitive approach by asking individuals their preferred terminology.
2 For an open-access trauma screening tool, visit the Trauma-Informed Care in Behavioral Health Services treatment improvement protocol (Exhibit 1.4-3: SLE Screening; CSAT, 2104, p. 107).

## References

Butler, L. D., Critelli, F. M., & Rinfrette, E. S. (2011). Trauma-informed care and mental health. *Directions in Psychiatry, 31,* 197–210.
Center for Substance Abuse Treatment. (2014). *Trauma-informed care in behavioral health services.* Substance Abuse and Mental Health Services Administration.
Centers for Disease Control and Prevention (2020, September 17). *Six guiding principles to a trauma-informed approach.* Centres for Preparedness and Response. www.cdc.gov/cpr/infographics/6_principles_trauma_info.htm
Deblinger, E., Pollio, E., Cooper, B., & Steer, R. A. (2020). Disseminating trauma-focused cognitive behavioral therapy with a systematic self-care approach to addressing secondary traumatic stress: PRACTICE What You Preach.

Community Mental Health Journal, 56(8), 1531–1543. https://doi.org/10.1007/s10597-020-00602-x

Fallot, R. D., & Harris, M. (2008). Trauma-informed approaches to systems of care. *Trauma Psychology Newsletter, 3*(1), 6–7.

Fallot, R. D., & Harris, M. (2009). *Creating cultures of trauma-informed care.* Washington DC: Community Connections.

Fiorillo, D., McLean, C., Pistorello, J., Hayes, S. C., & Follette, V. M. (2017). Evaluation of a web-based acceptance and commitment therapy program for women with trauma-related problems: A pilot study. *Journal of Contextual Behavioral Science, 6*(1), 104–113. https://doi.org/10.1016/j.jcbs.2016.11.003

Friman, P. C., Hayes, S. C., & Wilson, K. G. (1998). Why behavior analysts should study emotion: The example of anxiety. *Journal of Applied Behavior Analysis, 31*(1), 137–156. https://doi.org/10.1901/jaba.1998.31-137

Goodman, L. A., Sullivan, C. M., Serrata, J., Perilla, J., Wilson, J. M., Fauci, J. E., & DiGiovanni, C. D. (2016). Development and validation of the trauma-informed practice scales. *Journal of Community Psychology, 44*(6), 747–764. https://doi.org/10.1002/jcop.21799

Harris, M., & Fallot, R. D. (2001). Envisioning a trauma-informed service system: A vital paradigm shift. *New Directions for Mental Health Services, 2001*(89), 3–22. https://doi.org/10.1002/yd.23320018903

Kirmayer, L. J., Lemelson, R., & Barad, M. (2008). *Understanding trauma: Integrating biological, clinical and cultural perspectives.* Cambridge University Press. https://doi.org/10.1017/CBO9780511500008

Prather, W., & Golden, J. (2009). A behavioral perspective of childhood trauma and attachment issues: Toward alternative treatment approaches for children with a history of abuse. *The International Journal of Behavioral Consultation and Therapy, 5*(1), 56–74. https://doi.org/10.1037/h0100872

Rajaraman, A., Austin, J. L., Gover, H. C., Cammilleri, A. P., Donnelly, D. R., & Hanley, G. P. (2022). Toward trauma-informed applications of behavior analysis. *Journal of Applied Behavior Analysis, 55*(1), 40–61. https://doi.org/10.1002/jaba.881

Reeves, E. (2015). A synthesis of the literature on trauma-informed care. *Issues in Mental Health Nursing, 36*(9), 698–709. https://doi.org/10.3109/01612840.2015.1025319

Rich, A. J., DiGregorio, N., & Strassle, C. (2021). Trauma-informed care in the context of intellectual and developmental disability services: Perceptions of service providers. *Journal of Intellectual Disabilities, 25*(4), 603–618. https://doi.org/10.1177/1744629520918086

Sayed, S., Iacoviello, B. M., & Charney, D. S. (2015). Risk factors for the development of psychopathology following trauma. *Current Psychiatry Reports, 17*(8), 1–7. https://doi.org/10.1007/s11920-015-0612-y

Statistics Canada. (2022, May 20). *Survey on mental health and stressful events, August to December 2021.* www150.statcan.gc.ca/n1/daily-quotidien/220520/dq220520b-eng.htm

Storey, C., McDowell, C., & Leslie, J. C. (2017). Evaluating the efficacy of the Headsprout reading program with children who have spent time in care. *Behavioral Interventions, 32*(3), 285–293. https://doi.org/10.1002/bin.147

Substance Abuse and Mental Health Services Administration. (2014). *SAMHSA's concept of trauma and guidance for a trauma-informed approach.* HHS Publication No. (SMA) 14–4884. Substance Abuse and Mental Health Services Administration. https://ncsacw.acf.hhs.gov/userfiles/files/SAMHSA_Trauma.pdf

Van der Kolk, B. A. (1994). The body keeps the score: Memory and the evolving psychobiology of posttraumatic stress. *Harvard Review of Psychiatry, 1*(5), 253–265. https://doi.org/10.3109/10673229409017088

# 6.6 People Skills of Behavior Analysts as Co-designers with Autistic Adults
Applied Behavior Analysis Integrated into Virtual Reality Game-Based Intervention Supports

*Javier Alejandro Rojas*

## Introduction

Autistic adults[1] have highlighted their unmet needs in professional support and assistive technologies (Choi, 2021; & PHAC, 2020) to develop the skills required to succeed in post-secondary education (Ha et al., 2021; & Taylor et al., 2020), find and maintain jobs (Accardi & Duhaime, 2013; Smith et al., 2014, 2015; & Strickland et al., 2013), and develop skills for independent living in their communities (Dudley & Emery, 2014; Ha et al., 2021; Maich et al., 2020; & Stoddart, 2013).

Over the last two decades, serious videogames have been demonstrated as affordable, accessible, customizable, and effective technology based intervention support to meet the diverse learning needs of autistic individuals (Mazurek et al., 2015; & Tsikinas & Xinogalos, 2018, 2020). Serious videogames are defined as digital games designed and applied for utility purposes (e.g., to teach practical, academic, and on-the-job skills) in addition to their engaging mechanisms for entertainment (Alvarez et al., 2019; & Tsikinas & Xinogalos, 2018). Using serious videogames through virtual reality technologies can meet their learning needs if autistic individuals engage actively throughout intervention programs (Bouck et al., 2014; Plaisted, 2001; Smith et al., 2014, 2015; & Vasquez, 2015) by providing them with in-game choice opportunities.

Using an autistic adults-centered design method to create these in-game choice opportunities also creates opportunities for behavior analysts to develop their people skills. Throughout the co-design of virtual reality-game-based technologies, this method is oriented to support the active role of users as co-designers of their preferred choice opportunities: what they want to do, why they want to do it, and how they can best do it (Jerald, 2016). Behavior analysts may develop interpersonal and intrapersonal skills to facilitate and support co-design experiences in virtual reality tailored to teach choice-related skills so autistic adults can choose the order in which to complete tasks and pick the preferred activities, materials, and reinforcers through play-testing or virtual reality game-based workshops. In terms of intervention goals, when developmentally disabled individuals can choose games and rewards, they engage in goal-directed tasks more often than do individuals who have not been given such choice opportunities (Bannerman et al., 1990).

The following three sections are oriented to illustrate how behavior analysts may develop people skills by co-designing with autistic adults the integration of applied behavior analysis techniques into virtual reality-game-based supports for intervention programs. In the first section, some examples are provided to show how behavior analysts may develop problem-solving skills such as a design sense to effectively integrate evidence-based behavioral interventions used with autistic adults by balancing in-game teaching and learning strategies. The second section is oriented to illustrate how behavior analysts may develop leadership and teamwork skills as artists to craft and explore with autistic adults their preferred and most engaging game mechanisms that motivate their active participation throughout virtual reality game-based intervention programs. The last section is oriented to give examples of how behavior analysts may develop innovation and troubleshooting skills as co-programmers of evaluation functions to define, measure, and intervene in in-game target behaviors.

## Problem-Solving People Skills to Balance Behavioral in-Game Teaching and Learning Strategies

Behavior analysts may develop a problem-solving approach to design game-based technology supports using their analytical skills to break down and integrate evidence-based intervention techniques into effective in-game teaching and learning strategies. For instance, the integration of antecedent-based strategies and consequence-based interventions into in-game teaching and learning mechanisms to provide opportunities for choice in virtual reality game-based environments. Behavior analysts may integrate antecedent-based techniques, including task interspersal, task analysis, and prompting (Alberto & Troutman, 2013; Cooper et al., 2020; Maich et al., 2020; Martin & Pear, 2014; & Miltenberger, 2016). They also may integrate consequence-based interventions, including reinforcement and extinction (Alberto & Troutman, 2013; Cooper et al., 2020; Maich et al., 2020; Martin & Pear, 2014; & Miltenberger, 2016).

However, behavior analysts must develop design sense skills to balance the most effective antecedent-behavior-consequence interventions (NAC,

2015) integrated into in-game teaching and learning settings and the autistic adults needs and preferences. For instance, behavior analysts may use their design-sense skills to balance virtual reality features and autistic adults' needs, considering how they use their core skills in real-life environments. That is, their strengths and needs by using core cognitive, motor and sensory skills to develop behavioral skills throughout their lifespan in communication, social, and daily living settings (Dudley & Emery, 2014). Using design sense skills may allow behavior analysts to balance in-game learning goals and individual dispositions by scaffolding autistic adults to generalize the taught skills through social validation in post-secondary education (Ha et al., 2021; Lombardo & Baron-Cohen, 2011; Manett & Stoddart, 2012; O'Reilly et al., 2011; & Taylor et al., 2020), vocation and employment (Ha et al., 2021; MCSS, 2018; & Taylor et al., 2020), and independent life domains (Ha et al., 2021).

Suppose the goal is to engage autistic adults through in-game learning. In that case, behavior analysts must develop inspiration, imagination, and mind-mapping skills to design plenty of in-game choice opportunities to integrate them into immersive virtual reality environments' teaching and learning mechanisms. The inspiration, imagination and mind-mapping skills may help behavior analysts solve how to design preferred in-game choice opportunities to engage autistic adults in learning through virtual reality game-based intervention programs. Player engagement in serious videogames involves flow, immersion, presence, and arousal (Kiili et al., 2012; & Michailidis et al., 2018). Thus, the design of virtual reality game experiences should be appealing by crafting the players' preferred aesthetics.

When players achieve engagement in learning, they can attend to relevant environmental stimuli (Levin et al., 2021). This in-game teaching strategy can support decision-making skills training (Levin et al., 2021). Decision-making training can help autistic adults to practice tasks in everyday routines (Levin et al., 2021), for example, by considering how anxiety can impact such skills learning (Adams & Malone, 2021) to be addressed through in-game learning mechanisms. Behavior analysts also may use their mind-mapping skills to design in-game choice opportunities for learning decision-making skills by creatively targeting behavioral difficulties (e.g., turn-taking or eye contact) or supporting cognitive skills—for example, awareness of others' thoughts or emotions (Komeda, 2021).

**Communication, Leadership, and Teamwork Skills to Co-design Game Mechanics with Autistic Adults**

Communication, leadership, and teamwork people skills may allow behavior analysts to integrate and balance the behavioral teaching and learning techniques into engaging game mechanics to empower and motivate autistic adults throughout virtual reality-game-based intervention programs. Game mechanics are defined as the pieces a gamer can control, actions a player can take, and rules that govern the game environment (Zubek, 2020). For instance, behavior analysts can lead and facilitate the co-exploration and co-discovery with autistic adults of game mechanics integrating real-world rules that govern virtual reality environments to simulate real-life situations in a café or on a bus. Behavior analysts may co-define with autistic adults the game environment pieces (e.g., objects, people, and places arranged in such a game environment) to scaffold autistic players to learn practical skills to perform actions oriented to such goals as interpreting social scenes meaningfully, communicating, and behaving appropriately in daily living settings.

Communication, leadership and teamwork skills are crucial to behavior analysts for empowering autistic adults by facilitating an active role in iteratively co-creating game mechanics in virtual reality environments based on their preferences (Jerald, 2016). Behavior analysts may facilitate and lead virtual reality workshops (Knight et al., 2021) for autistic adults by using their artistic sense, brainstorming, and imagination skills as a strategy for balancing learning and game mechanics by co-designing game conceptualization (Arnab & Clarke, 2015). The conceptualization of the game mechanics is crafted iteratively in participatory design settings through virtual workshops driven by artistic expressions (Alvarez et al., 2019; Arnab & Clarke, 2015; Cobb et al., 2015; Dong et al., 2007; Macklin & Sharp, 2016; Malinverni et al., 2016; & Newell & Monk, 2007). Iteratively means repeating the rapid four-step technology co-development cycle with autistic adults to fail early and frequently until achieving the necessary improvement for game balance and effective integration of learning mechanics into game mechanics: conceptualization, prototyping, play-testing, and data-driven adjustment.

Behavior analysts may also use communication skills to explore and discover the data-driven adjustments required when interacting with autistic adults through virtual reality technologies. For instance, autistic adults might provide feedback through diverse expressions regarding the balance in the virtual reality affordances experienced. Affordances are defined as the relationships between users' needs, strengths, and preferences and the technology features they use (Jerald, 2016). Behavior analysts may use different communication strategies to obtain feedback about the interactive relationship between the game environment and the agent. Balancing these virtual reality affordances is essential to increase players' engagement, attention and playability by designing game mechanics in virtual reality environments that motivate them through rich experiences. For instance, behavior analysts may correct such balances by reducing useless redundancy—for example, avoiding repeating game mechanics, providing hints at the right time, and using the appropriate interface and interaction mode) (Catalano et al., 2014).

## Innovation and Troubleshooting Skills to Co-Program the Game Architecture with Autistic Adults

Behavior analysts may use analysis, design, logical reasoning, and troubleshooting skills to co-design and co-program algorithms organized in the game architecture for implementing artificial intelligence (AI) game agents that make decisions to control the in-game learning and game mechanics. Game architecture is defined as the programmed code organized into functions (Nystrom, 2014). These functions are a list of instructions computationally written to perform tasks such as running the game mechanisms, including game environment physics and pieces (objects) attributes, 3D graphics, and audio (Nystrom, 2014). Behavior analysts may develop logical reasoning skills to co-program such functions.

They also may develop troubleshooting skills to test and improve the programmed instructions within such functions to efficiently and comprehensively collect, retrieve, calculate, and display data of dynamic and ongoing in-game target behavior changes. For instance, behavior analysts may use their logical reasoning and troubleshooting skills to computationally co-program and test evaluation functions oriented to track in-game learning by effectively assessing in-game action steps followed by autistic adults. Logical reasoning skills would be advantageous for behavior analysts to systematically co-program evaluation functions in the game architecture oriented to follow up on autistic adults' in-game action execution. These functions may follow up on in-game action through tasks, activities, or challenge completions (Lameras et al., 2016). Such evaluation functions might be programmed for screening, operationalizing in quantitative dimensions the target behavior oriented to game action goals, pinpointing the target behavior as a dependent variable, monitoring assessment progress continually, and following up the game-based intervention results. Analysis and insight skills may also allow behavior analysis to break down in-game action dimensions to integrate such continuous measurement procedures used in applied behavior analysis into the game architecture, for instance, by integrating temporal dimensions (e.g., frequency of occurrence, event recording, duration, latency recording), and intensity (LeBlanc et al., 2015).

Behavior analysts may also use inspiration, innovation and reframing skills to consider from diverse perspectives how these evaluation functions would measure the in-game action dimensions regarding the target behavior considering the autistic adults' learning needs and strengths. In that case, behavior analysts must use their observation and analysis skills as co-programmers with autistic adults. The goal would be to balance learning mechanisms and fun by, for instance, co-programming feedback progress indicators such as social, cognitive, affective, motivational, and progress feedback toward players' mastery (Lameras et al., 2016).

Analysis and design skills as co-programmers may allow behavior analysts to design novel algorithms to be programmed as AI agents based on models of decision-making or training steps used in applied behavior analysis. An algorithm is defined as a computational procedure implemented into one or several functions within the game architecture to solve a well-specified problem by running the solution in one and the complete set of instances where such a solution is required (Skiena, 2021). For instance, behavior analysts may develop analytical and logical reasoning skills to co-design and implement AI algorithms as procedural steps computationally programmed to emulate human-like decisions and problem-solving. These AI algorithms might be inspired by the clinical decision-making models used by behavior analysts (LeBlanc et al., 2015) to co-program evaluation functions oriented to select the most appropriate game action measurement procedures. They may also use the six-step behavioral skills training (BST) (Parsons et al., 2012), a suitable algorithmic procedure to be implemented as AI game agents designed to train behavior analysts in virtual reality game-based environments for learning.

Behavior analysts may use troubleshooting and experimenting skills to co-design and co-program these AI algorithms by exploring and testing dataset patterns of in-game action as quantitative dimensions of target behavior measurements. Troubleshooting and experimenting skills may allow behavior analysts to find the best solution to AI agents learning from data patterns to create new algorithms and operate on their own, progressively becoming more efficient by adapting themselves to the new in-game assessment and intervention data.

Finally, troubleshooting and experimenting skills may allow behavior analysts to collect data through virtual reality technology combined with neurotechnologies (e.g., electroencephalography (EEG) signals, eye-, face, and motion tracking). They may balance creativity and logic to broaden and integrate multi-modal data inputs to measure target behaviors with higher accuracy, using real-time reinforcement in computer-simulated game-based intervention. For instance, eye-tracking systems may allow behavior analysts to change environmental stimuli properties in real-time to measure eye avoidance, scanning patterns, or detect whole/part focus (Tanaka & Sung, 2016). At the same time, they may accurately reinforce autistic adults by performing face-processing tasks in social interactions (Tanaka & Sung, 2016).

## Conclusion

Behavior analysts may develop people skills to co-design, and co-program engaging virtual reality game-based supports for intervention with autistic adults. These people skills may also allow behavior analysts to empower autistic adults using a user-centered design approach. Such people skills may enable behavior analysts to integrate applied behavior analysis principles and techniques into accessible, affordable, comprehensive, and effective technology-based supports for co-developing intervention solutions reaching underrepresented autistic communities globally. Behavioral technology is oriented to developing an ecosystem database of teaching gambits and a learning-management system to coordinate better services for autistic communities (Keenan et al., 2019). In particular, virtual reality technologies offer an immersive 360-degree look at in game-based reactive environments to explore and co-design multiple integrations into game mechanics, promoting essential people skills that behavior analysts may develop as co-designers, co-programmers, change agents, and leaders.

## Note

1 In this chapter, the author uses the identity-first language (e.g., "autistic adults") chosen autonomously by some autistic communities. They exert agency by choosing the identity-first language since it expresses that autism is not a detachable or negative condition but a life-span identification (Dunn & Andrews, 2015).

## References

Accardi, C., & Duhaime, S. (2013). Diversity in Ontario's youth and adults with autism spectrum disorders: Finding and keeping employment. *Autism Ontario*. www.autismontario.com/sites/default/files/2020-04/Finding%20and%20Keeping%20Employment_0.pdf

Adams, D., Malone, S. (2021). The impact of anxiety on decision making in individuals with intellectual and developmental disabilities or a diagnosis on the autism spectrum. In I. Khemka, & L. Hickson (Eds.), *Decision Making by Individuals with Intellectual and Developmental Disabilities* (pp. 173–196). Springer. https://doi.org/10.1007/978-3-030-74675-9_8.

Alberto, P., & Troutman, A. C. (2013). *Applied behavior analysis for teachers* (9th edn.). Pearson.

Alvarez, J., Irrmann, O., Djaouti, D., Taly, A., Rampnoux, O. & Sauvé, L. (2019). Design games and game design: relations between design, codesign and serious games in adult education. In S. Leleu-Merviel, D. Schmitt, & P. Useille (Eds.), *From UXD to LivXD: Lliving experience design* (pp. 229–253). Wiley.

Arnab, S., & Clarke, S. (2015). Towards a trans-disciplinary methodology for a game-based intervention development process. *British Journal of Educational Technology, 48*(2), 279–312. https://doi.org/10.1111/bjet.12377.

Bannerman, D. J., Sheldon, J. B., Sherman, J. A., & Harchick, A. E. (1990). Balancing the right to habilitation with the right to personal liberties: The rights of people with developmental disabilities to eat too many doughnuts and take a nap. *Journal of Applied Behavior Analysis, 23*, 79–89. https://doi.org/10.1901/jaba.1990.23-79.

Behavior Analyst Certification Board. (2017). *BCBA/BCaBA task list (5th ed.)*. Author. www.bacb.com/wp-content/uploads/2020/05/170113-BCBA-BCaBA-task-list-5th-ed-.pdf.

Behavior Analyst Certification Board. (2020). *Ethics code for behavior analysts.* https://bacb.com/wp-content/ethics-code-for-behavior-analysts/.

Bouck, E. C., Satsangi, R., Doughty, T. T., & Courtney, W. T. (2014). Virtual and concrete manipulatives: a comparison of approaches for solving mathematics problems for students with autism spectrum disorder. *Journal of Autism and Developmental Disorders, 44*(1), 180–193. https://doi.org/10.1007/s10803-013-1863-2.

Catalano, C. E., Luccini, A. M., & Mortara, M. (2014). Guidelines for an effective design of serious games. *International Journal of Serious Games, 1*(1). https://doi.org/10.17083/ijsg.v1i1.8.

Choi, R. (2021, October 27). *Canadian survey on disability.* Statistics Canada. www150.statcan.gc.ca/n1/en/pub/89-654-x/89-654-x2021002-eng.pdf?st=pzJEBhvj.

Cobb, S., Hawkins, T., Millen, L. & Wilson, J. (2015). Design and development of 3D interactive environments for special educational needs. In K. S. Hale & K. M. Stanney (Eds.), *Handbook of Virtual Environments: Design, implementation, and applications* (pp. 1075–1107). CRC Press.

Cooper, J. O., Heron, T. E., & Heward, W. L. (2020). *Applied behavior analysis* (3rd ed.). Pearson Education.

Dong, H., Nicolle, C., Brown, R. & Cassim, J. (2007). Designer-orientated user research methods. In R. Coleman, J. Clarkson, H. Dong, & J. Cassim (Eds.), *Design for Inclusivity: A practical guide to accessible, innovative and user-centred design* (pp. 131–148). Gower.

Dudley, C., & Emery, J. C. H. (2014). The value of caregiver time: costs of support and care for individuals living with autism spectrum disorder. *School of Public Policy Research Papers, 1*(7), 1–48.

Dunn, D. S., & Andrews, E. E. (2015). Person-first and identity first language: Developing psychologists' cultural competence using disability language. *American Psychologist, 70*(3), 255–264. https://doi.org/10.1037/a0038636.

Ha, K., Ziegert, A., Gorman, M., Hochberg, M., Morrison, A., Nowell, S., & Ramminge, T. (2021). *Life Journey through Autism: A guide for transition to adulthood.* Organization for Autism Research. https://researchautism.org/resources/a-guide-for-transition-to-adulthood/.

Jerald, J. (2016). *The VR Book: Human-centered design for virtual reality.* Association for Computing Machinery.

Keenan, M., Presti, G., & Dillenburger, K. (2019). Technology and behavior analysis in higher education. *European Journal of Behavior Analysis, 21*(1), 26–54. https://doi.org/10.1080/15021149.2019.1651569.

Kiili, K., de Freitas, S., Arnab, S., & Lainema, T. (2012). The design principles for flow experience in educational games. *Procedia Computer Science, 15,* 78–91.

Knight, I., West, J., Matthews, E., Kabir, T., Lambe, S., Waite, F., & Freeman, D. (2021). Participatory design to create a VR therapy for psychosis. *Design for Health, 5*(1), 98–119. https://doi.org/10.1080/24735132.2021.1885889.

Komeda, H. (2021). Cognitive, emotional, and moral decision making in adolescents and adults with autism spectrum disorder. In I. Khemka, & L. Hickson (Eds.), *Decision Making by Individuals with Intellectual and Developmental Disabilities* (pp. 353–374). Springer.

Lameras, P., Arnab, S., Dunwell, I., Stewart, C., Clarke, S., & Petridis, P. (2016). Essential features of serious games design in higher education: Linking learning attributes to game mechanics. *British Journal of Educational Technology, 48*(4), 972–994. https://doi.org/10.1111/bjet.12467.

LeBlanc, Raetz, P. B., Sellers, T. P., & Carr, J. E. (2015). A proposed model for selecting measurement procedures for the assessment and treatment of problem behavior. *Behavior Analysis in Practice, 9*(1), 77–83. https://doi.org/10.1007/s40617-015-0063-2.

Levin, I. P., Gaeth, G. J., Levin, A. M., & Chen, S. (2021). Social functioning and decision making: From group to individual differences across the autism spectrum. In I. Khemka, & L. Hickson (Eds.), *Decision making by individuals with intellectual and developmental disabilities* (pp. 333–351). Springer.

Lombardo, M. V., & Baron-Cohen, S. (2011). The role of the self in mindblindness in autism. *Consciousness and cognition, 20*(1), 130–140. https://doi.org/10.1016/j.concog.2010.09.006.

Macklin, C., & Sharp, J. (2016). *Games, design and play: A detailed approach to iterative game design.* Addison-Wesley.

Maich, K., Penney, S., Alves, K., & Hall, C. (2020). *Autism spectrum disorders in the Canadian Context: An introduction.* Canadian Scholars.

Michailidis, L., Balaguer-Ballester, E., & He, X. (2018). Flow and immersion in video games: the aftermath of a conceptual challenge. *Frontiers in Psychology, 9,* 1682. https://doi.org/10.3389/fpsyg.2018.01682.

Malinverni, L., Mora-Guiard, J. & Pares, N. (2016). Towards methods for evaluating and communicating participatory design: A multimodal approach. *International Journal of Human-Computer Studies, 94,* 53–63. https://doi.org/10.1016/j.ijhcs.2016.03.004.

Manett, J., & Stoddart, K. (2012). *Facing the Challenges of Post-Secondary Education: Strategies for individuals with autism spectrum disorder (ASD).* Autism Ontario. www.autismontario.com/sites/default/files/2020-05/Facing%20the%20Challenges%20of%20Post-Secondary%20Education.pdf.

Martin, G., & Pear, J. (2014). *Behavior modification* (10th ed.). Pearson Education.

Mazurek, M., Engelhardt, C., & Clark, K. (2015). Video games from the perspective of adults with autism spectrum disorder. *Computers in Human Behavior, 51,* 122–130. https://doi.org/10.1016/j.chb.2015.04.062.

Miltenberger, R. G. (2016). *Behavior modification: Principles and procedures* (6th ed.). Cengage Learning.

National Autism Center. (2015). *National standards project findings and conclusions: Phase 2*. Author.

Newell, A. & Monk, A. (2007). Involving older people in design. In R. Coleman, J. Clarkson, H. Dong, & J. Cassim (Eds.), *Design for inclusivity: A practical guide to accessible, innovative and user-centred design* (pp. 111–130). Gower.

Nystrom, R. (2014). *Game programming patterns*. Genever Benning.

Ontario Ministry of Community and Social Services. (2018). *Social assistance*. www.mcss.gov.on.ca/en/mcss/programs/social/directives/odsp/es/1_1_ODSP_ESDirectives.aspx.

O'Reilly, M., Little, A., Falcomata, T., Fragale, C., Green, V., Sigafoos, J., Lancioni, G., Edrisinha, C., & Choi, H. (2011). Interactive social skills. In J. K. Luiselli (Ed.), *Teaching and behavior support for children and adults with Autism spectrum disorder: A practitioner's guide* (pp. 104–110). Oxford University Press.

Parsons, M. B., Rollyson, J. H., & Reid, D. H. (2012). Evidence-based staff training: a guide for practitioners. *Behavior Analysis in Practice, 5*(2), 2–11. https://doi.org/10.1007/BF03391819.

Plaisted, K. (2001). Reduced generalization in autism: An alternative to weak central coherence. In A. Jacob, T. Charman, N. Yirmiya, & P. Zelaz (Eds.), *The development of autism: Perspectives from theory and research*. Lawrence Erlbaum Associates.

Public Health Agency of Canada. (2020, May 6). *Infographic: Autism spectrum disorder highlights from the Canadian survey on disability*. Government of Canada. www.canada.ca/en/public-health/services/publications/diseases-conditions/infographic-autism-spectrum-disorder-highlights-canadian-survey-disability.html.

Skiena, S. S. (2021). *Algorithm design manual*. Springer Nature.

Smith, M. J., Fleming, M. F., Wright, M. A., Losh, M., Humm, L. B., Olsen, D., & Bell, M. D. (2015). Brief report: vocational outcomes for young adults with autism spectrum disorders at six months after virtual reality job interview training. *Journal of Autism and Developmental Disorders, 45*(10), 3364–3369. https://doi.org/10.1007/s10803-015-2470-1.

Smith, M. J., Ginger, E. J., Wright, K., Wright, M. A., Taylor, J. L., Humm, L. B., Olsen, D. E., Bell, M. D., & Fleming, M. F. (2014). Virtual reality job interview training in adults with autism spectrum disorder. *Journal of Autism and Developmental Disorders, 44*(10), 2450–2463. https://doi.org/10.1007/s10803-014-2113-y.

Stoddart, K. P., Burke, L., Muskat, B., Manett, J., Duhaime, S., Accardi, C., Riosa, P. B., & Bradley, E. (2013, February 5). *Diversity in Ontario's youth and adults with ASD: Complex needs in unprepared systems*. www.community-networks.ca/wp-content/uploads/2015/11/march-14-2013-final-diversity-in-ontarios-youth-and-adults-with-asds-february-2013.pdf.

Strickland, D. C., Coles, C. D., & Southern, L. B. (2013). JobTIPS: a transition to employment program for individuals with autism spectrum disorders. *Journal of Autism and Developmental Disorders, 43*(10), 2472–2483. https://doi.org/10.1007/s10803-013-1800-4.

Tanaka, J. W., & Sung, A. (2016). The "eye avoidance" hypothesis of autism face processing. *Journal of Autism and Developmental Disorders, 46*(5), 1538–1552. https://doi.org/10.1007/s10803-013-1976-7.

Taylor, J., Wehman, P., & Pitonyak, V. (2020). Individual planning and community transition planning: Focus on inclusion. In P. Wehman (Ed.), *Essentials of transition planning* (2nd ed.). Brookes Publishing.

Tsikinas, S., & Xinogalos, S. (2018). Studying the effects of computer serious games on people with intellectual disabilities or autism spectrum disorder: A systematic literature review. *Journal of Computer Assisted Learning, 35*(1), 61–73. https://doi.org/10.1111/jcal.12311.

Tsikinas, S., & Xinogalos, S. (2020). Towards a serious games design framework for people with intellectual disability or autism spectrum disorder. *Education and Information Technologies, 25*, 3405–3425. https://doi.org/10.1007/s10639-020-10124-4.

Vasquez III, E., Nagendran, A., Welch, G., Marino, M., Hughes, D., Koch, A., & Delisio, L. (2015). Virtual learning environments for students with disabilities: A review and analysis of the empirical literature and two case studies. *Rural Special Education Quarterly, 34*(3), 26–32. https://doi.org/10.1177/875687051503400306.

Zubek, R. (2020). *Elements of game design*. The MIT Press.

# Index

ableism 18, 21, 267
acceptance 35, 64, 79, 96–7, 104–5
accessibility *see* accessible
accessible/accessibility 58, 66, 93, 172–7, 200–1, 262–72, 283–7, 299–304
ACT 8–12, 33–43, 50, 59, 106–7, 112–13, 132–3
active learning 200
active listening 52–5, 72, 97, 110–11, 164, 175, 293
adherence 52, 58–9, 65, 109–12, 115–18, 129–32, 137, 161, 170, 185, 292
advocacy/advocate 90, 148, 203–4, 207–12
agenda 51, 129, 170–5, 206–7
anti-oppressive practice 110–19
artificial intelligence 320–2
assessment 18–21, 59–61, 102, 134, 139, 144–6, 155–6, 186–8, 258–9, 280–3, 310–12, 320–1
assistive technology 316
autism 88–93, 131–2, 220–5, 263–7, 316–22
autistic *see* autism

behavior momentum 135
bias (types: pathology bias, confirmatory bias, hind-sight bias, over-confidence, racial or cultural bias, ableism) 7–9, 16–24, 68, 81, 123, 174, 252–7, 271
body language 53–5, 62–4, 169, 175, 198–9, 240, 292–3, 295
bracketing 218–19

BST 70, 110–12, 122, 231–3, 299, 321
burnout 6–9, 27–9, 66

capacity/capacity building 58, 99–102, 162–3, 265, 267, 280–1, 286–7
caregivers *see* mediators
chaining 135, 223
client ideas 65, 124
collaboration 128–32, 136, 138–40, 148, 151–5, 174–5, 211–12
communication 53–5, 77–9, 90–1, 111, 117, 148, 154–6, 159–65, 231–2, 265–6, 318–19
compassion/compassionate 10–12, 23–4, 64–6, 109–11, 124–5, 162, 177, 182–3, 230, 285; empathy 58–9, 62, 64–5, 91, 110–11, 149, 230, 244, 295
concerns 66, 99
concordance 129, 225
conflicts 152–3, 170
conversation *see* communication
corrective feedback *see* concerns
critical disability studies framework 89–90
criticism 71, 101, 211, 231
culture 34, 53–4, 68, 77–86, 133, 140, 225, 248–9, 252–4, 268; cultural identity 77; cultural diversity 21, 77, 83–4

data 16–18, 20–2, 104–5, 113–15, 136, 147–9, 152–3, 155–6, 176, 184–8, 201, 257, 319–21
decision-making 24, 29–30, 66–7, 154–5, 230, 267–9, 318, 321

disability 21–2, 88–93, 117, 225, 262–7
discriminative stimuli 135
diversity 16, 21, 80–1, 83–5, 232, 266–8
documentation 148, 152, 164, 240
documents *see* written materials
dress/dressing *see* professionalism
DSM-5 90–1

embedded curriculum approach 221–2
empathy *see* compassion
employee selection *see* hiring
environment 63–4, 78–9, 90–2, 105–6, 184–6, 220–2, 249, 270, 283–4, 293, 313
ethics 123–4, 148, 159–60, 183–4, 230, 263, 284–6, 295–6
evidence-based interventions 110, 130, 155, 186, 317
extinction 58, 133, 137, 156, 185–6, 317
eye contact 53, 92–3, 121–3, 199, 253, 292–3, 318

feedback 96–107, 121, 240, 243–50, 321; degree feedback 244
fidelity 23, 59, 103, 109–10, 115–19, 128–9, 131–2
function 19–21, 30–2, 65, 71, 136, 193, 253, 312
functional assessment *see* assessment

genuiness 59, 61, 63–4, 70
goals 4–5, 8, 18, 22, 33, 62–7, 80–1, 113, 128–31, 134–8, 144–9, 154–5, 171–2, 187–8, 207, 224–5, 238, 281, 283, 317–20

headings 303–4
health literacy 263–4
hiring 252–3, 255–7

identity-first language 88
imaginative free variations 218–19
instruction *see* teaching
interdisciplinary 143–50, 155–6
interpersonal skills 4, 6–8, 50, 72, 149, 289, 292–3, 316
interview 253–9; behavioural interview 256, 258

intrapersonal skills 5–10, 50, 72, 183, 257, 316

job crafting 31–2

language *see* verbal cues
layout 270–1, 302
learning history 58, 101, 107, 133, 160, 193, 308
line length 302
lists 300, 302

mediators 109–11, 113–15, 118–19, 122–4, 184–8, 263–4, 299–300
meetings 69, 128, 162–4, 169–78
mentorship 84, 161, 166, 286–7
mindfulness 9, 59, 153, 162
modelling 69, 112, 177, 186, 220, 229
motivating operation 98, 104, 106, 131, 134, 139, 186, 301; What's in it for me? 193–4, 301
motivational interviewing 130
movement education 219–20

neurodiversity paradigm 90, 93
non-judgemental 63–4
non verbal cues 22, 54, 56, 70, 162, 199, 254, 292, 297

pairing *see* reinforcement
paraphrase 55–7, 61
parent coaching 128–9, 139, 280
pausing 56, 197
people skills *see* interpersonal skills
performance feedback *see* feedback
person-first language 88
phenomenology 217
policy 203–4, 211, 267–8, 271
powerpoint 192–201, 304
present *see* mindfulness
presenter 192–201
problem-solving 294–5
professionalism 172–3
prompting 103, 117, 135–6, 317
psychological flexibility 8, 9, 59
punishment 4, 58, 82, 106, 137, 160, 165, 186, 231, 300, 313

questions 23, 56, 60, 66, 97, 110–15, 135, 139, 162, 164, 171, 173–6,

185, 188, 197–8, 238–9, 243–4, 253–5, 257–8, 311

rapport 20, 22–3, 51, 54–7, 60–6, 68–70, 100, 109–24, 131, 134, 161, 164–5, 173, 230, 140, 285–6, 293–4, 310, 313
realization 310
recognition 311
reinforcement 4, 17, 29, 60, 68–70, 103–5, 133–7, 145–7, 153, 156, 165–6, 185, 231, 237–8, 244, 247, 283–4, 300, 317, 321
relational leaderships 231–4
relationship *see* therapeutic relationship
resistance 206, 311–12
responding 311
response acquisition 136
response effort 136, 140, 146, 152, 300
role play *see* modelling
rule-governed behaviours 4, 132, 137

schedule 135–7, 162–3, 165, 170
self-awareness 5–9, 19, 81, 99
self-care 5–7, 29–33, 186
self-compassion *see* compassion
self-disclosure 61–2
self-regulation 5–8, 156, 165
shaping 22, 96, 100, 103, 135, 236
small talk 99, 171–3, 258
space and silence *see* pausing
station based pedagogy 220
stigma 266–267
structured interview *see* interview
success *see* reinforcement
summarizing 57

supervisee 81, 97, 103–4, 160, 175–6, 230–5, 239–40, 245, 247–9
supervision philosophy 237–9
supervisor 230–41

tacting feelings 60–1
task analysis 170–1
teaching 105, 109, 129–31, 161, 291–2; teaching 111–12
team 151–2
technology 291
telehealth 5–6, 22, 28, 96–100, 103–6, 166, 230–41, 280–7, 289–97
therapeutic relationship *see* reinforcement
therapeutic relationships 7, 50, 54, 59–61, 71, 129, 134, 139–40
training *see* teaching
trauma-informed 308–13
treatment adherence *see* adherence
treatment fidelity *see* fidelity
typeface 303

universal design 200–1, 270
unstructured interview *see* interview

value 38–41, 79, 83–5, 106, 205–8, 236–7, 301
verbal cues 54, 57, 198
virtual 169–71, 232, 280–7, 289–97
voice *see* verbal cues
vulnerable 11, 23, 63

waitlists 266
websites 271
white space 302
written materials 270–1, 299–305

Made in the USA
Columbia, SC
14 January 2025

51760865R00193